AF522211

ENCYCLOPEDIA OF BIOCHEMISTRY

ENCYCLOPEDIA OF BIOCHEMISTRY

Vol. 4

By

Sananda Chatterjee

DAE, DCA, CMCNet

Scientific Consultant

Institute for Natural Sciences

Kolkata

(West Bengal)

DISCOVERY PUBLISHING HOUSE PVT. LTD.

NEW DELHI-110 002

Published by:
Namit Wasan
DISCOVERY PUBLISHING HOUSE PVT. LTD.
4383/4B, Ansari Road, Darya Ganj
New Delhi-110 002 (India)
Phone : +91-11-23279245; 23253475; 43596065
E-mail : discoverybooksindia@gmail.com
discoverypublishinghouse@gmail.com
namitwasan9@gmail.com
web : www.discoverypublishinggroup.com

Reprinted: 2020
First Edition: 2012

ISBN: 978-81-8356-876-0 (Set)

Encyclopedia of Biochemistry

Printed at:
Infinity Imaging Systems
Delhi

Preface

It is always an ultimate outlook to give students the best information in a good package. So I took up the job of writing a series of books in chemistry starting from a very introductory level to this advance level of biochemistry.

Biochemistry, to express something about it, in one word, would be difficult for me. It should be ones part of life; the reactions of protein, amino-acids, the DNA RNA—double helix pattern. The nynhydrin reactions are all the wonders of biochemistry—man's anato-biochemical linkages.

Students who want to become doctors or medical practitioners must be a good biochemist first because biochemistry is the theory behind and the medicine is the practical application.

The effects and side effects of the particular medicine, whether it is allergic or non-allergic, to the patients are well understood by the doctors but are taught in biochemistry.

The chemical functions of the different human organs and systems; the kidney, the respiratory system, the blood, the chemical functions of hormones, vitamins, the liver etc are all well covered by this book.

The subject of this book is therefore deeply intended for those upcoming doctors who keep themselves for the real service of the mankind.

The object of this book is not only meant for the doctors but it is the last bangle of that long garland which has started all along from the introduction chapter of The Architecture of Chemistry, The Mans Scientific Attitude.

Biochemistry also comes in use in the field of agriculture, the biochemical configuration of the soil. The biochemical configuration of the plants to be cultivated, the application of the pesticides, and the relation of the pesticides with the human population are well studied under biochemistry.

The Agricultural Scientist, the Pathological Scientist, Forensic Scientist, is definitely inter-joined with each other by a single buffer of the Biochemistry.

This book is dedicated to those promising scientific readers and interested students of biochemistry and advanced chemical science, who wants to make a fruitful attempt in the field of modern scientific world.

The author will be grateful and surely thankful to see the best utilization of this book in any of the field of application.

Sananda Chatterjee

Contents

The Biochemistry of Human Systems

Introduction

In this chapter we are going to discuss the biochemical reactions in human systems. The process of Neuro-Communications, The Process of Digestions, The Process of Excretion, the chemistry and composition of Urine, and the chemistry of Respiration has been conversed.

THE CONSTITUTION OF HUMAN BODY A GENERAL DISCUSSION

The chemical composition of the body represents the entire physical and chemical basis of protoplasm and of the living processes which occur in it. Much discussion relative to it has occurred in the chapters on lipids, carbohydrates, and proteins.

The earlier work on the composition of tissues had to do primarily with the quantities of the various elements and of lipids, proteins, carbohydrates, minerals, water, etc., present. With the development of better techniques of isolation and analysis, it became possible to demonstrate the presence of a large number of different substances in tissues which are important in their functioning or as the products of their metabolism. Because tissues contain enzymes which may alter cell constituents after death, and particularly after mechanical disintegration, often it is necessary to proceed with great care in order to minimize these changes. In some instances, as in the isolation of glycogen, the tissues are excised quickly and heated in order to stop enzyme action. At times a specific enzyme poison may be added to prevent decomposition of a substance to be isolated; an example is the addition of sodium fluoride to muscle hash to prevent the enzymatic decomposition of hexose diphosphate. Another valuable procedure consists in freezing the tissue with liquid air or carbon dioxide snow, followed by powdering the frozen tissue and drying it at low temperature (lyophile process). In this way the cellular enzymes are inhibited by the low temperature; and after the powdered tissue is dry, its various constituents, including some enzymes, may be preserved for a long time. This lyophilized product may be used for the isolation of tissue constituents by extraction with appropriate solvents and other procedures. The composition of tissues is dynamic and it changes continuously both physically and biochemically.

The composition of any tissue is representation of balanced progress of synthesis and decomposition and also between processes that bring substances to the tissues as foods and those that remove substances from tissues as excretory products.

Numerous important tissue constituents are present in very small amounts at any given time, though the total amounts formed and decomposed during the course of a day may be of considerable amount.

When the several substances are formed and decomposed in sequence or in a process or cycle, the rate of process is slow. The least reactive substances accumulate in the largest amount and the most reactive substance accumulates less. Many of the active components of tissue are present in minute amounts. Their chemical instability and low concentration of ten cause detection and isolation to be difficult.

As we are acquainted with that knowledge that the six main constituents of human body are carbon, oxygen, calcium, phosphorous hydrogen and nitrogen, which comprises about 99% and the remaining 1% is sodium chloride magnesium iron, and traces of iodine. The Fig. 9.1 and 9.2 represents the composition of the human body.

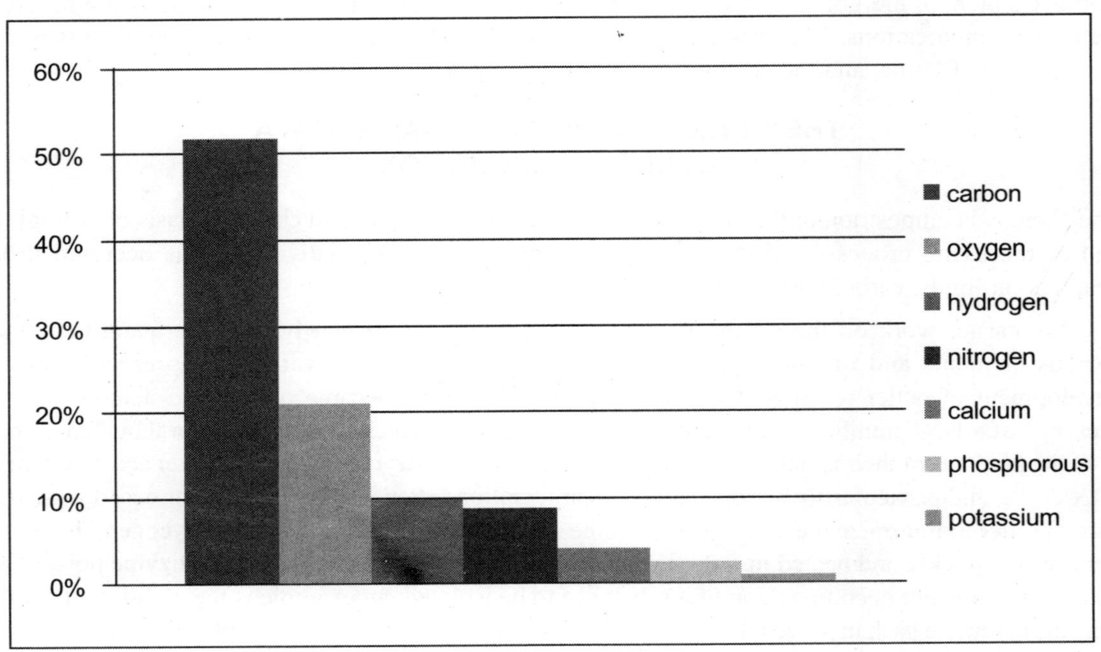

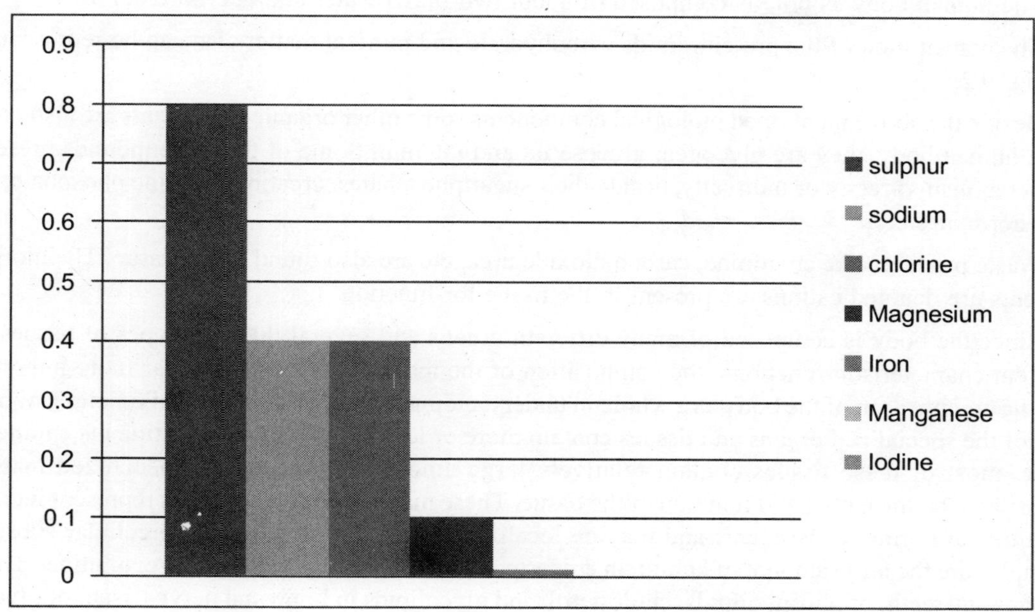

Fig. 9.1 : The Compositions of Human Body

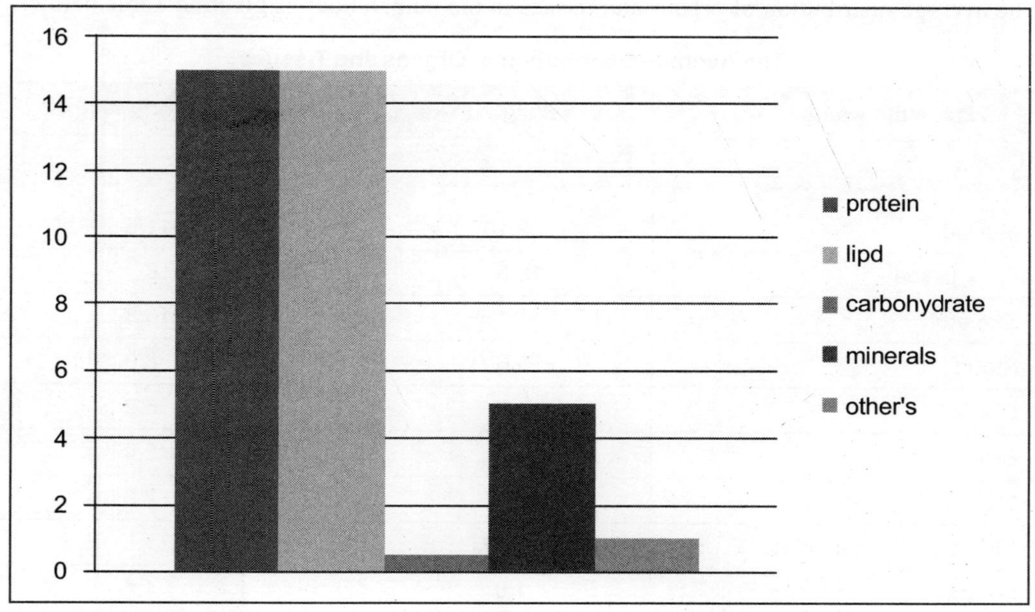

Fig. 9.2 : The Compositions of Human Body

The human body as hole is composed of about two-thirds water and one-third solids.

The constitution of the protein, lipids carbohydrate and mineral matter also can be represented in the Fig. 9.2.

Beside the above mentioned biological components some other organic ingredients are also present in the human body, they are glycogen, glucose fat and vitamin. Some of these compounds present in the metabolism directly or indirectly, beside their sugar phosphates, creatine, creatine phosphate, ATP, ADP hormones etc.

Waste products like creatinine, carbon dioxide urea, etc are also found in the tissue. The inorganic elements are denoted cations are present in the tissue for function.

Since the body is composed of many different organs and several different types of tissues with different characteristic functions, the composition of the individual tissues is of far more importance than the composition of the body as a whole in understanding body organization and function. Whereas most of the specialized organs and tissues contain more or less of most of the constituents enumerated above, most of these tissues contain relatively large amounts of secondary specialized materials necessitated by the nature and function of the tissue. These materials mayor may not represent increased quantities of normal constituents and may be localized within the cells or as intercellular substance. Examples are the large amount of keratin in epidermal tissue; of collagen in cartilage; of mineral matter in bones and teeth; of phospholipids, cholesterol, and glycolipids in brain and nerve tissue; of glycogen in liver; of hemoglobin in red blood cells; of fat in adipose tissue; of hormones in endocrine glands; and of the contractile protein myosin in muscle.

The average distribution of organs and tissues in the human body is given in Table below:

The Average Distribution of Organs and Tissues

Tissue or Organ	Weight in kg	Per Cent of Body Weight
1	2	3
Muscle	29.10	41.5
Fatty tissue	12.6	18
Skeleton	11.1	15.8
Blood	5.6	8.0
Skin	4.8	6.9
Brain	1.7	2.4
Liver	1.6	2.3
Stomch and intestines	1.27	1.8
Lungs	1.0	1.4
Heart	0.33	0.47

...(Contd.)

1	2	3
Kidneys	0.27	0.38
Spleen	0.13	0.18
Carebrospinal fluid	0.13	0.18
Pancreas	0.10	0.14
Salivary glands	0.05	0.07
Testicles	0.03	0.04
Thyroid	0.03	0.04
Thymus	0.016	0.02
Adrenals	0.007	0.01
Parathyroids	0.0004	0.006
Pituitary	0.003	0.004

The gross composition of tissues is generally designated by the contents of water, solids, proteins, lipids, carbohydrates, and extractives. The extractives include the water-soluble substances removed from tissues by treatment with hot water. They are composed of inorganic and a large variety of organic substances.

The composition of a definite type of tissue, such as mammalian striated muscle, shows some species variations but on the whole is characteristic. A young rapidly growing tissue generally contains more water and less total lipids and mineral matter than a more adult tissue. Many other variations in composition are also characteristic of tissue age.

The state of nutrition of an animal may markedly influence the composition of tissues. The glycogen content of the liver of a poorly nourished animal may amount to 1 per cent or less, whereas the liver glycogen of this same animal on a high carbohydrate diet may be as high as 15 per cent. The glycogen content of the muscles and other tissues shows similar but less spectacular variations. The very great variation in the fat content of tissues with the nutritive state is well known. The proteins of the blood may become severely reduced as the result of inadequate dietary proteins. The mineral and water contents of tissues also vary with the intake of these materials. The composition of tissues may be drastically altered by pathologic conditions. Liver and muscle glycogen are very low in uncontrolled severe diabetes mellitus, whereas liver glycogen is very high in glycogenosis or von Gierke's disease. Very large amounts out may be deposited in the liver in pathologic conditions such as acute yellow atrophy. In the xanthomatoses or lipoidoses which appear to be congenital diseases of childhood there is accumulation of excess lipid in large cells of the reticuloendothelial system. The type of lipid in the cells varies. It is glycolipid (kerasin) in Gaucher's disease, phospholipid in Nieman-Pick's disease, and cholesterol and its esters in Schuller-Christian syndrome. In cases of progressive muscle dystrophy the muscles may show a marked decreltse in creatine, adenosine triphosphate, and other extractives.

Abnormal deposition of calcium salts may occur in tissues in cases of pathologic calcification, and the mineral content of bone may be depleted in conditions of hyperparathyroidism. Pathologic variations in the composition of blood are numerous and form the basis of much of our modern clinical diagnosis.

Tables below give the gross composition of the adult human body. While the gross composition o tissues may bring out a few points of interest, it is the detailed molecular composition which affords insight into the finer structures and physiologic functions of tissues. This has not been completely determined for any tissue. Since blood is the easiest of the tissues to obtain and analyze, our knowledge of its composition is more nearly complete than that of any other tissue.

The Average Distribution of Gross Composition of the Avarage Human Body

	Total Body	Water Water	Ether Extract	Crude Protein (N×6.25)	Ash	Ca	P
Skin	6.33	57.71	14.23	27.33	0.62	0.0034	0.070
Skeleton	17.58	28.17	25.04	19.71	26.62	10.68	4.61
Teeth	0.08	5.00*		23.00*	67.95	25.05	12.09
Straited muscle	39.76	70.09	6.60	21.94	1.01	0.0066	0.156
Brain, spinal cord, nerve trunks	2.99	75.09	12.35	11.50	1.37	0.0147	0.299
Liver	2.34	71.58	3.11	22.24	1.35	0.0133	0.303
Heart	0.52	62.95	16.58	17.48	0.61	0.0058	0.144
Lungs$	3.30	77.28	1.32	19.20	1.03	0.0090	0.132
Spleen*	0.11	78.69	1.19	17.81	1.13	0.0089	0.217
Kidneys	0.51	70.58	7.18	19.28	0.87	0.0057	0.188
Pancreas*	0.14	73.08	13.08	12.69	0.93	0.0143	0.155
Alimentary tract	1.86	77.40	9.17	12.77	0.53	0.0140	0.098
Adipose tissue	11.37	23.02	71.57	5.85	0.20	0.0078	0.031
Remaining tissue, solid	11.43	59.29	22.47	17.28	0.85	0.0257	0.088
Remaining tisue, liquid	0.59	81.45	2.55	13.58	0.82	0.0044	0.104
Bile, content of bladder and alimentary tract	0.99	—	—	—	—	—	—
Gaur abd bauks	9.10	—	—	—	—	—	—
Average (weighted) all tissues	98.91	55.74	19.66	18.82	5.49	1.928	0.936
Total composition whole body weighing 53.80 kilos	100.00	55.13	19.44	18.62	5.43	1.907	0.925
Composition fat-free body		69.38	—	23.43	6.83	2.400	1.164

*Chemical composition assumed.
$Congested with blood.

The Mineral Distribution of the Avarage Human Body

Tissue	Ca	P	Mg	B p.p.m.*	Co p.p.m.*	Be p.p.m.*	Sr p.p.m.*
Skin	0.014	0.075	0.0075	0.39	0.019	0	
Skeleton	10.83	5.0	0.18	0.74	0.028	0	55
Teeth	25.46	13.24	0.62				
Straited muscle	0.014	0.13	0.019	0.28	0.007	0	0
Nerve tissue	0.028	0.26	0.01	0.05	0.015		
Liver	0.01	0.17	0.0065	0.09	0.095	0.006	0
Heart	0.017	0.11	0.014	0.39	0.034		
Lungs	0.025	0.17	0.0058	0.14	0.038	0.011	0
Spleen	0.010	0.169	0.0124	0.08			
Kidney	0.039	0.16	0.02	0.031	0.034	0	
Alimentary tract	0.021	0.10	0.017	0.024	0.021	0	0
Adipose tissue	0.06	0.044	0.0061	0.10	0.011		
Remaining tissue (solid)	0.12	0.15	0.012	0.19	0.033		
Remaining tisue (liquid)	0.018	0.11	0.0046		0.015		
Whole body	1.76	0.94	0.04				
Fat free whole body	2.11	1.12	0.048				
Bone ash	38.41	17.75	0.636				

*Parts per million.

Much has been learned of the composition of muscle which is of importance in understanding muscle contraction. The composition of brain and nerve tissue and its relation to function are much less well understood. While the liver and kidneys and other glandular organs are of primary importance, our knowledge of their compositions is very incomplete.

Under the following subsections the different systems of human body has will be discussed, regarding their biochemical aspects and their metabolic functions, here we are going to discuss the neurotransmitter, their biochemistry, the biochemistry of respiration, the biochemistry of digestion and excretion also the biochemical composition of urine.

THE BIOCHEMISTRY OF NERVOUS SYSTEM

Introduction

Humans, like all living organisms, can respond to their environment. Humans have two complimentary control systems to do this: the nervous system and the endocrine (hormonal) system. The human nervous system controls everything from breathing and producing digestive enzymes, to memory and intelligence. Under the next subsections we are going to discuss the biochemistry of the nervous system, the neurotransmitters, the process of communications between the brain and the organs etc.

THE NEURON

A neuron also known as a neurone or nerve cell is an electrically excitable cell that processes and transmits information by electrochemical signaling, via connections with other cells called synapses. Neurons are the core components of the nervous system, which includes the brain, spinal cord, and peripheral ganglia. A number of specialized types of neurons exist: sensory neurons respond to touch, sound, light and numerous other stimuli affecting cells of the sensory organs that then send signals to the spinal cord and brain. Motor neurons receive signals from the brain and spinal cord and cause muscle contractions and affect glands. Interneurons connect neurons to other neurons within the same region of the brain or spinal cord.

A typical neuron possesses a cell body (often called soma), dendrites, and an axon. Dendrites are filaments of protoplasm that extrude from the cell body, often extending for hundreds of microns and branching multiple times, giving rise to a complex "dendritic tree". An axon is a special protoplasmic filament that arises from the cell body at a site called the axon hillock and travels through the body, often for a great distance. The cell body of a neuron frequently gives rise to multiple dendrites, but never to more than one axon, although the axon may branch hundreds of times before it terminates. At the majority of synapses, signals are sent from the axon of one neuron to a dendrite of another. There are, however, many exceptions to these rules: neurons that lack dendrites, neurons that have no axon, synapses that connect an axon to another axon or a dendrite to another dendrite, etc.

All neurons are electrically excitable, maintaining voltage gradients across their membranes by means of metabolically driven ion pumps, which combine with ion channels embedded in the membrane to generate intracellular-versus-extracellular concentration differences of ions such as sodium, potassium, chloride, and calcium. Changes in the cross-membrane voltage can alter the function of voltage-dependent ion channels. If the voltage changes by a large enough amount, an all-or-none electrochemical pulse called an action potential is generated, which travels rapidly along the cell's axon, and activates synaptic connections with other cells when it arrives.

Neurons do not undergo cell division, and usually cannot be replaced after being lost. In most cases they are generated by special types of stem cells, although astrocytes (a type of glial cell) have been observed to turn into neurons as they are sometimes pluripotent.

A neuron is a special type of cell that is found in the bodies of most animals (all members of the group Eumetazoa, to be precise—this excludes only sponges and a few other very simple animals). The features that define a neuron are electrical excitability and the presence of synapses, which are complex membrane junctions used to transmit signals to other cells. The body's neurons, plus the glial cells that give them structural and metabolic support, together constitute the nervous system. In vertebrates, the majority of neurons belong to the central nervous system, but some reside in peripheral ganglia, and many sensory neurons are situated in sensory organs such as the retina and cochlea.

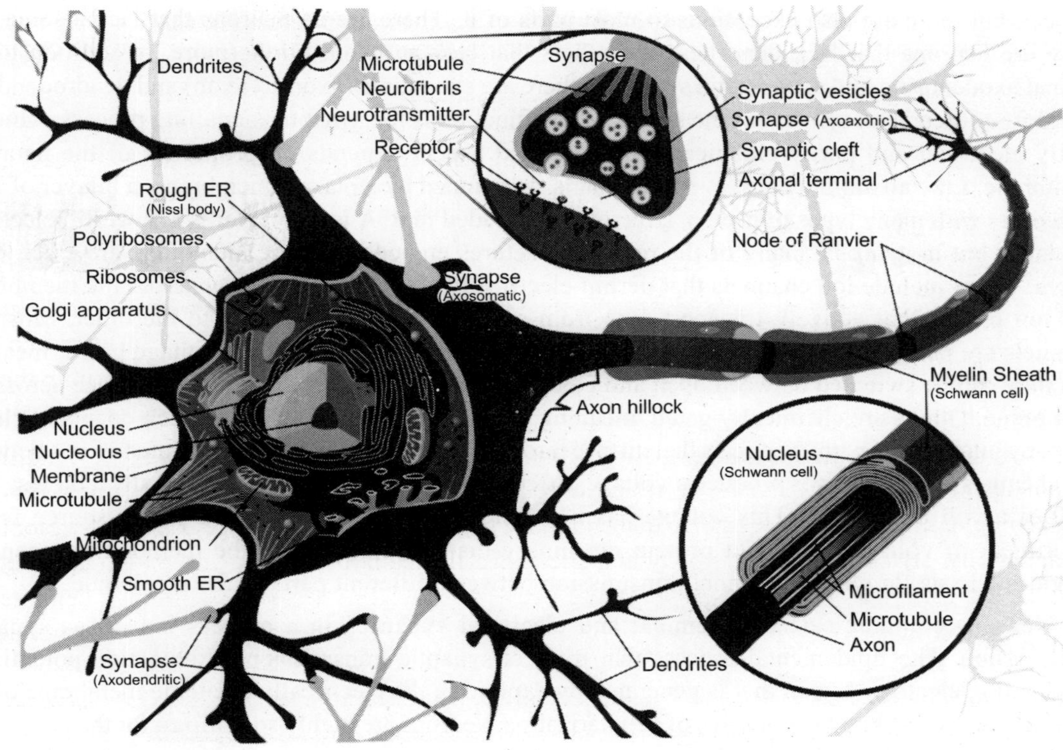

Fig. 9.3 : The Description of A Nerve Cell

Although neurons are very diverse and there are exceptions to nearly every rule, it is convenient to begin with a schematic description of the structure and function of a "typical" neuron. A typical neuron Humans, like all living organisms, can respond to their environment. Humans have two complimentary control systems to do this: the nervous system and the endocrine (hormonal) system. The human nervous system controls everything from breathing and producing digestive enzymes, to memory and intelligence is divided into three parts: the soma or cell body, dendrites, and axon. The soma is usually compact; the axon and dendrites are filaments that extrude from it. Dendrites typically branch profusely, getting thinner with each branching, and extending their farthest branches a few hundred microns from the soma. The axon leaves the soma at a swelling called the axon hillock, and can extend for great distances, giving rise to hundreds of branches. Unlike dendrites, an axon usually maintains the same diameter as it extends. The soma may give rise to numerous dendrites, but never to more than one axon. Synaptic signals from other neurons are received by the soma and dendrites; signals to other neurons are transmitted by the axon. A typical synapse, then, is a contact between the axon of one neuron and a dendrite or soma of another. Synaptic signals may be excitatory or inhibitory. If the net excitation received by a neuron over a short period of time is large enough, the neuron generates a brief pulse called an action potential, which originates at the soma and propagates rapidly along the axon, activating synapses onto other neurons as it goes. Many neurons fit the foregoing schema in every respect, but there are also exceptions to most parts of it. There are no neurons that lack a soma, but there are neurons that lack dendrites, and others that lack an axon. Furthermore, in addition to the typical axodendritic and axosomatic synapses, there are axoaxonic (axon-to-axon) and dendrodendritic (dendrite-to-dendrite) synapses. The key to neural function is the synaptic signalling process, which is partly electrical and partly chemical. The electrical aspect depends on properties of the neuron's membrane. Like all animal cells, every neuron is surrounded by a plasma membrane, a bilayer of lipid molecules with many types of protein structures embedded in it. A lipid bilayer is a powerful electrical insulator, but in neurons, many of the protein structures embedded in the membrane are electrically active. These include ion channels that permit electrically charged ions to flow across the membrane, and ion pumps that actively transport ions from one side of the membrane to the other. Most ion channels are permeable only to specific types of ions. Some ion channels are voltage gated, meaning that they can be switched between open and closed states by altering the voltage difference across the membrane. Others are chemically gated, meaning that they can be switched between open and closed states by interactions with chemicals that diffuse through the extracellular fluid. The interactions between ion channels and ion pumps produce a voltage difference across the membrane, typically a bit less than 1/10 of a volt at baseline. This voltage has two functions: first, it provides a power source for an assortment of voltage-dependent protein machinery that is embedded in the membrane; second, it provides a basis for electrical signal transmission between different parts of the membrane.

Neurons communicate by chemical and electrical synapses in a process known as synaptic transmission. The fundamental process that triggers synaptic transmission is the action potential, a propagating electrical signal that is generated by exploiting the electrically excitable membrane of the neuron. This is also known as a wave of depolarization.Neurons are highly specialized for the processing and transmission of cellular signals. Given the diversity of functions performed by neurons in different parts of the nervous system, there is, as expected, a wide variety in the shape, size, and electrochemical properties of neurons. For instance, the soma of a neuron can vary from 4 to 100 micrometers in diameter.

- The soma is the central part of the neuron. It contains the nucleus of the cell, and therefore is where most protein synthesis occurs. The nucleus ranges from 3 to 18 micrometers in diameter.
- The dendrites of a neuron are cellular extensions with many branches, and metaphorically this overall shape and structure is referred to as a dendritic tree. This is where the majority of input to the neuron occurs.
- The axon is a finer, cable-like projection which can extend tens, hundreds, or even tens of thousands of times the diameter of the soma in length. The axon carries nerve signals away from the soma (and also carries some types of information back to it). Many neurons have only one axon, but this axon may—and usually will—undergo extensive branching, enabling communication with many target cells. The part of the axon where it emerges from the soma is called the axon hillock. Besides being an anatomical structure, the axon hillock is also the part of the neuron that has the greatest density of voltage-dependent sodium channels. This makes it the most easily-excited part of the neuron and the spike initiation zone for the axon: in neurological terms it has the most negative action potential threshold. While the axon and axon

hillock are generally involved in information outflow, this region can also receive input from other neurons.

- The axon terminal contains synapses, specialized structures where neurotransmitter chemicals are released in order to communicate with target neurons.

Although the canonical view of the neuron attributes dedicated functions to its various anatomical components, dendrites and axons often act in ways contrary to their so-called main function.

Axons and dendrites in the central nervous system are typically only about one micrometer thick, while some in the peripheral nervous system are much thicker. The soma is usually about 10–25 micrometers in diameter and often is not much larger than the cell nucleus it contains. The longest axon of a human motoneuron can be over a metre long, reaching from the base of the spine to the toes. Sensory neurons have axons that run from the toes to the dorsal columns, over 1.5 metres in adults. Giraffes have single axons several meters in length running along the entire length of their necks. Much of what is known about axonal function comes from studying the squid giant axon, an ideal experimental preparation because of its relatively immense size (0.5–1 millimetres thick, several centimeters long).

Fully differentiated neurons are permanently amitotic; however, recent research shows that additional neurons throughout the brain can originate from neural stem cells found throughout the brain but in particularly high concentrations in the subventricular zone and subgranular zone through the process of neurogenesis.

Nerve cell bodies stained with basophilic dyes show numerous microscopic clumps of **Nissl substance** (named after German psychiatrist and neuropathologist Franz Nissl, 1860–1919), which consists of rough endoplasmic reticulum and associated ribosomal RNA. The prominence of the Nissl substance can be explained by the fact that nerve cells are metabolically very active, and hence are involved in large amounts of protein synthesis.

The cell body of a neuron is supported by a complex meshwork of structural proteins called **neurofilaments**, which are assembled into larger **neurofibrils**. Some neurons also contain pigment granules, such as **neuromelanin** (a brownish-black pigment, byproduct of synthesis of catecholamines) and **lipofuscin** (yellowish-brown pigment that accumulates with age).

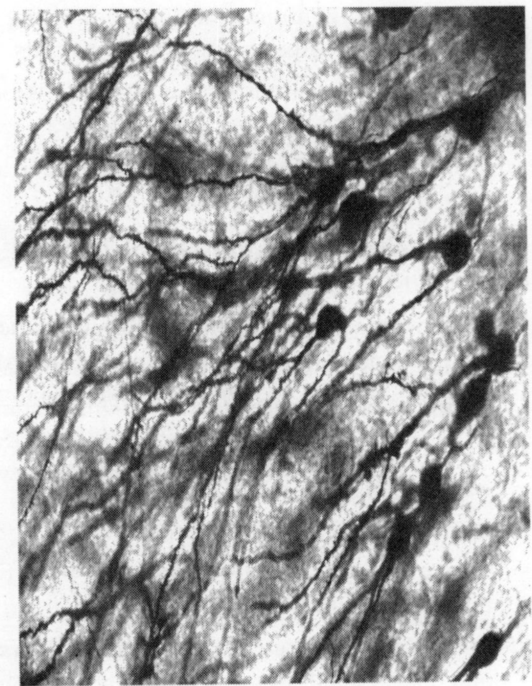

Fig. 9.4 : The Section of Human Brain under the Microscope Showing the Magnified Image of Nerve Cells

There are different internal structural characteristics between axons and dendrites. Typical axons almost never contain ribosomes, except some in the initial segment. Dendrites contain granular endoplasmic reticulum or ribosomes, with diminishing amounts with distance from the cell body.

Neurons exist in a number of different shapes and sizes and can be classified by their morphology and function. The anatomist Camillo Golgi grouped neurons into two types; type I with long axons used to move signals over long distances and type II with short axons, which can often be confused with dendrites. Type I cells can be further divided by where the cell body or soma is located. The basic morphology of type I neurons, represented by spinal motor neurons, consists of a cell body called the soma and a long thin axon which is covered by the myelin sheath. Around the cell body is a branching dendritic tree that receives signals from other neurons. The end of the axon has branching terminals (axon terminal) that release neurotransmitters into a gap called the synaptic cleft between the terminals and the dendrites of the next neuron.

Most neurons can be anatomically characterized as:

- Unipolar or pseudounipolar: dendrite and axon emerging from same process.
- Bipolar: axon and single dendrite on opposite ends of the soma.
- Multipolar: more than two dendrites:
  - Golgi I: neurons with long-projecting axonal processes; examples are pyramidal cells, Purkinje cells, and anterior horn cells.
  - Golgi II: neurons whose axonal process projects locally; the best example is the granule cell.

Furthermore, some unique neuronal types can be identified according to their location in the nervous system and distinct shape. Some examples are:

- Basket cells, neurons with dilated and knotty dendrites in the cerebellum.
- Betz cells, large motor neurons.
- Medium spiny neurons, most neurons in the corpus striatum.
- Purkinje cells, huge neurons in the cerebellum, a type of Golgi I multipolar neuron.
- Pyramidal cells, neurons with triangular soma, a type of Golgi I.
- Renshaw cells, neurons with both ends linked to alpha motor neurons.
- Granule cells, a type of Golgi II neuron.
- anterior horn cells, motoneurons located in the spinal cord.
- Afferent neurons convey information from tissues and organs into the central nervous system and are sometimes also called sensory neurons.
- Efferent neurons transmit signals from the central nervous system to the effector cells and are sometimes called motor neurons.
- Interneurons connect neurons within specific regions of the central nervous system.

Afferent and efferent can also refer generally to neurons which, respectively, bring information to or send information from the brain region.

A neuron affects other neurons by releasing a neurotransmitter that binds to chemical receptors. The effect upon the target neuron is determined not by the source neuron or by the neurotransmitter, but by the type of receptor that is activated. A neurotransmitter can be thought of as a key, and a receptor as a lock: the same type of key can here be used to open many different types of locks. Receptors can be classified broadly as *excitatory* (causing an increase in firing rate), *inhibitory* (causing a decrease in firing rate), or *modulatory* (causing long-lasting effects not directly related to firing rate).

In fact, however, the two most common neurotransmitters in the brain, glutamate and GABA, have actions that are largely consistent. Glutamate acts on several different types of receptors, but most of them have effects that are excitatory. Similarly GABA acts on several different types of receptors, but all of them have effects (in adult animals, at least) that are inhibitory. Because of this consistency, it is common for neuroscientists to simplify the terminology by referring to cells that release glutamate as "excitatory neurons," and cells that release GABA as "inhibitory neurons." Since well over 90% of the neurons in the brain release either glutamate or GABA, these labels encompass the great majority of neurons. There are also other types of neurons that have consistent effects on their targets, for example "excitatory" motor neurons in the spinal cord that release acetylcholine, and "inhibitory" spinal neurons that release glycine.

The distinction between excitatory and inhibitory neurotransmitters is not absolute, however. Rather, it depends on the class of chemical receptors present on the target neuron. In principle, a single neuron, releasing a single neurotransmitter, can have excitatory effects on some targets, inhibitory effects on others, and modulatory effects on other still. For example, photoreceptors in the retina constantly release the neurotransmitter glutamate in the absence of light. So-called OFF bipolar cells are, like most neurons, excited by the released glutamate. However, neighboring target neurons called ON bipolar cells are instead *inhibited* by glutamate, because they lack the typical ionotropic glutamate receptors and instead express a class of inhibitory metabotropic glutamate receptors.[5] When light is present, the photoreceptors cease releasing glutamate, which relieves the ON bipolar cells from inhibition, activating them; this simultaneously removes the excitation from the OFF bipolar cells, silencing them.

Neurons can be classified according to their electrophysiological characteristics:

- **Tonic or regular spiking**. Some neurons are typically constantly (or tonically) active. Example: interneurons in neurostriatum.
- **Phasic or bursting**. Neurons that fire in bursts are called phasic.
- **Fast spiking**. Some neurons are notable for their fast firing rates, for example some types of cortical inhibitory interneurons, cells in globus pallidus, retinal ganglion cells.
- **Thin-spike**. Action potentials of some neurons are more narrow compared to the others. For example, interneurons in the prefrontal cortex are thin-spike neurons.

Neurons differ in the type of neurotransmitter they manufacture. Some examples are

- cholinergic neurons - acetylcholine

  Acetylcholine is released from presynaptic neurons into the synaptic cleft. It acts as a ligang for Nicotinic Acetylcholine receptors, which are ligand gated Na^+ ion channels. Ligand binding opens the channel causing depolarization and increases the probability of an action potential firing, occcuring once the threshold is reached.

- GABAergic neurons-gamma aminobutyric acid

  GABA is one of two neuroinhibitors in the CNS, the other being Glycine. GABA has a homologous function to ACh, gating anion channels that allow Cl- ions to enter the post synaptic neuron. Cl- causes hyperpolarization within the neuron, decreasing the probability of an action potential firing as the voltage becomes more negative (recall that for an action potential to fire, a positive voltage threshold must be reached).

- glutamatergic neurons-glutamate

  Glutamate is one of two primary neuroexcitors, the other being Aspartate (not Aspartame). Glutamate receptors are one of four chategories, three of which are ion channels and one of which is a G-protein coupled receptor (often referred to as GPCR). 1-AMPA and Kainate receptors (really two different receptors) both function as Na+ cation channels mediating fast excitatory synaptic transmission 2-NMDA receptors are another ction channel, instead for Ca^{++}. The function of NMDA receptors is dependant on Glycine binding to a mediator binding spot on the channel pore. NMDA receptors will not function without both ligands present-antagonist drugs for Glycine will cause NMDA receptors to malfunction. 3-Metabotropic receptor, a GPCR that modulates synaptic transmission Glutamate can cause excitotoxicity when blood flow to the brain is interrupted, resulting in brain damage. When blood flow is suppressed, glutamate is released from presynaptic neurons causing NMDA and AMPA receptor activation moreso than would normally be the case outside of stress conditions, leading to elevated Ca^{++} and Na^{+} entering the post synaptic neuron and cell damage.

- dopaminergic neurons-dopamine

  Dopamine is a neurotransmitter that acts on a GPCR. Dopamine is connected to mood and behaviour, and modulates post synaptic neurotransmission. Loss of dopamine neurons has been linked to Parkinson's disease, schizophrenia, and ADD.

- serotonergic neurons-serotonin

  Serotonin, full name 5-Hydroxytriptamine, can act as excitatory or inhibitory. Of the four 5HT receptor classes, 3 are GPCR and 1 is ligand gated cation channel. Serotonin is synthesized from tryptophan by tryptophan hydroxylase, and then further by aromatic acid decarboxylase. A lack of 5HT at postsynaptic neurons has been linked to depression. Drugs that are antagonistic for reabsorption by presynaptic neurons are used for treatment, such as Prozac and Zoloft.

Neurons communicate with one another via synapses, where the axon terminal or *en passant* boutons (terminals located along the length of the axon) of one cell impinges upon another neuron's dendrite, soma or, less commonly, axon. Neurons such as Purkinje cells in the cerebellum can have over 1000 dendritic branches, making connections with tens of thousands of other cells; other neurons, such as the magnocellular neurons of the supraoptic nucleus, have only one or two dendrites, each of which receives thousands of synapses. Synapses can be excitatory or inhibitory and will either increase or decrease activity in the target neuron. Some neurons also communicate via electrical synapses, which are direct, electrically-conductive junctions between cells.

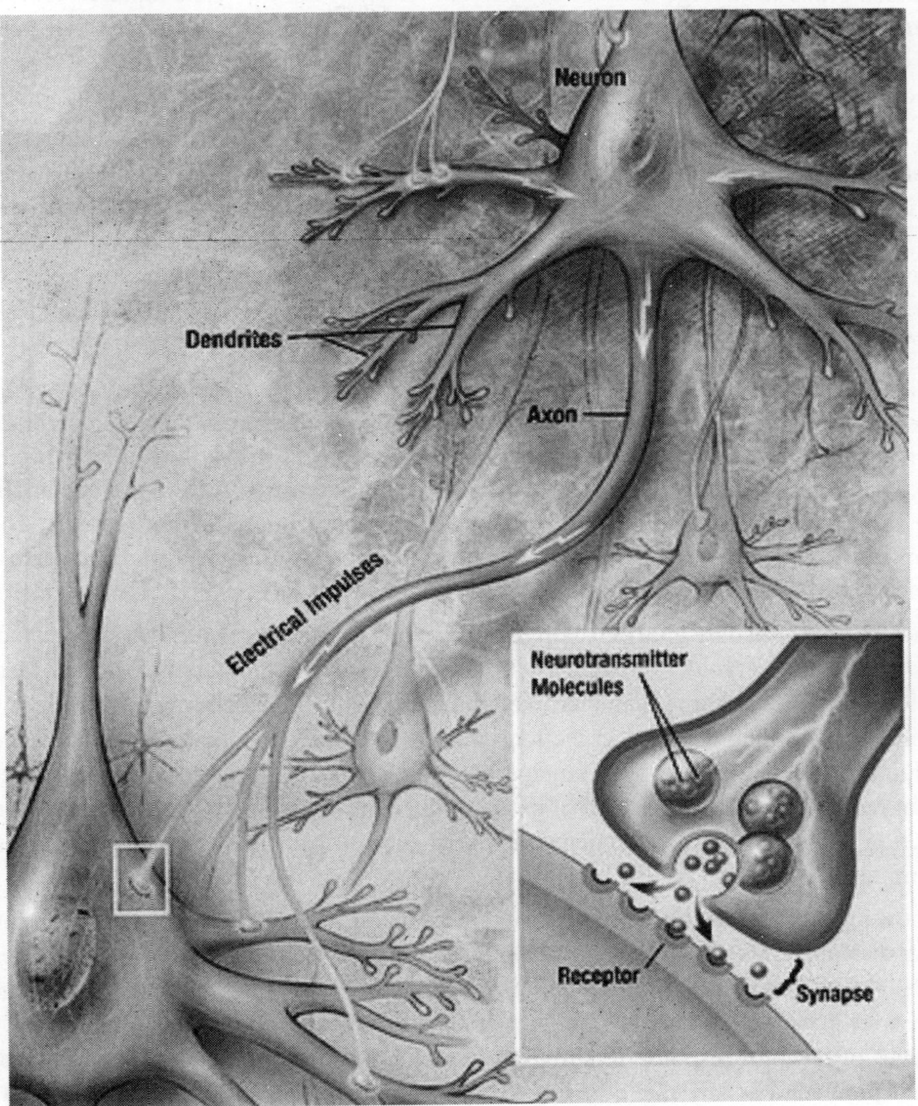

Fig. 9.5 : A Signal Propagating Down an Axon to the Cell Body and Dendrites of the Next Cell.

In a chemical synapse, the process of synaptic transmission is as follows: when an action potential reaches the axon terminal, it opens voltage-gated calcium channels, allowing calcium ions to enter the terminal. Calcium causes synaptic vesicles filled with neurotransmitter molecules to fuse with the membrane, releasing their contents into the synaptic cleft. The neurotransmitters diffuse across the synaptic cleft and activate receptors on the postsynaptic neuron.

The human brain has a huge number of synapses. Each of the 10^{11} (one hundred billion) neurons has on average 7,000 synaptic connections to other neurons. It has been estimated that the brain of a three-year-old child has about 10^{15} synapses (1 quadrillion). This number declines with age, stabilizing by adulthood. Estimates vary for an adult, ranging from 10^{14} to 5×10^{14} synapses (100 to 500 trillion).

In 1937, John Zachary Young suggested that the squid giant axon could be used to study neuronal electrical properties. Being larger than but similar in nature to human neurons, squid cells were easier to study. By inserting electrodes into the giant squid axons, accurate measurements were made of the membrane potential. The cell membrane of the axon and soma contain voltage-gated ion channels which allow the neuron to generate and propagate an electrical signal (an action potential). These signals are generated and propagated by charge-carrying ions including sodium (Na^+), potassium (K^+), chloride (Cl^-), and calcium (Ca^{2+}).

There are several stimuli that can activate a neuron leading to electrical activity, including pressure, stretch, chemical transmitters, and changes of the electric potential across the cell membrane. Stimuli cause specific ion-channels within the cell membrane to open, leading to a flow of ions through the cell membrane, changing the membrane potential. Thin neurons and axons require less metabolic expense to produce and carry action potentials, but thicker axons convey impulses more rapidly. To minimize metabolic expense while maintaining rapid conduction, many neurons have insulating sheaths of myelin around their axons.

The sheaths are formed by glial cells: oligodendrocytes in the central nervous system and Schwann cells in the peripheral nervous system. The sheath enables action potentials to travel faster than in unmyelinated axons of the same diameter, whilst using less energy. The myelin sheath in peripheral nerves normally runs along the axon in sections about 1 mm long, punctuated by unsheathed nodes of Ranvier which contain a high density of voltage-gated ion channels. Multiple sclerosis is a neurological disorder that results from demyelination of axons in the central nervous system. Some neurons do not generate action potentials, but instead generate a graded electrical signal, which in turn causes graded neurotransmitter release.

Such nonspiking neurons tend to be sensory neurons or interneurons, because they cannot carry signals long distances. The conduction of nerve impulses is an example of an all-or-none response. In other words, if a neuron responds at all, then it must respond completely. The greater the intensity of stimulation does not produce a stronger signal but can produce *more* impulses per second. There are different types of receptor response to stimulus, slowly adapting or tonic receptors respond to steady stimulus and produce a steady rate of firing. These tonic receptors most often respond to increased intensity of stimulus by increasing their firing frequency, usually as a power function of stimulus plotted against impulses per second.

This can be likened to an intrinsic property of light where to get greater intensity of a specific frequency (colour) there has to be more photons, as the photons can't become "stronger" for a specific frequency. There are a number of other receptor types that are called quickly adapting or phasic receptors, where firing decreases or stops with steady stimulus; examples include: skin when touched by an object causes the neurons to fire, but if the object maintains even pressure against the skin the neurons stop firing. The neurons of the skin and muscles that are responsive to pressure and vibration have filtering accessory structures that aid their function. The pacinian corpuscle is one such structure; it has concentric layers like an onion which form around the axon terminal. When pressure is applied and the corpuscle is deformed, mechanical stimulus is transferred to the axon, which fires. If the pressure is steady there is no more stimulus, thus typically these neurons respond with a transient depolarization during the initial deformation and again when the pressure is removed which cause the corpuscle to change shape again. Other types of adaptation are important in extending the function of a number of other neurons.

THE NEURO-COMMUNICATIONS

The communication method of the nervous system is termed as neuro-communication. Regarding this some chemical elements are used called neuro-transmitters. The process of communication in the nerve cells is described below:

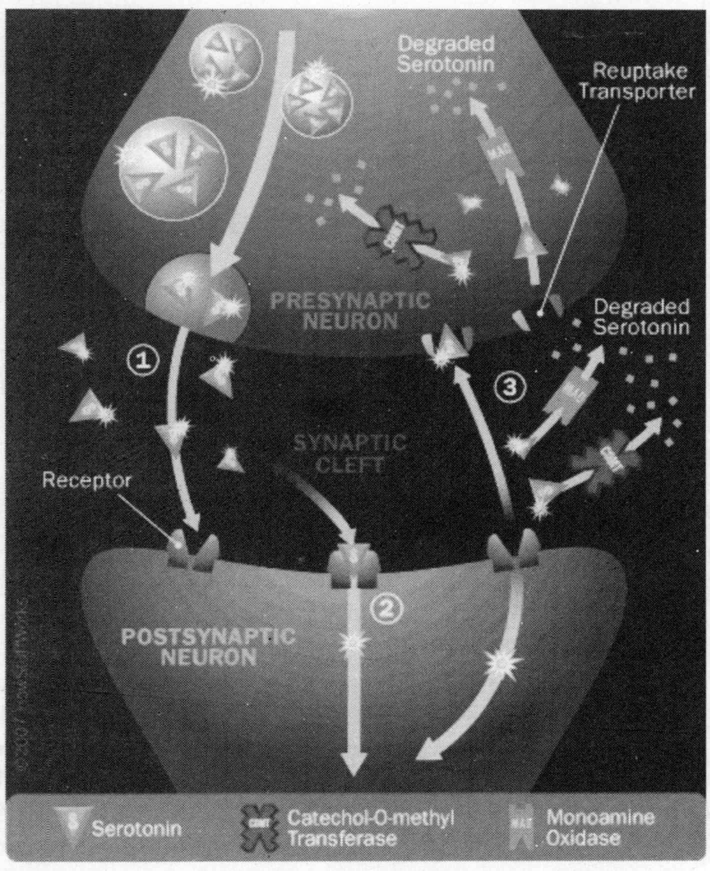

Fig. 9.6 : The Function of Neuro-transmitters

1. The presynaptic cell (sending cell) makes **serotonin** (5-hydroxytryptamine, 5HT) from the amino acid tryptophan and packages it in vesicles in its end terminals.
2. An electrochemical nerve signal passes down the presynaptic cell into its end terminals.
3. The nerve signal stimulates the vesicles containing serotonin to fuse with the cell membrane and dump serotonin into the synaptic cleft.
4. Serotonin passes across the synaptic cleft, binds with special proteins called **receptors** on the membrane of the postsynaptic cell (receiving cell) and sets up a new electrochemical signal in that cell (the signal can stimulate or inhibit the postsynaptic cell). Serotonin fits with its receptor like a lock and key.
5. The remaining serotonin molecules in the cleft and those released by the receptors after use get destroyed by enzymes in the cleft [monoamine oxidase (MAO) and catechol-o-methyl transferase (COMT)]. Some get taken up by specific transporters on the presynaptic cell (**reuptake**). In the presynaptic cell, the absorbed serotonin molecules get destroyed by MAO and COMT. This enables the nerve signal to be turned "off."

A similar process occurs for norepinephrine, which is also implicated in mood, emotions and MDD. Serotonin, norepinephrine and dopamine are chemically similar and belong to a class of neurotransmitters called **monoamine neurotransmitters**. Because these chemicals are structurally similar, they are all recognized by the enzymes MAO and COMT.

THE NEURO-TRANSMITTERS

Neurotransmitter, chemical made by neurons, or nerve cells. Neurons send out neurotransmitters as chemical signals to activate or inhibit the function of neighboring cells.

Within the central nervous system, which consists of the brain and the spinal cord, neurotransmitters pass from neuron to neuron. In the peripheral nervous system, which is made up of the nerves that run from the central nervous system to the rest of the body, the chemical signals pass between a neuron and an adjacent muscle or gland cell.

Types of Neurotransmitters

Nine chemical compounds-belonging to three chemical families-are widely recognized as neurotransmitters. In addition, certain other body chemicals, including adenosine, histamine, enkephalins, endorphins, and epinephrine, have neurotransmitterlike properties. Experts believe that there are many more neurotransmitters as yet undiscovered.

The first of the three families is composed of amines, a group of compounds containing molecules of carbon, hydrogen, and nitrogen. Among the amine neurotransmitters are acetylcholine, norepinephrine, dopamine, and serotonin. Acetylcholine is the most widely used neurotransmitter in the body, and neurons that leave the central nervous system (for example, those running to skeletal muscle) use acetylcholine as their neurotransmitter; neurons that run to the heart, blood vessels, and other organs may use acetylcholine or norepinephrine. Dopamine is involved in the movement of muscles, and it controls the secretion of the pituitary hormone prolactin, which triggers milk production in nursing mothers.

Dopamine also plays a major role in the obtaining and retaining of the male erection. The drug Apomorphine is used as a sexual dysfunction drug by stimulating the production of dopamine and is also a powerful human growth hormone stimulator. The second neurotransmitter family is composed of amino acids, organic compounds containing both an amino group (NH2) and a carboxylic acid group (COOH). Amino acids that serve as neurotransmitters include glycine, glutamic and aspartic acids, and gamma-amino butyric acid (GABA). Glutamic acid and GABA are the most abundant neurotransmitters within the central nervous system, and especially in the cerebral cortex, which is largely responsible for such higher brain functions as thought and interpreting sensations.

The third neurotransmitter family is composed of peptides, which are compounds that contain at least 2, and sometimes as many as 100 amino acids. Peptide neurotransmitters are poorly understood, but scientists know that the peptide neurotransmitter called substance P influences the sensation of pain.

In general, each neuron uses only a single compound as its neurotransmitter. However, some neurons outside the central nervous system are able to release both an amine and a peptide neurotransmitter.

Neurotransmitters are manufactured from precursor compounds like amino acids, glucose, and the dietary amine called choline. Neurons modify the structure of these precursor compounds in a series of reactions with enzymes. Neurotransmitters that come from amino acids include serotonin, which is derived from tryptophan; dopamine and norepinephrine, which are derived from tyrosine; and glycine, which is derived from threonine. Among the neurotransmitters made from glucose are glutamate, aspartate, and GABA. Choline serves as the precursor for acetylcholine.

FUNCTIONS OF NEURO-TRANSMITTERS

Neurotransmitters are released into a microscopic gap, called a synapse, that separates the transmitting neuron from the cell receiving the chemical signal. The cell that generates the signal is called the presynaptic cell, while the receiving cell is termed the postsynaptic cell.

After their release into the synapse, neurotransmitters combine chemically with highly specific protein molecules, termed receptors, that are embedded in the surface membranes of the postsynaptic cell. When this combination occurs, the voltage, or electrical force, of the postsynaptic cell is either increased (excited) or decreased (inhibited).When a neuron is in its resting state, its voltage is about -70 millivolts. An excitatory neurotransmitter alters the membrane of the postsynaptic neuron, making it possible for ions (electrically charged molecules) to move back and forth across the neuron's membranes. This flow of ions makes the neuron's voltage rise toward zero. If enough excitatory receptors have been activated, the postsynaptic neuron responds by firing, generating a nerve impulse that causes its own neurotransmitter to be released into the next synapse. An inhibitory neurotransmitter causes different ions to pass back and forth across the postsynaptic neuron's membrane, lowering the nerve cell's voltage to -80 or -90 millivolts. The drop in voltage makes it less likely that the postsynaptic cell will fire.

If the postsynaptic cell is a muscle cell rather than a neuron, an excitatory neurotransmitter will cause the muscle to contract. If the postsynaptic cell is a gland cell, an excitatory neurotransmitter will cause the cell to secrete its contents.While most neurotransmitters interact with their receptors to create new electrical nerve impulses that energize or inhibit the adjoining cell, some neurotransmitter interactions do not generate or suppress nerve impulses. Instead, they interact with a second type of receptor that changes the internal chemistry of the postsynaptic cell by either causing or blocking the formation of chemicals called second messenger molecules. These second messengers regulate the postsynaptic cell's biochemical processes and enable it to conduct the maintenance necessary to continue synthesizing neurotransmitters and conducting nerve impulses. Examples of second messengers, which are formed and entirely contained within the postsynaptic cell, include cyclic adenosine monophosphate, diacylglycerol, and inositol phosphates.Once neurotransmitters have been secreted into synapses and have passed on their chemical signals, the presynaptic neuron clears the synapse of neurotransmitter molecules. For example, acetylcholine is broken down by the enzyme acetylcholinesterase into choline and acetate. Neurotransmitters like dopamine, serotonin, and GABA are removed by a physical process called reuptake. In reuptake, a protein in the presynaptic membrane acts as a sort of sponge, causing the neurotransmitters to reenter the presynaptic neuron, where they can be broken down by enzymes or repackaged for reuse.

ROLE OF NEURO-TRANSMITTERS IN DISEASES

Neurotransmitters are known to be involved in a number of disorders, including Alzheimer's disease. Victims of Alzheimer's disease suffer from loss of intellectual capacity, disintegration of personality, mental confusion, hallucinations, and aggressive-even violent-behavior. These symptoms are the result of progressive degeneration in many types of neurons in the brain. Forgetfulness, one of the earliest symptoms of Alzheimer's disease, is partly caused by the destruction of neurons that normally release the neurotransmitter acetylcholine. Medications that increase brain levels of acetylcholine have helped restore short-term memory and reduce mood swings in some Alzheimer's patients.

Neurotransmitters also play a role in Parkinson's disease, which slowly attacks the nervous system, causing symptoms that worsen over time. Fatigue, mental confusion, a masklike facial expression, stooping posture, shuffling gait, and problems with eating and speaking are among the difficulties suffered by Parkinson's victims. These symptoms have been partly linked to the deterioration and eventual death of neurons that run from the base of the brain to the basal ganglia, a collection of nerve cells that manufacture the neurotransmitter dopamine. The reasons why such neurons die are yet to be understood, but the related symptoms can be alleviated. L-dopa, or levodopa, widely used to treat Parkinson's disease, acts as a supplementary precursor for dopamine. It causes the surviving neurons in the basal ganglia to increase their production of dopamine, thereby compensating to some extent for the disabled neurons.Many other effective drugs have been shown to act by influencing neurotransmitter behavior. Some drugs work by interfering with the interactions between neurotransmitters and intestinal receptors. For example, belladonna decreases intestinal cramps in such disorders as irritable bowel syndrome by blocking acetylcholine from combining with receptors. This process reduces nerve signals to the bowel wall, which prevents painful spasms.Other drugs block the reuptake process. One well-known example is the drug fluoxetine (Prozac), which blocks the reuptake of serotonin. Serotonin then remains in the synapse for a longer time, and its ability to act as a signal is prolonged, which contributes to the relief of depression and the control of obsessive-compulsive behaviors.

SYNAPSE: MEETING POINT BETWEEN NEURONS

Since neurons form a network of electrical activities, they somehow have to be interconnected. When a nerve signal, or impulse, reaches the ends of its axon, it has traveled as an action potential, or a pulse of electricity. However, there is no cellular continuity between one neuron and the next; there is a gap called synapse. The membranes of the sending and receiving cells are separated from each other by the fluid-filled synaptic gap. The signal cannot leap across the gap electrically. So, special chemicals called neurotransmitters have this role. They are released by presynaptic sending membrane and seep across the gap tp receptors on the receiving neuron's postsynaptic membrane. The binding of neurotranmitters to these receptors has the effect to allowing ions (charged particles) to pass in and out of the receiving cell, as we have seen in the paper about neural conduction.

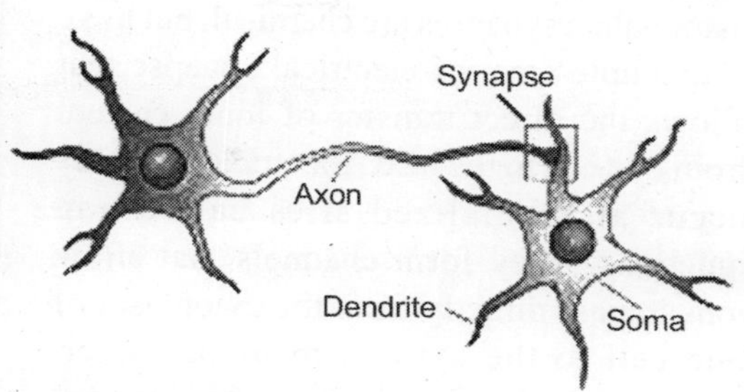

Fig. 9.7 : A Detal Synapse

Synapse

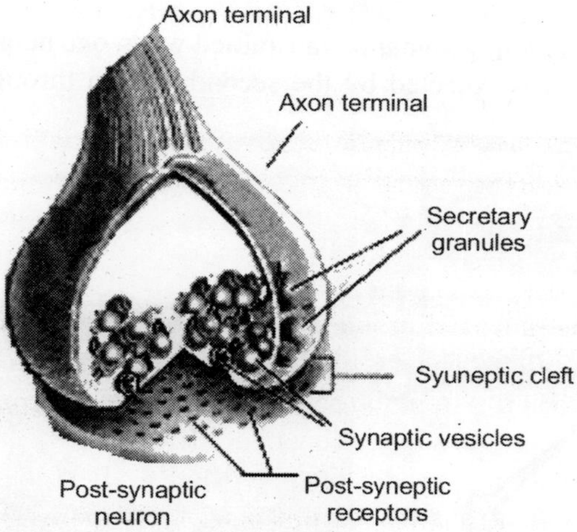

Axon terminal
Axon terminal
Secretary granules
Syuneptic cleft
Synaptic vesicles
Post-synaptic neuron
Post-syneptic receptors

Fig. 9.8 : The detail communication action between neurons through synapse

The normal direction of information flow is from the axon terminal to the target neuron; thus, the axon terminal is said to be presynaptic (carries information towards a synapse) and the target neuron is said to be postsynaptic (carries information from a synapse).

TYPES OF SYNAPSES

The typical and most frequent type of synapse is the one in which the axon of one neuron connects to a second neuron by usually making contact with one of its dendrites or with the cell body. There are two ways in which this can happen: the electrical and the chemical synapses. Most mammalian synapses are chemical, but there is a simple form of electrical synapse that allows the direct transfer of ionic current from one cell to the next. Electrical synapses occur at specialized sites called gap junctions. They form channels that allow ions to pass directly from the cytoplasm of one cell to the cytoplasm of the other. Transmission at electrical synapses is very fast, thus, an action potential in the presynaptic neuron can produce almost instantaneously, an action potential in the postsynaptic neuron. Electrical synapses in mammalian CNS, are mainly found in specialized locations where normal function requires that the activity of neighboing neurons be highly synchronized. Although gap junctions are relatively rare between adult mammalian neurons, they are very common in a large variety of non-neural cells, including smooth cardiac muscle cells, epithelial cells, some glandular cells, glia, etc. They are also common in many invertebrates.

> **Synapse.** As an electrical impulse travels down the "tail" of the cell, called the axon and arrives at its terminal, it triggers vesicles containing a neuro-transmitter to move toward the terminal membrane. The vesicles fuse with the terminal membrane to release their contents. Once inside the synaptic cleft (the space between the 2 neurons) the neuro-transmitter can bind to receptors (specific proteins) on the membrane of a neighbouring neuron.

THE CHEMICAL SYNAPSE

In this type of synapse the incoming signal is transmitted when one neuron releases a neurotransmitter into the synaptic cleft which is detected by the second neuron through the activation of receptors

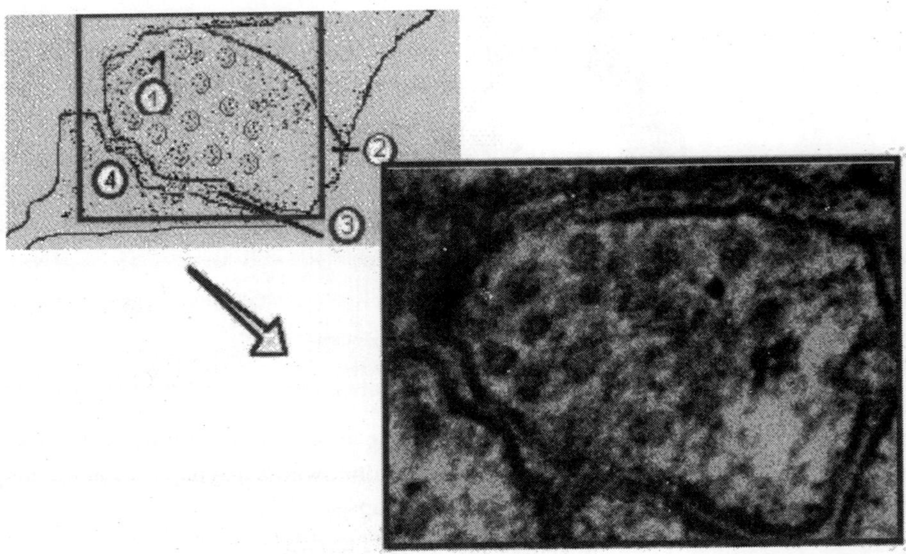

Fig. 9.9 : The Diagram and micrography of a synapse of the neuromuscular junction. 1-Synaptic vesicles; 2- Presynaptic neuron (terminal axon); 3- Synaptic cleft; 4- Postsynaptic neuron

placed opposite to the release site. Neurotransmitters are chemicals made by neurons and used by them to transmit signals to the other neurons or non-neuronal cells (e.g., skeletal muscle; myocardium, pineal glandular cells) that they innervate. The chemical binding of the neurotransmitter to the receptors causes a series of physiological changes in the second neuron which constitutes the signal. Usually the release from the first neuron (called presynaptic) is caused by a series of intracellular events evoked by a depolarization of its membrane, and almost invariably when an action potential takes place.

THE TRIGGER OF THE NEURO-TRANSMITTER

Some mechanism must exist whereby the action potential causes the transmitter stored in synaptic vesicles to be expelled into the cleft. The action potential stimulates the influx of Ca^{2+}, which causes synaptic vesicles to attach to the release sites, fuse with the plasma membrane and expel their supply of transmitter. The transmitter diffuses to the target cell, where it binds to a receptor protein on the external surface of the cell membrane. After a brief period the transmitter dissociates from the receptor and the response is terminated. In order to prevent the transmitter from rebinding to the receptor and repeating the cycle, the transmitter is either destroyed by degradative action of an enzime or it is taken up, usually into the presynaptic ending. Each neuron can produce only one kind of transmitter.

CATEGORIES OF CHEMICAL SYNAPSES

There are two types of chemical synapses, according to the effect it causes on the postsynaptic element:

An impulse arriving in the presynaptic terminal causes the release of neurotransmitter (Fig. 9.10A). The molecules bind to transmitter-gated ion channels in the postsynaptic membrane. If Na+ enters the postsynaptic cell through the open channels, the membrane will become depolarized (Fig. 9.10B). The molecules bind to transmitter-gated ion channels in the postsynaptic membrane. If Cl- enters the possynaptic cell through the open channels, the membrane will become hyperpolarized. The resulting change in membrane potential, as recorded by a microeletrode in the cell is saw in figure below (Generation of an EPSP and IPSP).

Excitatory synapses: They cause an excitatory electrical change in the postsynaptic potential (EPSP). This happens when the net effect of transmitter release is to depolarize the membrane, bringing it nearer to the electrical threshold for firing an action potential. This effect typically is mediated by the opening of membrane channels (kind of pores which traverses the cell membranes) for sodium and calcium ions.

Inhibitory synapses: They cause an inhibitory postsynaptic potential (IPSP), because the net effect of transmitter release is now to hyperpolarize the membrane, making it more difficult to reach the electrical threshold potential. This type of inhibitory synapse works by opening different ion channels in the membrane: typical chloride (Cl-) or potassium (K+) channels.

In Fig. 9.11, the recording of the transmembrane electrical potential in function of time shows that there is a gradual upward deflection of the trace when an excitatory synapse is activated (EPSP). The flux of ions causes a depolarization, i.e, the membrane becomes less polarized. Remember that usually the exterior face of the membrane is negative in relation of the interior, and that the resting potential of the postsynaptic membrane is around -70 millivolts. Any depolarization decreases this value, making it less negative, therefore causing an upward deflection (closer to the zero level).

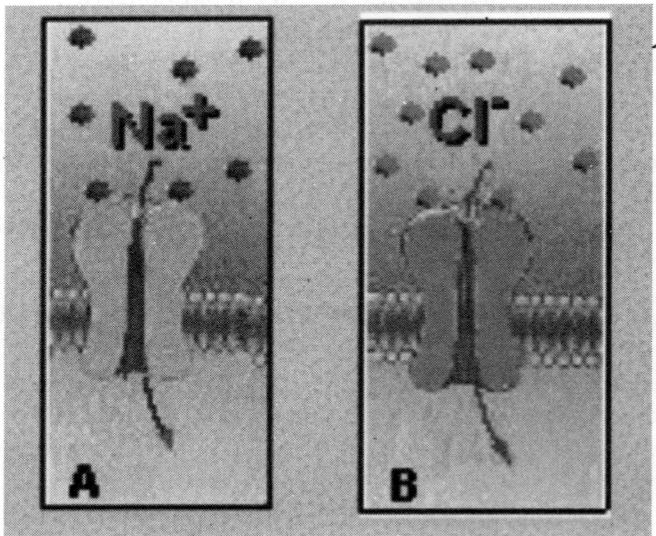

Fig. 9.10 : The Action of Na^+ Cl^- in the Nerve Cell

The recording of the membrane potential for an inhibitory postsynaptic potential (IPSP: in green) shows a hyperpolarization, i.e., a downward deflection in the tracing, because it becomes more negative than the resting potential.

A single neural cell usually has hundreds or thousands of chemical excitatory and inhibitory synapses arriving at its dendrites or cell body. EPSPs and IPSPs are summed up algebraically, so that the resulting curve may lean toward a net depolarization or hyperpolarization. If net depolarization reaches the threshold level, the postsynaptic cell fires action potentials (Fig. 9.11).

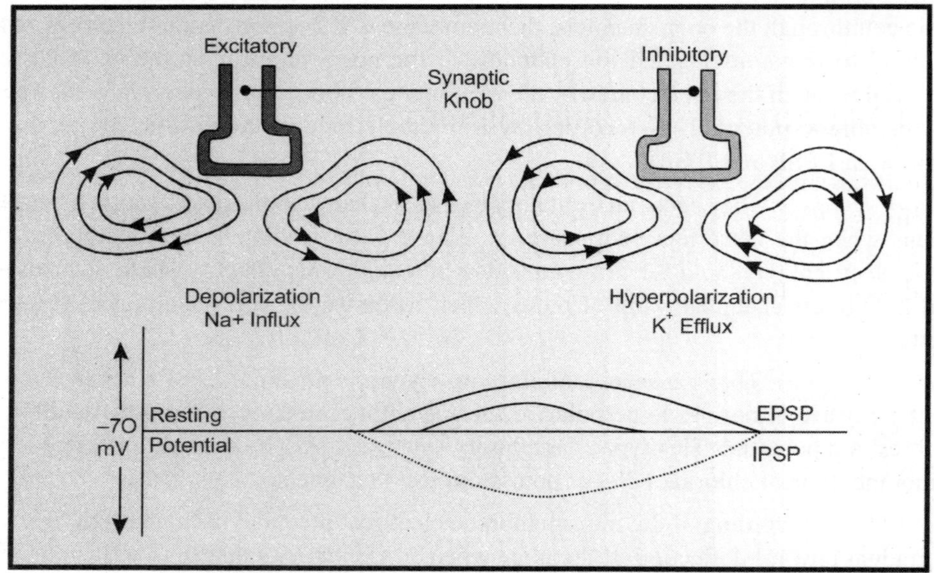

Fig. 9.11 : The Generation of an EPSP and IPSP

SYNAPSES IN THE CENTRAL NERVOUS SYSTEM

Different types of synapse may be distinguished by which part of the neuron is postsynaptic to the axon terminal. If the postsynaptic membrane is on a dendrite, the synapse is said to be axo-dendritic. If the postsynaptic membrane is on the cell body, the synapse is said to be axosomatic. In some cases the postsynaptic membrane is on another axon, and these synapses are called axoaxonic. In certain specialized neurons, dendrites actually form synapses with one another; these are called dendrodendritic synapses.

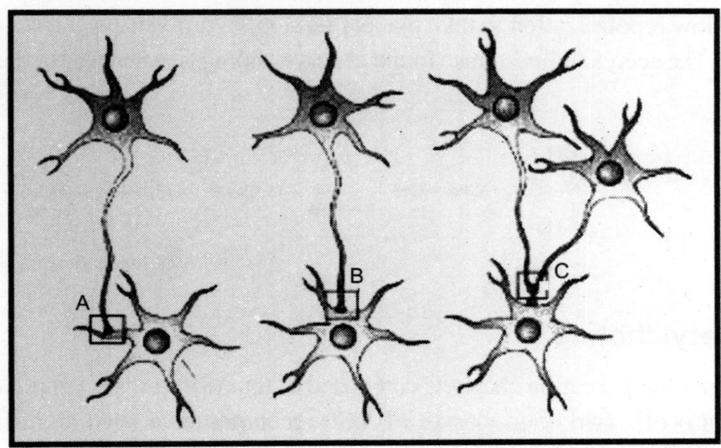

Fig. 9.12 : The Synaptic Arrangements in the CNS. A. An axodendritic synapse, B. An axosomatic synapse. C. An axoxonic synapse

A NEUROTRANSMITTERS: MESSENGERS OF THE BRAIN

Chemically, neurotransmitters are relatively small and simple molecules. Different types of cells secrete different neurotransmitters. Each brain chemical works in widely spread but fairly specific brain locations and may have a different effect according to where it is activated. Some 60 neurotransmitters have been identified, and they fall mainly into one of four classes:

(1) cholines; of which acetylcholine is the most important one;

(2) biogenic amines: serotonin, histamine, and the catecholamines - dopamine and norepinephrine

(3) amino acids - glutamate and aspartate are well known excitatory transmitters, while gamma-aminobutyric acid (GABA), glycine and taurine are inhibitory neurotransmitters.

(4) neuropeptides,- these are formed by longer chains of amino acids (like a small protein molecule). Over 50 of them are known to occur in the brain, and many of them have been implied in the modulation or transmission of neural information.

IMPORTANT NEUROTRANSMITTERS AND THEIR FUNCTION

Acetylcholine

Acetylcholine (ACh) is a simple molecule synthesized from choline and acetyl-CoA through the action of choline acetyltransferase. Neurons that synthesize and release ACh are termed **cholinergic neurons.** When an action potential reaches the terminal button of a presynaptic neuron a voltage-gated calcium channel is opened. The influx of calcium ions, Ca^{2+}, stimulates the exocytosis of presynaptic vesicles containing ACh, which is thereby released into the synaptic cleft. Once released, ACh must be removed rapidly in order to allow repolarization to take place; this step, hydrolysis, is carried out by the enzyme, acetylcholinesterase. The acetylcholinesterase found at nerve endings is anchored to the plasma membrane through a glycolipid.

$$\underset{\text{AcetylCoA}}{CoA\text{—}S\text{—}C(=O)\text{—}CH_3} + \underset{\text{Choline}}{H_3C\text{—}\overset{+}{N}(CH_3)_2\text{—}CH_2\text{—}CH_2\text{—}OH} \longrightarrow \underset{\text{Acetylcholine (Ach)}}{H_3C\text{—}N(CH_3)_2\text{—}CH_2\text{—}CH_2\text{—}O\text{—}C(=O)\text{—}CH_3} + SH\text{—}CoA$$

Synthesis of Acetylcholine

ACh receptors are ligand-gated cation channels composed of four different polypeptide subunits arranged in the form $[(\alpha 2)(\beta)(\gamma)(\delta)]$. Two main classes of ACh receptors have been identified on the basis of their responsiveness to the toadstool alkaloid, muscarine, and to nicotine, respectively: the **muscarinic receptors** and the **nicotinic receptors**. Both receptor classes are abundant in the human brain. Nicotinic receptors are further divided into those found at neuromuscular junctions and those found at neuronal synapses. The activation of ACh receptors by the binding of ACh leads to an influx of Na^+ into the cell and an efflux of K^+, resulting in a depolarization of the postsynaptic neuron and the initiation of a new action potential.

Cholinergic Agonists and Antagonists

Numerous compounds have been identified that act as either agonists or antagonists of cholinergic neurons. The principal action of cholinergic agonists is the excitation or inhibition of autonomic effector cells that are innervated by postganglionic parasympathetic neurons and as such are referred to as **parasympathomimetic** agents. The cholinergic agonists include choline esters (such as ACh itself) as well as protein- or alkaloid-based compounds. Several naturally occurring compounds have been shown to affect cholinergic neurons, either positively or negatively.

The responses of cholinergic neurons can also be enhanced by administration of cholinesterase (ChE) inhibitors. ChE inhibitors have been used as components of nerve gases but also have significant medical application in the treatment of disorders such as glaucoma and myasthenia gravis as well as in terminating the effects of neuromuscular blocking agents such as atropine.

Natural Cholinergic Agonist and Antagonists

	Source of Compound	Mode of Action
Agonists		
Nicotine	alkaloid prevalent in the tobacco plant	activates nicotinic class of ACh receptors, locks the channel open
Muscarine	alkaloid produced by *Amanita muscaria* mushrooms	activates muscarinic class of ACh receptors
α-Latrotoxin	protein produced by the black widow spider	induces massive ACh release, possibly by acting as a Ca^{2+} ionophore
Antagonists		
atropine (and related compound Scopolamine)	alkaloid produced by the deadly nightshade, *Atropa belladonna*	blocks ACh actions only at muscarinic receptors
Botulinus toxin	eight proteins produced by *Clostridium botulinum*	inhibits the release of ACh
α-Bungarotoxin	protein produced by *Bungarus* genus of snakes	prevents ACh receptor channel opening
d-Tubocurarine	active ingredient of curare	prevents ACh receptor channel opening at motor end-plate

Catecholamines

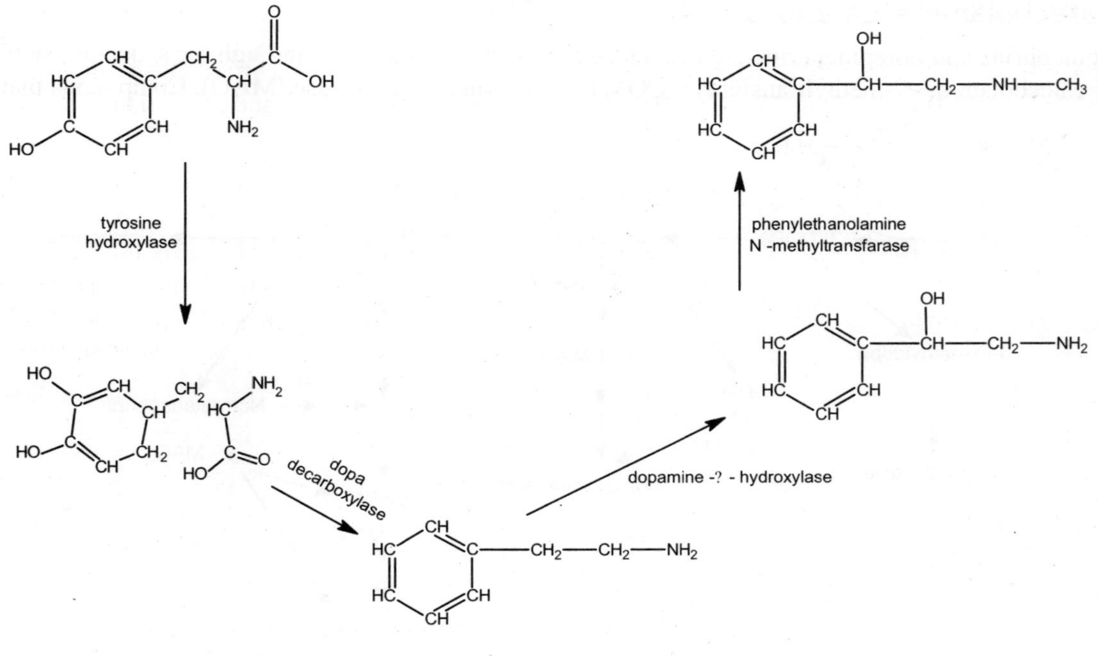

tyrosine hydroxylase

dopa decarboxylase

dopamine -? - hydroxylase

phenylethanolamine N -methyltransfarase

The principal catecholamines are **norepinephrine**, **epinephrine** and **dopamine**. These compounds are formed from phenylalanine and tyrosine. Tyrosine is produced in the liver from phenylalanine through the action of phenylalanine hydroxylase. The tyrosine is then transported to catecholamine-secreting neurons where a series of reactions convert it to dopamine, to norepinephrine and finally to epinephrine

Synthesis of the Catecholamines from Tyrosine

Catecholamines exhibit peripheral nervous system excitatory and inhibitory effects as well as actions in the CNS such as respiratory stimulation and an increase in psychomotor activity. The excitatory effects are exerted upon smooth muscle cells of the vessels that supply blood to the skin and mucous membranes. Cardiac function is also subject to excitatory effects, which lead to an increase in heart rate and in the force of contraction. Inhibitory effects, by contrast, are exerted upon smooth muscle cells in the wall of the gut, the bronchial tree of the lungs, and the vessels that supply blood to skeletal muscle.

In addition to their effects as neurotransmitters, norepinephrine and epinephrine can influence the rate of metabolism. This influence works both by modulating endocrine function such as insulin secretion and by increasing the rate of glycogenolysis and fatty acid mobilization.

The catecholamines bind to two different classes of receptors termed the α- and β-adrenergic receptors. The catecholamines therefore are also known as **adrenergic neurotransmitters**; neurons that secrete them are **adrenergic neurons**. Norepinephrine-secreting neurons are **noradrenergic**. The adrenergic receptors are classical serpentine receptors that couple to intracellular G-proteins. Some of the norepinephrine released from presynaptic noradrenergic neurons recycled in the presynaptic neuron by a reuptake mechanism.

Catecholamine Catabolism

Epinephrine and norepinephrine are catabolized to inactive compounds through the sequential actions of catecholamine-*O*-methyltransferase (COMT) and monoamine oxidase (MAO). Compounds that

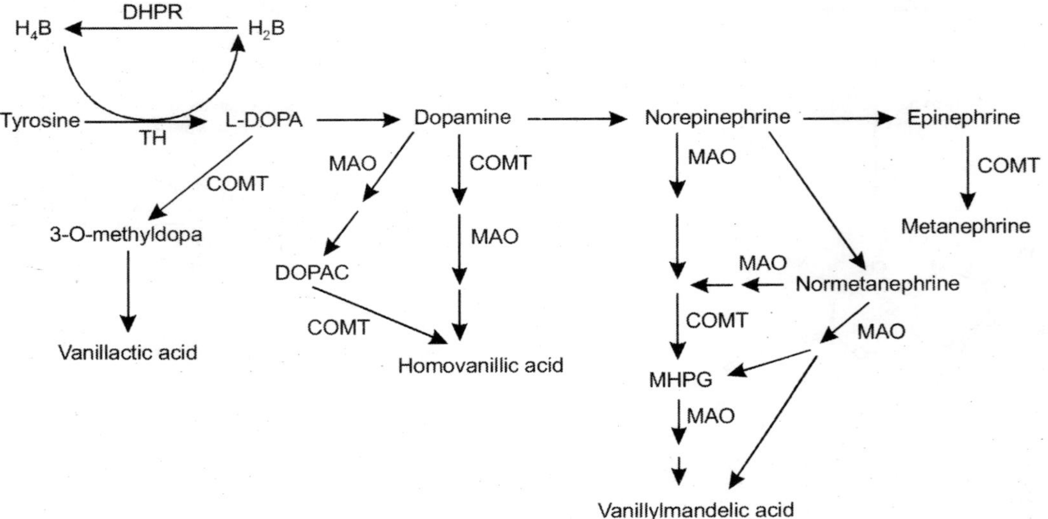

inhibit the action of MAO have been shown to have beneficial effects in the treatment of clinical depression, even when tricyclic antidepressants are ineffective. The utility of MAO inhibitors was discovered serendipitously when patients treated for tuberculosis with isoniazid showed signs of an improvement in mood; isoniazid was subsequently found to work by inhibiting MAO.

Serotonin

Serotonin (5-hydroxytryptamine, 5HT) is formed by the hydroxylation and decarboxylation of tryptophan

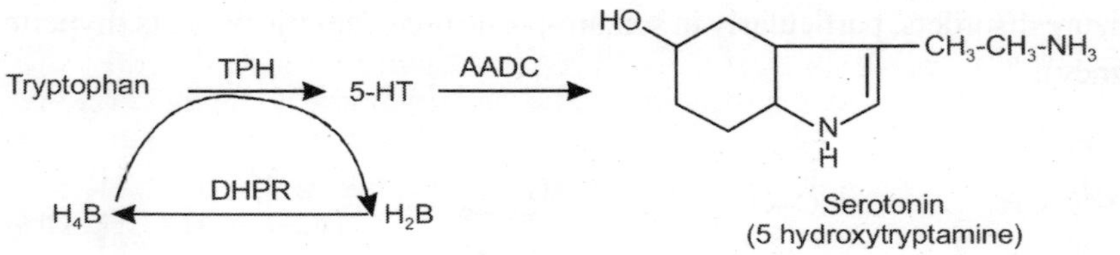

Tryptophan —TPH→ 5-HT —AADC→ HO– … –CH_3-CH_3-NH_3

H_4B ←DHPR— H_2B

Serotonin
(5 hydroxytryptamine)

The greatest concentration of 5HT (90%) is found in the enterochromaffin cells of the gastrointestinal tract. Most of the remainder of the body's 5HT is found in platelets and the CNS. The effects of 5HT are felt most prominently in the cardiovascular system, with additional effects in the respiratory system and the intestines. Vasoconstriction is a classic response to the administration of 5HT.

Neurons that secrete 5HT are termed **serotonergic**. Following the release of 5HT, a portion is taken back up by the presynaptic serotonergic neuron in a manner similar to that of the reuptake of norepinephrine.

The function of serotonin is exerted upon its interaction with specific receptors. Several serotonin receptors have been cloned and are identified as $5HT_1$, $5HT_2$, $5HT_3$, $5HT_4$, $5HT_5$, $5HT_6$, and $5HT_7$. Within the $5HT_1$ group there are subtypes $5HT_{1A}$, $5HT_{1B}$, $5HT_{1D}$, $5HT_{1E}$, and $5HT_{1F}$. There are three $5HT_2$ subtypes, $5HT_{2A}$, $5HT_{2B}$, and $5HT_{2C}$ as well as two $5HT_5$ subtypes, $5HT_{5a}$ and $5HT_{5B}$. Most of these receptors are coupled to G-proteins that affect the activities of either adenylate cyclase or phospholipase Cγ. The $5HT_3$ class of receptors are ion channels.

Some serotonin receptors are presynaptic and others postsynaptic. The $5HT_{2A}$ receptors mediate platelet aggregation and smooth muscle contraction. The $5HT_{2C}$ receptors are suspected in control of food intake as mice lacking this gene become obese from increased food intake and are also subject to fatal seizures. The $5HT_3$ receptors are present in the gastrointestinal tract and are related to vomiting. Also present in the gastrointestinal tract are $5HT_4$ receptors where they function in secretion and peristalsis. The $5HT_6$ and $5HT_7$ receptors are distributed throughout the limbic system of the brain and the $5HT_6$ receptors have high affinity for antidepressant drugs.

GABA

Several amino acids have distinct excitatory or inhibitory effects upon the nervous system. The amino acid derivative, γ-aminobutyrate, also called 4-aminobutyrate, (GABA) is a well-known inhibitor of presynaptic transmission in the CNS, and also in the retina. Neurons that secrete GABA are termed **GABAergic**.

The formation of GABA occurs by the decarboxylation of glutamate catalyzed by glutamate decarboxylase (GAD). GAD is present in many nerve endings of the brain as well as in the β-cells of the pancreas. The activity of GAD requires pyridoxal phosphate (PLP) as a cofactor. PLP is generated from the B_6 vitamins (pyridoxine, pyridoxal, and pyridoxamine) through the action of pyridoxal kinase. Pyridoxal kinase itself requires zinc for activation. A deficiency in zinc or defects in pyridoxal kinase can lead to seizure disorders, particularly in seizure-prone preeclamptic patients (hypertensive condition in late pregnancy).

$$HO{-}\overset{O}{\overset{\|}{C}}{-}CH_2{-}\underset{NH_2}{\underset{|}{C}}H_2{-}CH{-}\overset{O}{\overset{\|}{C}}{-}OH \xrightarrow[Co_2]{GAD} HO{-}\overset{O}{\overset{\|}{C}}{-}CH_2{-}CH_2{-}CH_2{-}NH_2$$

Glutamate GABA

GABA Synthesis

GABA exerts its effects by binding to two distinct receptors, GABA-A and GABA-B. The GABA-A receptors form a Cl^- channel. The binding of GABA to GABA-A receptors increases the Cl^- conductance of presynaptic neurons. The anxiolytic drugs of the benzodiazepine family exert their soothing effects by potentiating the responses of GABA-A receptors to GABA binding. The GABA-B receptors are coupled to an intracellular G-protein and act by increasing conductance of an associated K^+ channel.

Pathway for serotonin synthesis from tryptophan. Abbreviations: THP = tryptophan hydroxylase, DHPR = dihydropteridine reductase, H_2B = dihydrobiopterin, H_4B = tetrahyrobiopterin, 5-HT = 5-hydroxytryptophan, AADC = aromatic L-amino acid decarboxylase.

THE CERIBRO SPINAL FLUID

Cerebrospinal fluid (CSF), *Liquor cerebrospinalis*, is a clear bodily fluid that occupies the subarachnoid space and the ventricular system around and inside the brain. In essence, the brain "floats" in it.

The CSF occupies the space between the arachnoid mater (the middle layer of the brain cover, meninges), and the pia mater (the layer of the meninges closest to the brain). It constitutes the content of all intra-cerebral (inside the brain, cerebrum) ventricles, cisterns, and sulci (singular sulcus), as well as the central canal of the spinal cord. It acts as a "cushion" or buffer for the cortex, providing a basic mechanical and immunological protection to the brain inside the skull. It is produced in the choroid plexus.

Circulation

CSF is produced in the brain by modified ependymal cells in the choroid plexus (approx. 50-70%), and the remainder is formed around blood vessels and along ventricular walls. It circulates from the lateral ventricles to the foramen of Monro, third ventricle, aqueduct of Sylvius, fourth ventricle, foramina of Magendie and Luschka; subarachnoid space over brain and spinal cord; reabsorption into venous sinus blood via arachnoid granulations.

It had been thought that CSF returns to the vascular system by entering the dural venous sinuses via the arachnoid granulations or villi. However, some have suggested that CSF flow along the cranial nerves and spinal nerve roots allow it into the lymphatic channels; this flow may play a substantial role in CSF reabsorbtion, in particular in the neonate, in which arachnoid granulations are sparsely distributed. The flow of CSF to the nasal submucosal lymphatic channels through the cribiform plaque seems to be specially important.

Amount and Constitution

Reference Ranges in CSF

Substance	Lower limit	Upper limit	Unit
Glucose	50	80	ng/dL
Protein	15	40 - 45	mg/dL
RBCs	n/a	0 / negative	cells/μL
WBCs	0[3]	3	cells/μL

The CSF is produced at a rate of 500 ml/day. Since the brain can contain only from 135 to 150 ml, large amounts are drained primarily into the blood through arachnoid granulations in the superior sagittal sinus. Thus the CSF turns over about 3.7 times a day. This continuous flow into the venous system dilutes the concentration of larger, lipoinsoluble molecules penetrating the brain and CSF.

The CSF contains approximately 0.3% plasma proteins, or approximately 15 to 40 mg/dL, depending on sampling site. CSF pressure ranges from 80 to 100 mmH2O (780-980 Pa or 4.4-7.3 mmHg) in newborns, and < 200 mmH20 (1.94 kPa) in normal children and adults, with most variations due to coughing or internal compression of jugular veins in the neck.

There are quantitative differences in the distributions of a number of proteins in the CSF. In general, globular proteins and albumin are in lower concentration in ventricular CSF compared to lumbar or cisternal fluid.

Functions

CSF serves four primary purposes:

1. *Buoyancy*—The actual mass of the human brain is about 1400 grams; however the net weight of the brain suspended in the CSF is equivalent to a mass of 25 grams. The brain therefore exists in neutral buoyancy, which allows the brain to maintain its density without being impaired by its own weight, which would cut off blood supply and kill neurons in the lower sections without CSF.
2. *Protection*—CSF protects the brain tissue from injury when jolted or hit. In certain situations such as auto accidents or sports injuries, the CSF cannot protect the brain from forced contact with the skull case, causing hemorrhaging, brain damage, and sometimes death.
3. *Chemical Stability*—CSF flows throughout the inner ventricular system in the brain and is absorbed back into the bloodstream, rinsing the metabolic waste from the central nervous system through the blood-brain barrier. This allows for homeostatic regulation of the distribution of neuroendocrine factors, to which slight changes can cause problems or damage to the nervous system. For example, high glycine concentration disrupts temperature and blood pressure control, and high CSF pH causes dizziness and syncope.
4. *Prevention of brain ischemia*—The prevention of brain ischemia is made by decreasing the amount of CSF in the limited space inside the skull. This decreases total intracranial pressure and facilitates blood perfusion

The cerebrospinal fluid **(CSF)** is produced from arterial blood by the choroid plexuses of the lateral and fourth ventricles by a combined process of diffusion, pinocytosis and active transfer. A small amount is also produced by ependymal cells. The **choroid plexus** consists of tufts of capillaries with thin fenestrated endothelial cells. These are covered by modified ependymal cells with bulbous microvilli. The total volume of CSF in the adult is about 140 ml. The volume of the ventricles is about 25 ml. CSF is produced at a rate of 0.2 - 0.7 ml per minute or 600-700 ml per day. The circulation of CSF is aided by the pulsations of the choroid plexus and by the motion of the cilia of ependymal cells. CSF is absorbed across the arachnoid villi into the venous circulation. The arachnoid villi act as one-way valves between the subarachnoid space and the dural sinuses. The rate of absorption correlates with the CSF pressure. CSF acts as a cushion that protects the brain from shocks and supports the venous sinuses. It also plays an important role in the homeostasis and metabolism of the central nervous system.

CSF from the lumbar region contains **15 to 45 mg/dl protein** (lower in childen) and **50-80 mg/dl glucose** (two-thirds of blood glucose). Protein concentration in cisternal and ventricular CSF is lower. Normal CSF contains **0-5 mononuclear cells**. The CSF pressure, measured at lumbar puncture (LP), is 100-180 mm of H2O (8-15 mm Hg) with the patient lying on the side and 200-300 mm with the patient sitting up.

Analysis of CSF provides important information in the diagnosis of CNS disease. Commonly, CSF fluid is analyzed to detect infections of the CNS, to confirm a subarachnoid hemorrhage and to confirm the diagnosis of chronic inflammatory diseases of the CNS, such as multiple sclerosis.

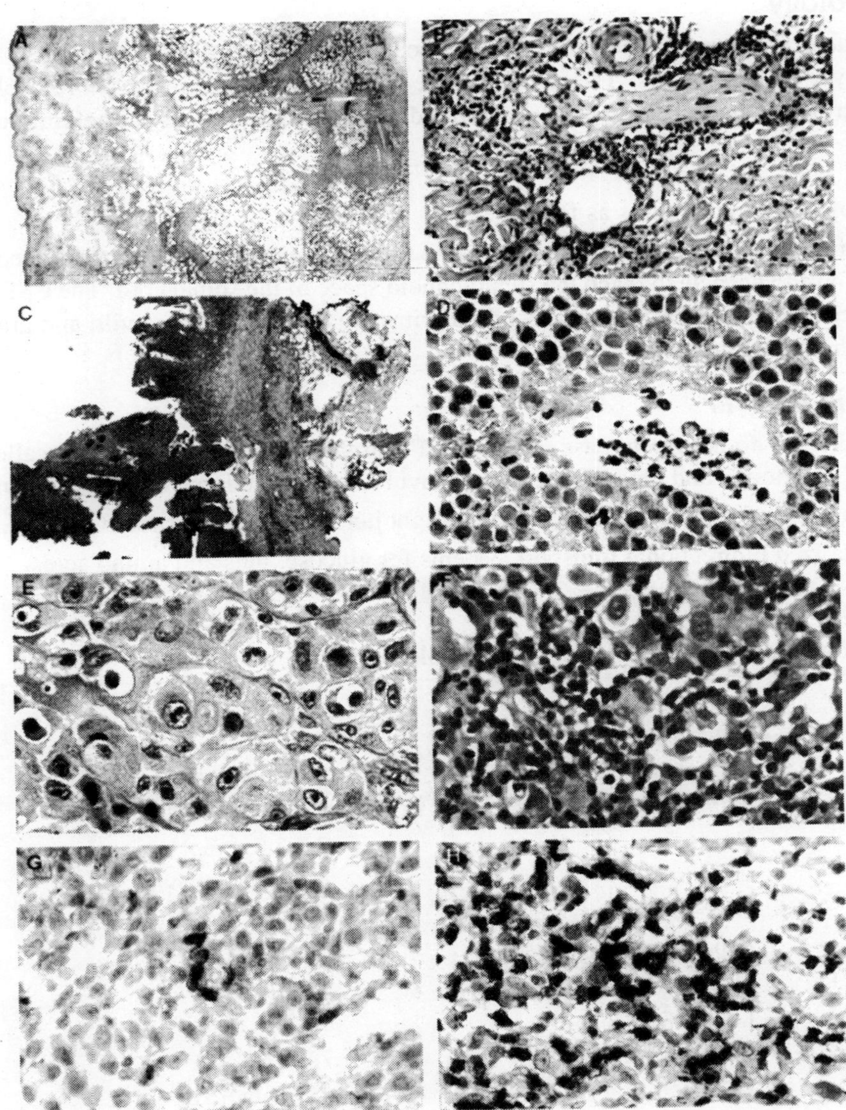

Fig. 9.13: (*A*) Injection Site of Irradiated GM-CSF Secreting Melanoma Cells.

Note the extensive inflammatory reaction throughout all layers of the skin and the marked fibrosis in the subcutaneous fat. (*B*) Injection site of irradiated nontransfected melanoma cells after vaccination. Note the prominent infiltrate composed primarily of lymphocytes and eosinophils. (*C*) Melanoma metastasis after vaccination showing extensive necrosis and fibrosis. (*D*) Vasculopathy in metastasis after vaccination. (*E*) Absence of infiltrate in metastasis pretreatment. (*F*) Diffuse infiltrate of T lymphocytes and plasma cells in metastasis after vaccination. (*G*) CD4-positive T cell reaction in metastasis after vaccination. (*H*) CD8 positive T cell reaction in metastasis after vaccination.

CSF Physiology

CSF is produced at the rate of 500 ml/day and occupies a volume of 135 ml (adults). Consequently, CSF turns over every 6 hours. Two-thirds of the CSF is produced in the choroid plexuses of the cerebral ventricles. One-third of the CSF is formed via fluid shifts across the cerebral capillaries.

Anatomy

The pattern of CSF circulation is as follows:

Fluid formed in the lateral ventricles flows into the third ventricle, through the Aqueduct of Sylvius and into the fourth ventricle to reach the subarachnoid space of the spinal cord. The CSF then migrates toward the cerebral sinuses. CSF is re-absorbed primarily by the arachnoid villi and granulations that project into the dural sinuses.

Blood-brain Barrier

Plasma constituents do not freely pass into the CSF and this phenomenon has been called the "blood-brain barrier". Anatomic and physiologic factors involved in maintaining the blood-brain barrier are:

(1) brain capillaries have anatomic barriers [tight junctions and fenestrated choroidal capillaries);

(2) specialized bidirectional transport systems for glucose, ions and amino acids.

CSF Composition

Reference values for CSF are as follows in the table below:

Protein	0.15-0.45 g/L
Electrolytes	
Osmolality	280-300 mmol/L
Sodium	135-150 mmol/L
Potassium	2.6-3.0 mmol/L
Chloride	115-1 30 mmol/L
Carbon dioxide	20-25 mmol/L
Calcium	1.00-1 .40
Magnesium	1.2-1.5 mmol/L
Lactate	1.1-2.4 mmol/L
pH	7.28-7.32
PCO_2	44-50 mm Hg
PO_2	40-44 mm Hg
Other constituents	
Creatinine	50-110 µmol/L
Glucose	2.8-4.4 mmol/L
Iron	0.2-0.4 umol/L
Phosphorus	0.4-0.6 µmol/L
Urea	3.0-6.5 mmol/L

A comparison of the composition of CSF and plasma reveals that:

1. CSF proteins are -1 % that of plasma;
2. CSF calcium levels are -50% that of plasma;
3. CSF glucose levels are -60% that of plasma;
4. CSF chloride and magnesium levels are higher than plasma;
5. CO_2 diffuses rapidly, and HCO_3 slowly, from the plasma into the CSF.

CSF function

The CSF has several functions:

1. it provides physical support for the brain;
2. it protects against acute changes in arterial and venous blood pressure;
3. it is a route of waste excretion, replacing in many ways the function of lymphatics [the brain has no lymphatics);
4. it is involved in intra-cerebral transport, ex. hypothalamic releasing factors; 5) it helps maintain the ionic homeostasis of the CNS.

Laboratory Investigation of CSF

Important information on CSF can be derived from the following parameters:

1. opening pressure;
2. gross appearance;
3. total and differential cell count;
4. bacterial culture and sensitivity;
5. protein and glucose;
6. analysis of immunoglobulins [detect chronic CNS inflammatory conditions);
7. cytology [detect malignant cells).

Sampling and CSF-pressure reading

After the needle has been inserted between the 3rd and 4th vertebrae, the CSF pressure is measured using a sterile manometer tube. Normal opening pressure is 90-180 mm Hg. Causes of increased opening pressure [>180 mm Hg) include congestive heart failure, meningeal inflammation, cerebral edema, mass lesions, or obstruction of the intracranial venous sinuses. If the CSF pressure is normal, 10-20 ml of fluid can be removed. Samples are collected sequentially in tubes contain 1-3 ml. The samples are processes as follows:

(1) tube #1 chemistry and immunology;

(2) tube #2 microbiological analysis;

(3) tube #3 cell count and differential.

Appearance

(1) clear and colourless (normal);

(2) *bright red* - This indicates the presence of blood. The appearance of fresh blood in all three tubes supports the diagnosis of a subarachnoid hemorrhage. A traumatic tap [blood vessel damaged as the needle enters the spinal canal) will be indicated by the presence of blood in the first sample, with the disappearance of blood in subsequent samples

(3) *xanthochromic (yellow)* - If there is no liver failure (jaundice can cause CSF to be yellow), xanthochromic CSF suggests that a subarachnoid hemorrhage has recently occurred [within days). The yellow color is due to bilirubin generated in the CNS by the breakdown of hemoglobin released from ABC's.

(4) *turbid* - This indicates the presence of white cells and is suggestive of a CNS infection.

Cell counts

CSF normally contains a small number of cells (usually lymphocytes and monocytes) and the total cell count is less than 5 cells/ cu mm. An increase in cell counts suggests either an infection of the CNS, or a number of CNS pathological conditions. The differential cell count provides further information regarding the possible cause of the CNS disease:

(1) *increased neutrophils* - This indicates a bacterial meningitis. Other causes of an increased neutrophil count include a cerebral abscess, following seizures and following CNS hemorrhage;

(2) *increased lymphocytes* - This indicates a viral meningitis. Lymphocyte counts are also elevated in meningitis, due to TB, syphilis, fungal and parasitic infections. Degenerative diseases of the CNS, such as multiple sclerosis, will also generate elevated lymphocyte counts.

(3) *"mixed reaction"* - Here, there is an increase in neutrophils, lymphocytes and plasma cells. This is characteristic of T8 meningitis, fungal meningitis and chronic bacterial meningitis;

(4) *increased plasma cells* - This is a feature of TB meningitis and chronic inflammatory disorders, ex. multiple sclerosis;

(5) *leukemic cells* - The presence of leukemic cells in the CSF indicates meningeal infiltration by leukemic cells. Leukemic cells typically appear in the CSF after several remissions have been achieved by chemotherapy. The blood-brain barrier prevents chemotherapeutic drugs from reaching the CNS and allows leukemic cells to escape treatment.

Cytology

Tumor cells can be detected by cytological examination of the CSF. Tumor cells can be from primary CNS tumors [ex. medulloblastoma], or metastatic CNS tumors [ex. lung, breast, GI tract, melanoma).

Biochemical Analysis

(1) *CSF Total protein*—The CSF normally contains less than 0.45 g/L protein. Increased levels may be found in: (*i*) infection; (*ii*) blood contamination; (*iii*) chronic inflammatory disorders

of the CNS (TB, syphilis, Guillain-Barre); (*iv*) Froin's syndrome. This syndrome is characterized by a blockage of the spinal canal [tumor, disc, infection) and the spinal fluid is xanthochromic with a very high protein content.

(2) *Electrophoretic separation* of CSF proteins and detection of CSF immunoglobulin CSF immunoglobulin can arise from three causes: (*i*) secondary to an increase in plasma immunoglobulin [ex. multiple myeloma); (*ii*) drainage/impairment of the blood-brain barrier; (*iii*) local synthesis in the CNS [ex. multiple sclerosis and other chronic inflammatory disorders of the CNS).

Multiple sclerosis [MS] is characterized pathologically by foci of demyelination in the white matter of the CNS. The diagnosis of MS is made by clinical criteria and laboratory investigations. The two most useful tests are magnetic resonance imaging and the CSF immunoglobulin. In 70% of MS patients, an increase in CSF immunoglobulin can be detected at some point in their clinical course. The increase in CSF immunoglobulin is characterized by oligoclonal pattern immunoglobulin synthesis. This pattern of synthesis suggests that several (but not many) clones of lymphocytes are proliferating and producing immunoglobulin at an increased rate. An oligoclonal pattern can be detected in 90%f MS patients. To detect oligoclonal immunoglobulin, CSF fluid is concentrated and the proteins separated by a high resolution electrophoretic gel. The proteins are stained and the oligoclonal bands (if present) will be detected in the gamma globulin region.

(3) *CSF glucose Low levels of CSF glucose suggest*: (*i*) infection (local metabolism by white cells); (*ii*) hypoglycemia. Increased levels are found in hypoglycemic states. CSF glucose is of limited diagnostic utility and the plasma glucose concentration must be known in order to interpret the CSF glucose level properly.

(4) CSF B_2 - transferrin levels [tau protein]

Occasionally, the laboratory is asked to determine if a nasal or ear fluid is CSF. CSF fluid contains a modified transferrin protein called B_2-transferrin (tau protein), which is not present in plasma or other fluids. The presence of B_2-transferrin in a fluid strongly suggests that the fluid is CSF.

Newer tests to detect CSF pathogen: Polymerase Chain Reaction Assay

The polymerase chain reaction (PCR) is a technique to rapidly amplify a defined region of DNA or RNA PCR has been used to detect the presence of bacterial pathogens (ex. syphilis. TB) and viral pathogens (ex. HIV) in the CSF.

THE BIOCHEMSITRY OF RESPIRATION

Breathing consists of two phases, inspiration and expiration. During inspiration, the diaphragm and the intercostal muscles contract. The diaphragm moves downwards increasing the volume of the thoracic (chest) cavity, and the intercostal muscles pull the ribs up expanding the rib cage and further increasing this volume. This increase of volume lowers the air pressure in the alveoli to below atmospheric pressure. Because air always flows from a region of high pressure to a region of lower pressure, it rushes in through the respiratory tract and into the alveoli. This is called negative pressure breathing, changing the pressure inside thelunsg relative to the pressure of the outside atmosphere. In contrast to inspiration, during expiration the diaphragm and intercostal muscles relax. This returns the thoracic cavity to it's original volume, increasing the air pressure in the lungs, and forcing the air out.

The Pathway of Respiration

- Air enters the nostrils
- passes through the nasopharynx,
- the oral pharynx
- through the glottis
- into the trachea
- into the right and left bronchi, which branches and rebranches into bronchioles, each of which terminates in a cluster of alveoli

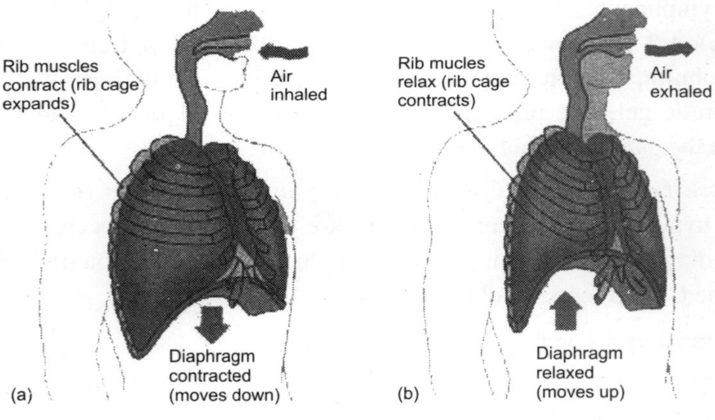

Fig. 9.14 : The Mechanism of Lungs

EXTERNAL RESPIRATION

When a breath is taken, air passes in through the nostrils, through the nasal passages, into the pharynx, through the larynx, down the trachea, into one of the main bronchi, then into smaller broncial tubules, through even smaller broncioles, and into a microscopic air sac called an alveolus. It is here that external respiration occurs. Simply put, it is the exchange of oxygen and carbon dioxide between the air and the blood in the lungs. Blood enters the lungs via the pulomanory arteries. It then proceeds through arterioles and into the alveolar capillaries. Oxygen and carbon dioxide are exchanged between blood and the air. This blood then flows out of the alveolar capillaries, through venuoles, and back to the heart via the pulmanory veins. For an explanation as to why gasses are exchanged here, see partial pressure.

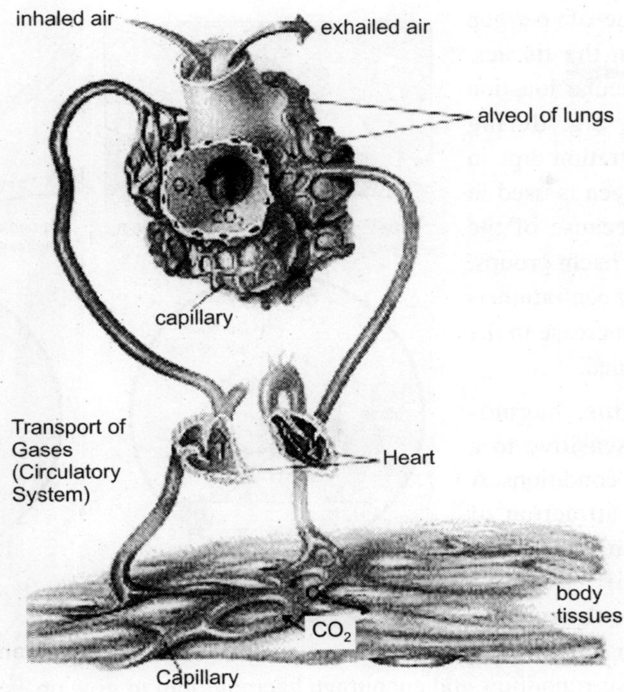

Fig. 9.15 : The Mechanism of Respiration

Gas Transport

If 100mL of plasma is exposed to an atmosphere with a pO_2 of 100mm Hg, only 0.3mL of oxygen would be absorbed. However, if 100mL of blood is exposed to the same atmosphere, about 19mL of oxygen would be absorbed. This is due to the presence of haemoglobin, the main means of oxygen transport in the body. The respiratory pigment haemoglobin is made up of an iron-containing porphyron, haem, combined with the protein globin. Each iron atom in haem is attached to four pyrole groups by covalent bonds. A fifth covalent bond of the iron is attached to the globin part of the molecule and the sixth covalent bond is avaiable for combination with oxygen. There are four iron atoms in each heamoglobin molecule and therefore four heam groups (Fig. 9.16).

Oxygen Transport

In the loading and unloading of oxygen, there is a cooperation between these four haem groups. When oxygen binds to one of the groups, the others change shape slighty and their attraction to oxygen increases. The loading of the first oxygen, results in the rapid loading of the next three (forming oxyhaemoglobin). At the other end, when one group unloads it's oxygen, the other three rapidly unload as their groups change shape again having less attraction for oxygen. This method of cooperative binding and release can be seen in the dissociation curve for haemoglobin. Over the range of oxygen concentrations where the curve has a steep slope, the slighest change in concentration will cause haemoglobin to load or unload a substantial amount of oxygen. Notice that the steep part of the curve corresponds to the range of oxygen concentrations found in the tissues. When the cells in a particular location begin to work harder, e.g. during exercise, oxygen cincentration dips in that location, as the oxygen is used in cellular respiration. Because of the cooperation between the haem groups, this slight change in concentration is enough to cause a large increase in the amount of oxygen unloaded.

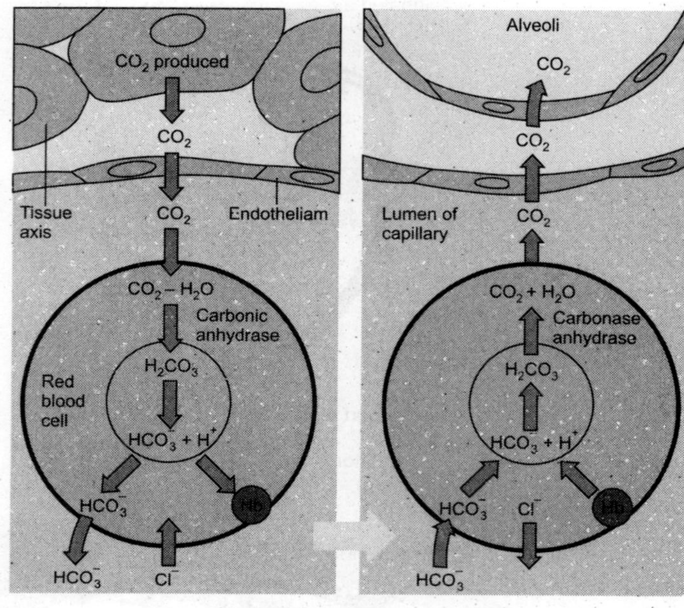

Fig. 9.16 : The Gas Transport

As with all proteins, haemoglobin's shape shift is sensitive to a variety of environmental conditions. A drop in pH lowers the attraction of haemoglobin to oxygen, an effect knownas the Bohr shift. Because carbon dioxide reacts with water to produce carbinic acid, an active tissue will lower the pH of it's surroundings and encourage haemoglobin to give up extra oxygen, to be used in cellular respiration. Haemoglobin a notable molecule for it's ability to tranport oxygens from regions of supply to regions of demand.

Carbon Dioxide Transport - Out of the carbon dioxide released from respiring cells, 7% dissolves into the plasma, 23% binds to the multiple amino groups of haemoglobin (Caroxyhaemoglobin), and 70% is carried as bicarbonate ions. Carbon dioxide created by respiring cells diffuses into the blood plasma and then into the red blood cells, where most of it is converted to bicarbonate ions. It first reacts with water forming carbonic acid, which then breaks down into H^+ and CO_3^-. Most of the hydrogen ions that are produced attach to haemoglobin or other proteins.

INTERNAL RESPIRATION

The body tissues need the oxygen and have to get rid of the carbon dioxide, so the blood carried throughout the body exchanges oxygen and carbon dioxide with the body's tissues. Internal respirtaion is basically the exchange of gasses between the blood in the capillaries and the body's cells.

TISSUE RESPIRATION

Tissue respiration is the release of energy, usually from glucose, in the tissues of all animals, green plants, fungi and bacteria. All these living things require energy for other processes such as growth, movement, sensitivity, and reproduction.

The most efficient form of respiration is aerobic respiration: this requires oxygen. When oxygen is not available, some organisms can respire anaerobically i.e. without air or oxygen. Yeast can respire in both ways. Yeast gets more energy from aerobic respiration, but when it runs out of oxygen it does not die. It can continue to respire anaerobically, but it does not get so much energy from the sugar. Yeast produces ethanol (alcohol) when it respires anaerobically and ultimately the ethanol will kill the yeast.

We can respire in both ways too. Normally we use oxygen, but when we are running in a race, we may not get enough oxygen into our blood, so our muscles start to respire anaerobically. Unlike yeast we produce lactic acid. Of course if we produced alcohol in our muscles it would make us drunk! Fine thing if you are running away from a predator and you end up drunk! Making lactic acid is not much better. Lactic acid causes cramp.

Glucose + Oxygen = Carbon Dioxide + Water + Energy

This word equation means: "sugar and oxygen are turned into carbon dioxide and water releasing energy". You must memorise the word equation (and the balanced chemical equation if you want a grade A, B or C).

Glucose = Carbon Dioxide + Ethanol + Energy

This word equation means: "glucose is turned into carbon dioxide and ethanol releasing energy". You must memorise this word equation.

Ventilating the Lung

This is the proper term for what many people call breathing. If you open a window, you can ventilate a room and get rid of all the nasty smells of BO, cigarette smoke, and cooking; but if you open your mouth it is not enough to ventilate your lungs. You must actually suck air into your lungs. We do this by contracting the diaphragm when we are sitting down and "breathing" gently, and by raising and expanding our rib cage when we are "breathing" deeply. Sucking air into the lungs is called **inhaling**. Inhaled air is mixed with the stale air already in our lungs; so although the air we inhale contains 20% Oxygen, the air in our alveoli (alveolar air) only contains 14% Oxygen.

Don't expect all the stale used air to come out just by opening your mouth, it must be pushed out. This happens when the diaphragm relaxes and the muscles of the abdomen (tummy) push the lungs up. The rib cage can also be pulled down and in. So this is how you **exhale.** Although alveolar air only contains 14% Oxygen, it gets mixed with rather fresher air in your trachea. This means that exhaled air may contain 16% Oxygen.

When I inhale and exhale as deeply as I possibly can, about 5.5 litres of air comes in and out.

Gaseous Exchange

All forms of respiration require some form of gaseous exchange. In aerobic respiration, Oxygen must enter our blood and Carbon Dioxide must leave the blood through our lungs. Gaseous exchange is **the exchange of Oxygen and Carbon Dioxide across a respiratory surface**. Many animals which live in water or very wet places use gills for gaseous exchange. Animals which live on dry land use lungs. Our lungs have an **enormous surface area** so that Oxygen can get into the blood quickly enough and Carbon Dioxide can get out of our blood quickly enough. Our lungs contain billions of very tiny sacs called alveoli. Each alveolus is microscopic; but if we took all the alveoli in someone's lungs and laid them flat side by side we would end up with a sheet the size of a tennis court. As well as having a very, very, very large surface area, the walls of out alveoli are **incredibly thin**, so the distance between the air in our lungs and the blood in our capillaries is very, very, very small.

So in your exam remember that respiratory surfaces:

- have a very large surface area and
- are very thin.

These two things allow the respiratory gases (Oxygen and Carbon Dioxide) to get in or out of the blood fast enough. If you don't believe this, find someone who has been smoking cigarettes for fifty years. They might have a disease called **emphysema**. What happens is that instead of having billions of very tiny alveoli, they have millions of larger ones; this means that the surface area of their lungs is not the size of a tennis court but just the size of a dining room table. People with emphysema get out of breath very quickly. Even getting out of the chair to change channels on the TV makes them puff and pant as though they had run the marathon. Perhaps that is why we need remote controls for our TVs and radios.

Gaseous exchange is also necessary for photosynthesis. Green plants do respire: at night time they exchange gases just as we do, Oxygen in and Carbon Dioxide out. In the daytime they do just the opposite. Carbon Dioxide enters a plant because it is needed for photosynthesis, and Oxygen leaves. This is still called gaseous exchange.

THE RESPIRATION IN THE ALVEOLI

The lungs and associated structures form an extremely efficient system with minimal water loss and maximum gas exchange. Air reaches the lungs through the nose where it is moistened and dust and germs are caught. From here the air passes through the throat or pharynx and then into the trachea (*wind pipe*). The trachea, which is kept open by C - shaped rings of cartilage, splits into two bronchi which lead directly into the right and left lungs. Bronchi then split into many bronchiole which all then eventually end in alveoli. The alveolar epithelium is covered with a thin layer of detergent fluid called surfactant, into which oxygen dissolves. Due to the ball like structure of the 700 million alveoli they create a total surface area of 70 square meters. Tiny blood capillaries surround the alveolus with only two cells width between them and a distance of about 0.3 micrometers, so as to maximise the amount of gaseous exchange. An alveolus is comprised of three epithelial cells which form its boundaries and secrete surfactant. The rate at which we breath is dependant on our activities and need for oxygen. During rest we breath using only the **tidal volume** of our lungs which amount to about half a litre of air. If we breath in deeply we can take in about 3 extra litres of air, called **inspiratory reserve volume**. If we breathe out more that usual we exhale the **expiratory reserve volume** of about a litre. Thus the total volume is the sum of the two reserve volumes and the tidal air, which adds up to between 4 and 6 litres. In addition there is also a residual volume of air (1,5 litres) that is never expelled which ensures that there is always air which waste can diffuse into, and oxygen can diffuse out off.

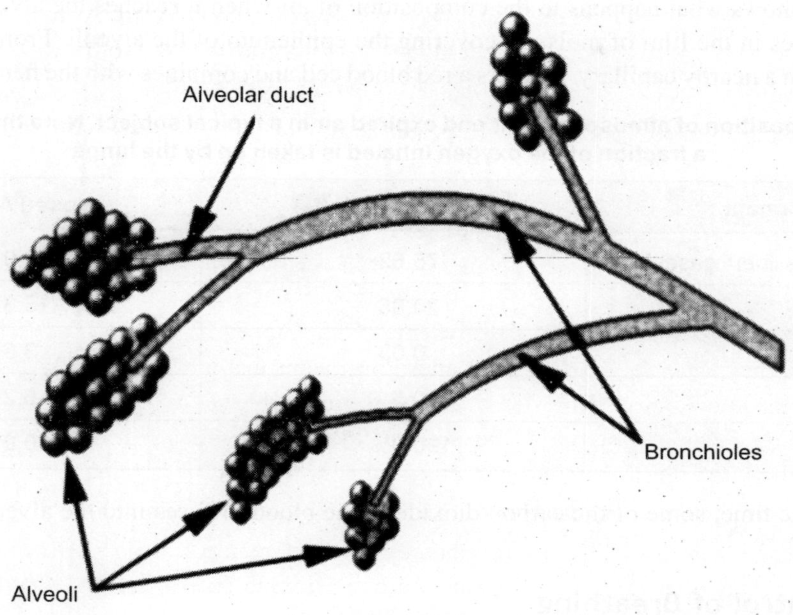

Fig. 9.17 : The Alveoli

Air is sucked into our lungs by a lower pressure which is induced by expanding the volume of the lungs. The lungs are enlarged by contracting the intercostal muscles, between the ribs, and by contracting (straightening) the diaphragm, which is underneath the lungs. This causes inspiration. Expiration on the other hand is cause by relaxing the above groups of muscles causing the chest area to decrease in size and force the air out. These actions are all controlled by the brain.

The rate of breathing is controlled by chemoreceptors in the neck, they test the blood's pH level. When carbon dioxide is in the blood it forms carbonic acid and thus by checking the pH of the blood the amount of carbon dioxide can accurately be ascertained. If there is too much carbon dioxide in the blood the chemoreceptors inform the brain which intern causes the rate of breathing to be stepped up.

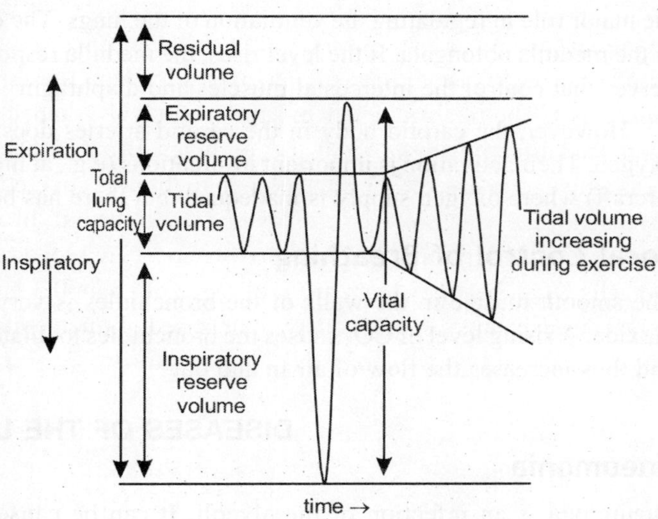

Fig. 9.18 : The Breathing Cycle

The table shows what happens to the composition of air when it reaches the alveoli. Some of the oxygen dissolves in the film of moisture covering the epithelium of the alveoli. From here it diffuses into the blood in a nearby capillary. It enters a red blood cell and combines with the hemoglobin therein.

Composition of atmospheric air and expired air in a typical subject. Note that only a fraction of the oxygen inhaled is taken up by the lungs

Component	Atmospheric Air (%)	Expired Air (%)
N_2 (plus inert gases)	78.62	74.9
O_2	20.85	15.3
CO_2	0.03	3.6
H_2O	0.5	6.2
	100.0%	100.0%

At the same time, some of the carbon dioxide in the blood diffuses into the alveoli from which it can be exhaled.

Central Control of Breathing

The rate of cellular respiration (and hence oxygen consumption and carbon dioxide production) varies with level of activity. Vigorous exercise can increase by 20-25 times the demand of the tissues for oxygen. This is met by increasing the rate and depth of breathing.

It is a rising concentration of carbon dioxide—not a declining concentration of oxygen—that plays the major role in regulating the ventilation of the lungs. The concentration of CO_2 is monitored by cells in the medulla oblongata. If the level rises, the medulla responds by increasing the activity of the motor nerves that control the intercostal muscles and diaphragm.

However, the carotid body in the carotid arteries does have receptors that respond to a drop in oxygen. Their activation is important in situations (e.g., at high altitude in the unpressurized cabin of an aircraft) where oxygen supply is inadequate but there has been no increase in the production of CO_2.

Local Control of Breathing

The smooth muscle in the walls of the bronchioles is very sensitive to the concentration of carbon dioxide. A rising level of CO_2 causes the bronchioles to dilate. This lowers the resistance in the airways and thus increases the flow of air in and out.

DISEASES OF THE LUNGS

Pneumonia

Pneumonia is an infection of the alveoli. It can be caused by many kinds of both bacteria (e.g., Streptococcus pneumoniae) and viruses. Tissue fluids accumulate in the alveoli reducing the surface area exposed to air. If enough alveoli are affected, the patient may need supplemental oxygen.

Asthma

In asthma, periodic constriction of the bronchi and bronchioles makes it more difficult to breathe in and, especially, out. Attacks of asthma can be

- triggered by airborne irritants such as chemical fumes and cigarette smoke
- airborne particles to which the patient is **allergic**.

Emphysema

In this disorder, the delicate walls of the alveoli break down, reducing the gas exchange area of the lungs. The condition develops slowly and is seldom a direct cause of death. However, the gradual loss of gas exchange area forces the heart to pump ever-larger volumes of blood to the lungs in order to satisfy the body's needs. The added strain can lead to heart failure.

The immediate cause of emphysema seems to be the release of proteolytic enzymes as part of the inflammatory process that follows irritation of the lungs. Most people avoid this kind of damage during infections, etc. by producing an enzyme inhibitor (a **serpin**) called **alpha-1 antitrypsin**. Those rare people who inherit two defective genes for alpha-1 antitrypsin are particularly susceptible to developing emphysema.

Chronic Bronchitis

Any irritant reaching the bronchi and bronchioles will stimulate an increased secretion of mucus. In chronic bronchitis the air passages become clogged with mucus, and this leads to a persistent cough. Chronic bronchitis is usually associated with cigarette smoking.

Chronic Obstructive Pulmonary Disease (COPD)

Irritation of the lungs can lead to asthma, emphysema, and chronic bronchitis. And, in fact, many people develop two or three of these together. This constellation is known as chronic obstructive pulmonary disease (COPD).

Among the causes of **COPD** are

- cigarette smoke (often)
- cystic fibrosis (rare)

Cystic fibrosis is a genetic disorder caused by inheriting two defective genes for the **c**ystic **f**ibrosis **t**ransmembrane conductance **r**egulator (**CFTR**), a transmembrane protein needed for the transport of Cl^- ions out of the epithelial cells of the lung thus enabling water to follow by osmosis. Diminished CFTR function reduces the water content of the fluid in the lungs making it more viscous and difficult for the ciliated cells to move it up out of the lungs. The accumulation of mucus plugs the airways interfering with breathing and causing a persistent cough. Cystic fibrosis is the most common inherited disease in the U.S. white population.

Lung Cancer

Lung cancer is the most common cancer and the most common cause of cancer deaths in U.S. males. Although more women develop breast cancer than lung cancer, since 1987 U.S. women have been dying in larger numbers from lung cancer than from breast cancer.

Lung cancer, like all cancer, is an uncontrolled proliferation of cells. There are several forms of lung cancer, but the most common (and most rapidly increasing) types are those involving the epithelial cells lining the bronchi and bronchioles.

Ordinarily, the lining of these airways consists of two layers of cells. Chronic exposure to irritants

- causes the number of layers to increase. This is especially apt to happen at forks where the bronchioles branch.
- The ciliated and mucus-secreting cells disappear and are replaced by a disorganized mass of cells with abnormal nuclei.
- If the process continues, the growing mass penetrates the underlying basement membrane.
- At this point, malignant cells can break away and be carried in lymph and blood to other parts of the body where they may lodge and continue to proliferate.
- It is this metastasis of the primary tumor that eventually kills the patient.

THE ARTIFICIAL RESPIRATION

Normally, Alveolar Ventilation is unconsciously regulated to maintain constant arterial blood gas tensions, despite variable levels of oxygen consumption and CO_2 production.

Many drugs and techniques used in anaesthesia interfere with control or mechanics of ventilation, and it is the Anaesthetists responsibility to ensure the adequacy of ventilation during the perioperative period.

Equipment related to ventilation is consequently of great importance to the Anaesthetist and Anaesthetic Technician. Correct use of the equipment relies on a good understanding of basic respiratory physiology as well as how the individual ventilator operates.

Basic Principles

Venous blood always has a lower PaO_2 (40 mmHg or 75% saturated or 15 ml O_2/100ml blood) and higher $PaCO_2$ (46 mmHg) than inspired gas (PiO_2 150 mmHg, $PiCO_2$ usually 0), so that there is normally a partial pressure gradient driving Oxygen in and CO_2 out of the pulmonary capillary blood.

Ventilation of the lungs with inspired gases results in mixing of the inspired gases with alveolar gas.

If there is no ventilation at all, there will be no replenishment of oxygen and no removal of CO_2, so the PAO_2 will fall and $PACO_2$ will rise towards the venous O_2 and CO_2 tensions.

If the ventilation is much greater than is needed, then the alveolar gas tensions will be much closer to inspired gas.

Definitions

Ventilation is the process by which Oxygen and CO_2 are transported to and from the Lungs.

Tidal Volume (VT) is the amount of gas expired per breath - typically 500ml at rest.

Deadspace Volume (VD) is the sum of the Anatomic Deadspace, due to the volume of the airways (typically 150ml), and Physiologic Deadspace, due to alveoli which are ventilated but not perfused (usually insignificant).

Minute Volume (VE) is the amount of gas expired per minute.

Alveolar Ventilation (VA) is the amount of gas which reaches functional respiratory units (ie, alveoli) per minute. VA = (Tidal Volume - Deadspace) × Respiratory rate

Lung Volumes

- FRC (Functional Residual Capacity) 2.2l.(supine)
- TLC (Total Lung Capacity) 6.2l.
- Maximum Inspiratory Volume 4.0l. above FRC.
- ERV (Expiratory Reserve Volume) 1.0l. below FRC.
- RV (Residual Volume) 1.2l.
- MVV (Maximal Voluntary Ventilation) 150 l/m.

Lung Mechanics Inspiration

An active process requiring musular effort; 75% diaphragmatic at rest; intercostals used on exertion.

Inspiratory effort causes:

- Fall in intrapleural pressure
- Fall in Alveolar pressure
- Pressure gradient from mouth to alveoli
- Gas flow down pressure gradient

Maximum inspiratory force sometimes used as an index of resp. effort; if < 20 cmH_2O most patients have difficulty

Expiration

Usually a passive process due to lung recoil:

- Relaxation of inspiratory muscles causes:
- Intrapleural pressure becomes less negative
- Alveolar pressure rises
- Pressure gradient from alveoli to mouth
- Gas flow down pressure gradient

Airway Resistance

- Limits gas flow down airways
- Due mostly to airway/ETT diameter (fourth power of radius)
- Normal response to increased resistance is increased effort
- GA's increase resistance and decrease response, causing hypoventilation

Intrapleural Pressure

- Normally -10cm H_2O, due to elastic recoil of lung opposed by chest wall.
- Becomes more negative on inspiration.

- Less at the dependent regions of the lung, reducing alveolar size.

Compliance

"Static" Compliance is a measure of the "stiffness" of lung and chest wall, typically 50 ml/cmH_2O in adults and proportionally less in kids. It is usually due equally to lung and chest wall compliances (100 ml/cmH_2O each).

Surfactant improves lung compliance, especially at low lung volumes; its absence as in ARDS, results in stiff lungs and a tendency for the alveoli to collapse and fill with fliud.

"Dynamic" compliance includes the extra pressure needed to overcome resistance to airflow, inertia of chest wall, and viscoelasticity of tissues.

Total compliance varies from person to person and from time to time. A ventilator with pressure limited inspiration will deliver varying tidal volumes during an anaesthetic and from patient to patient. Most modern anaesthesia ventilators are of the "Volume Preset" type to minimise this problem.

Work of Breathing

Work = Pressure × Volume

Respiratory work at rest or during exercise is seldom responsible for more than 5% of the total body work. Most of this is used to overcome the lung and chest wall stiffness during inspiration. Work to overcome airway resistance is usually very small, except during exercise or in athsmatics.

Patients with most respiratory diseases have increased respiratory workloads, which may be due to high respiratory rates, stiff lungs, or high airway resistances. When the patient becomes so exhausted that they can no longer keep up the workload, respiratory failure ensues. Anaesthetic machine tubing, one-way valves, and ETTs all increase total resistance and respiratory work, while drugs will diminish respiratory effort, so that the patient with poor respiratory function usually requires ventilating both during and after the operation.

CO_2 Elimination

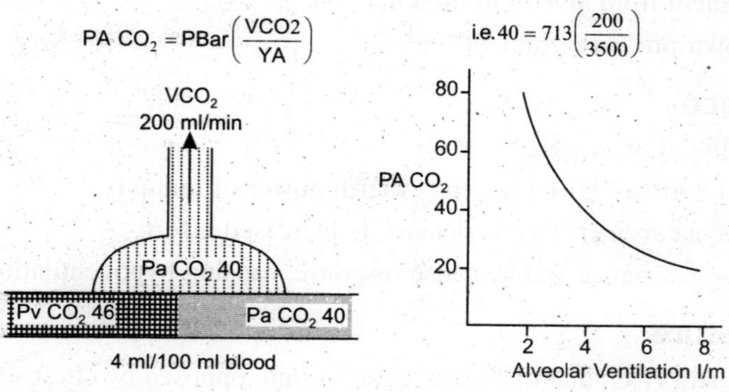

Fig. 9.19 : The Relation between the $PaCO_2$ and Alveolar Ventilation

Oxygen Transport

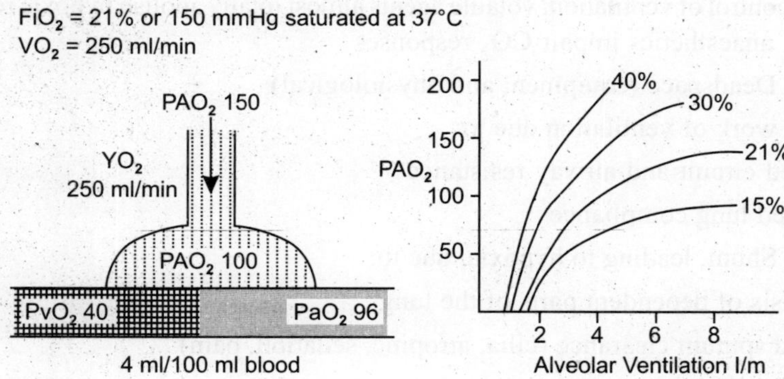

Fig. 9.20 : The Oxygen Transport during the Alveolar Ventilation

Effect of Shunts

Some venous blood passes through the lungs without equilibration with Alveolar gas. This "Venous Admixture" or "Shunt" subsequently mixes with oxygenated blood in the pulmonary veins, and has the effect of reducing PaO_2 and elevating $PaCO_2$.

While the slight rise in $PaCO_2$ can be overcome easily by increasing the ventilation to normal alveoli, the same is not true for PaO_2. For example, a 50% shunt needs 100% inspired oxygen to get a PaO_2 of about 60 mmHg, but only a doubling of ventilation for normocarbia.

This is because the normal alveoli can blow off lots more CO_2 than normal, but can never saturate the Hb any more than 100%.

Control of Ventilation

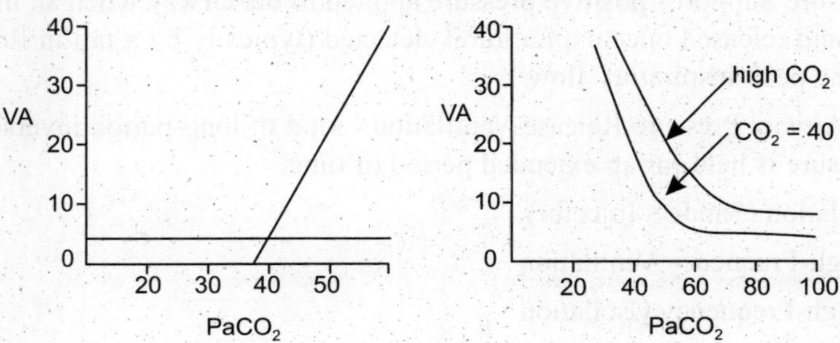

Fig. 9.21 : The Graphical Representation of the Control of the Artificial Ventilation

Effects of Anaesthesia

- Impaired control of ventilation; volatile agents almost totally abolish hypoxic responses, narcotics, sedatives, anaesthetics impair CO_2 responses
- Increased Deadspace (equipment and physiological)
- Increased work of ventilation due to:
  - Increased circuit and airway resistance
  - Decreased lung compliance
- Increased Shunt, leading to hypoxia, due to:
  - Atelectasis of dependent parts of the lung
  - Impaired sputum clearance (cilia, atropine, sedation, pain)
  - Decreased FRC

Ventilation

Classification

- Mouth-to-Mouth/mask/ett etc
- IPPV - "Conventional" Mechanical Ventilation
- PCV - Pressure Control Ventialtion
- IMV - Intermittent Mandatory (Volume) Ventilation
- MMV - Mandatory Minute Ventilation (amount of ventilation is automatically adjusted byt he ventilator to ensure a constant minute volume).
- SIMV — Synchronised IMV ("Assisted") - inspirations are brought forward in time (within a limited time window) if the patient makes a triggering effort.
- PRVC - Pressure Regulated Volume Controlled (volume preset pressure ventilation; machine alters pressure on a breath by breath basis to generate the tidal volume set by the user)
- BiPAP - Two-level CPAP (pt can breathe during inspiration and expiration)
- PS - Pressure Support - positive pressure applied to the airway when an inspiratory trigger is detected and released once expiration is detected (typically by a fall in flow rate to less than 25% of the peak inspiratory flow.
- APRV - Airway Pressure Release Ventilation - kind of long-period inverse BiPAP where the high pressure is held for an extended period of time.
- J et Ventilation (Sanders Injector)
- HFV - High-Frequency Ventilation
- HFO - High-Frequency Oscilation
- HFJV - High-Frequency Jet Ventilation
- PEEP - Positive End-Expiratory Pressure

- CPAP - Constant Positive Airway Pressure (It can breathe during expiration)
- NPV - Negative Pressure Ventilation
- TRIO - Tracheal Insufflation of Oxygen
- Apnoeic oxygenation

Effects of IPPV/PEEP

Respiratory:

- Decreased $PaCO_2$ due to increased Alveolar Ventilation
- Improved PaO_2
- Intrapleural Pressure less negative
- Work of breathing reduced
- Decreased lung water
- Optimum PEEP increases alveolar size, FRC, compliance, etc.
- Hazards associated with incubation, paralysis or sedation, equipment failure.

Cardiovascular:

- Pressure gradient for venous return decreased whenever intrathoracic pressure rises
- CVP and peripheral venous pressure rise
- Reduced RV filling and increased RV afterload; opposite effects on LV
- May cause fall in Cardiac Output, particularly in hypovolaemic patients, causing reflex increase in contractility, heart rate, MVO_2, vasoconstriction to augment venous pressure, reduced mixed venous oxygen tension, which may worsen aterial PO_2

Renal

- Decreased renal function due to fall in Cardiac Output and Renal perfusion
- Increased ADH due to decreased central venous wall tension

Effects of CPAP/IMV or BiPAP

- Patient can breathe spontaneously; paralysis not always required
- Optimum CPAP and a low-resistance circuit reduces work of breathing
- Intrapleural pressure generally not as high as for IPPV so less depression of C.O.

But most IMV/CPAP systems do not maintain CPAP well and finding "Optimal" CPAP is difficult.

Ventilators

Classification

- TYPE OF VENTILATION

  - Positive/Negative

- OTHER CAPABILITIES
- PEEP/CPAP/IMV/MMV/HFV/HFO/HFJV etc
- CYCLING (reason inspiration commences)

  (*a*) Automatic

  - Time (Campbell, Bird)
  - Pressure
  - Other

  (*b*) Manual

  (*c*) Patient-Triggered
- INSPIRATION LIMIT
  - Automatic
  - Volume +/- Pressure limit (Bird with Bellows)
  - Time (Campbell)
  - Pressure (Bird)
  - Flow
  - Manual
- INSPIRATORY FLOW PATTERN
  - Constant (Bird with Air-Mix control closed)
  - Decelerating (Venturi-type, i.e. Campbell)
  - Programmable
  - Sinusoidal (Piston driven)
- CONTROL MECHANISM
  - Pneumatic +/- Magnetic (Bird)
  - Electronic (Servo)
  - Fluidic Logic (Campbell)
- PATIENT CIRCUIT
  - Single or Dual
- POWER REQUIREMENTS CONTROLS

Use in Anaesthesia

- Aim for normocarbia or slight hypocarbia
- Usually Volume preset IPPV devices
- Tidal volume and rate adjusted to suit patient (CO_2 analysers useful)

With CO_2 absorber ON:

- All inspired gas is free of CO_2

- Effective ventilation depends only on Ventilator settings
- Very low Fresh Gas Flows may be used in the circle circuit

With the CO_2 absorber OFF:

- Provided that the ventilator settings deliver normal alveolar ventilation, the effective ventilation depends on Fresh Gas Flow.
- CO_2 rebreathing occurs.

Hazards

- Disconnection from circuit
- Failure to deliver ventilation
- Barotrauma

Monitoring Ventilation

- Colour of the Patient
- Watching the chest move
- Precordial/Oesophageal Stethoscope +/- telemetry
- Listening to sound of ventilator
- Measurement of Circuit Parameters, such as pressure or tidal volume
- Measurement of Patient Parameters, such as $ETCO_2$, SpO_2, chest wall impedance, etc

Humidification

Physics

- Vapour - Gas Phase of a liquid below boiling point
- Aerosol/Mist - suspension of fine droplets of a liquid in a gas
- Absolute Humidity - amount of water vapour per unit of gas (mg/l)
- Relative Humidity - Absolute humidity of the sample as a % of the absolute humidity of fully saturated gas at the same temperature

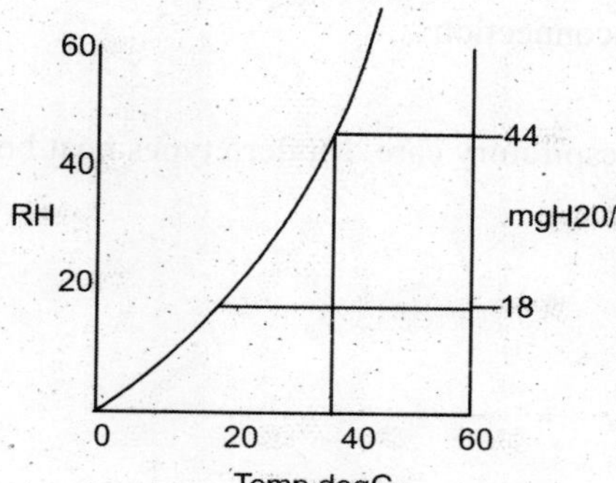

Note
(1) Exponential rise in absolute humidity with increasing temperature.
(2) At 37 degC fully set is 44 mg/l
(3) At 20 degC fully set is 18 mg/l

Measurement of Humidity

- Dew Point Hygrometer
- Hair Hygrometer
- Wet/Dry thermometer
- Humidity Sensors
- Measurement of water used by humidifier

Physiology

The nose is a very efficient humidifier:

- 60% RH at the post-nasal space
- 5% RH in the mouth
- 100% at 37°C in the bronchi

Mouth-breathing is less efficient (60% RH in the upper trachea)

Heat and water loss through the nose is minimised by cooling on inspiration and warming on expiration.

Humidification is required to maintain of ciliary activity, prevent squamous epithelial changes (Mucosal changes in 2-3 hours), prevent dehydration and thickening of secretions and possible ETT obstrucion, minimise atelectasis and tracheitis, and to decrease heatloss

Methods

Anaesthetic Circuit Considerations

- Cylinder gas is completely dry, and tracheal intubation bypasses the nose
- Waters CO_2 absorber heats and humidifies gas very effectively
- Circle CO_2 absorbers are of slight benefit only
- Bain circuit allows some warming but very little humidification

Heat and Moisture Exchangers

- Relatively cheap
- 70%-80% effective humidification
- Increased deadspace, resistance, risk of disconnection

Humidifiers

Up to 100% humidification, essential for longterm respiratory care. Modern types heat both the water bath and patient hose to prevent rainout.

Disadvantages:

- Cost
- Potential for leaks, disconnection
- Drowning if tipped
- Source of infection

- Unreliable
- Airway burns
- Increased airway resistance

Equipment

Fisher & Paykel

- Water heated to 37°C , servo controlled hose heaters in newer units to prevent 'rain-out'.

Grant - Nicholas

- Water heated to 45°C , hose servo to 37°C. Inefficient at >10l/min flows

Bourns

- Basic Kettle type

Nebulisers

- Produce aerosols with humidity depending on temperature.
- Air is usually cooled by the droplets ->cold wet air
- Most useful for drug delivery

THE BIOCHEMISTRY OF DIGESTION

Introduction

Only a small portion of the food we eat is in the proper formation for absorption into the blood or the lymph system of the body. Such things as water, uncombined mineral salts and uncombined vitamins may not need digestive action prior to absorption. But the rest of the bulk food must undergo profound chemical changes and process before the resulting molecules can be absorbed. The proteins an protein split produces must be hydrolyzed to amino acids, the oligosaccharides and polysaccharides to hexose sugars and the fats to fatty acids and glycerol in parts at least. Other lipids are also hydrolyzed to smaller product molecules. Many reactions and enzymatic processes are involved in these changes to constitute the process of digestion.

A number of factors help the digestive process; Cooking of foods produces several changes. It softens certain tissue, breaks the cellulose covering around starch granules allowing a better access to digestive enzymes, partially hydrolyzes some food components, brings about coagulation of liquid proteins of eggs and other foods, and improves the flavour of some which may result in increased flow of saliva and gastric juice.

The ripening of fruits and vegetables and aging of many foods produce desirable chemical changes. Mastication greatly increases the surface area of food for better contact with the digestive juices.

Absorption from the intestine by removing digested molecules brings about by a mass law effect, more nearly complete enzyme action. The most important factor is of course the action of the enzymes and other specific components of the digestive juices.

The enzyme and other compounds important in digestion are separated by various glands of the body. The control of these secretions is nervous hormonal or both among these five secretions are known to be involved they are:

1. Saliva
2. Gastric Juice
3. Pancreatic Juice
4. Intestinal Juice
5. The Bile Juice

THE PROCESS OF DIGESTION

Where does food enter your body? Many would answer, "As soon as you put it into your mouth." Yet this is NOT where food actually enters your body. When food is in your mouth, or in your stomach or even in your intestines, it is still really outside of your body. The digestive tract or gut is really just a tube-like continuation of your body's external surface run through your body.

To enter the body, food must be absorbed across the epithelium that lines the digestive tract. Digestion prepares food for this journey. Digestion is a process that breaks or dissembles the things that you eat into molecules small enough to be absorbed by cells that line the digestive tract. Ultimately these small molecules enter the cytoplasm of every cell in your body, where their nutritive value is utilized.

The human digestive tract is about 9 metres (30 feet) in length. It consists of:

- mouth
- esophagus
- stomach
- small intestine
- large intestine
- anus

The human digestive tract is a model food processing plant. As food passes through, it is mixed with various fluids, churned and moved by the musculature of the tract, broken down by various enzymes and absorbed by cells that line the tract. Digestion begins in the mouth where food is cut and ground by teeth. This makes food easier to dissolve and affords a greater surface area for various fluids to act upon. While in the mouth, saliva is added through secretions of the salivary glands which initiates chemical digestion. Saliva contains enzymes that initiate the digestion of starch. Saliva also contains mucin, a major protein of mucus, which acts as a lubricant. Ever tried to swallow a dry pill without salivating a bit?

Mucin also bunches the food together into a mass called a bolus. During swallowing, the bolus is pushed into the esophagus, the tubular channel that leads to the stomach. During the swallowing process, openings to the respiratory and nasal passages are automatically closed to ensure that food is kept out of these places. In spite of your best efforts, you simply cannot breathe while swallowing.

The walls of the esophagus contain muscles that contract in successive waves, a process called peristalsis. Peristalsis constricts the esophagus and pushes the bolus of food through a sphincter into the stomach.

Once in the stomach, food gets churned into a paste-like material and is mixed with gastric juices. Eventually this is called chyme. Gastric juice is produced by secretory cells that are located in pits in the wall of the stomach. Hydrochloric acid, a strong acid, is one of the compounds that makes up the gastric juice. Hydrochloric acid lowers the pH of the stomach contents to perhaps 2.0, providing an extremely acidic environment that kills most microbes in food, including many that could cause human illness.

If the acidic contents leaks back into the esophagus, the irritation of the lining there we perceive as "heartburn". While most enzymatic digestion occurs in the small intestine, protein digestion begins in the stomach through the action of an enzyme called pepsin. Pepsin is secreted as an inactive material called pepsinogen. The acid environment of the stomach activates it, converting it to pepsin.

With all this digestion going on and the severe pH environment found within the stomach, why is the stomach itself not injured or digested? Because it protected by a fairly thick coating of alkaline mucus. When this protective lining is breached, one can develop gastric ulcers. For a long time, gastric ulcers have been treated with antacids. For many people, however, antacids were simply ineffective. Within the past several years, studies have shown that while stomach acidity may be a factor in stomach ulcers, the primary cause seems to be infection with a bacterium, *Helicobacter pylori*. Ulcer patients treated with antibiotics that kill these bacteria seem to enjoy a greater cure rate. What do you think of the possibility of a vaccine becoming available to prevent ulcers?

How is gastric secretion controlled? The first phase seems to involve stimulation by nerve impulses that reach the stomach from the brain as a result of the smell, taste or even the thought of food. When food actually enters the stomach, other signals are generated. One signal is carried by sensory neurons from the stomach to the brainstem which responds by sending impulses down autonomic nerve fibers which stimulates the digestive glands of the stomach to release their products. Another signal is a chemical message sent by the hormone gastrin which is released by endocrine cells in the stomach lining. The message is carried in blood vessels of the stomach wall to the stomach's glandular cells, triggering these cells to secrete gastric juices.

Only a few small molecules such as aspirin and alcohol are able to enter the bloodstream through the stomach wall. This explains the rapid onset of their effects. Other materials must wait for passage into the small intestine for absorption. Peristaltic waves moving along the wall of the stomach propel small quantities of chyme into the small intestine. The small intestine is about 7 meters (21 feet) in length and is a highly coiled muscular tube about 2.5 cm (1 inch) in diameter.It is here in the small intestine that the macromolecules we ingest as food are broken down into small, organic molecules such as simple sugars, amino acids and nucleotides. These are the things that are absorbed into the bloodstream.

The digestion of materials in the small intestine requires the help/participation of several major organs and their secretions. These include intestinal juice, pancreatic secretions, and emulsifying lipids from the liver and gall bladder. When the inflow of chyme stretches the intestinal wall, the action triggers a nerve response that causes the cells of the intestinal lining to secrete intestinal juices and mucus.

Under normal conditions, the intestinal lining secretes 2-3 liters of fluid each day. This fluid is needed to dissolve the molecules for digestion and to facilitate absorption across the intestinal epithelium. Cholera is a nasty human disease and results from a bacterial toxin that greatly increases the fluid released from the intestinal lining. When stimulated by cholera toxin, these cells can churn out over a liter of fluid each HOUR, most of which is simply lost through diarrhoea.

Death by cholera is death due to severe, acute dehydration. Babies, and the very young are truly challenged by cholera. Treatment is largely one of replacing the lost fluids. The pancreas is a gland that secretes digestive enzymes and releases them into the small intestine. The pancreas also released sodium bicarbonate, an alkaline material that neutralizes somewhat, the extremely acidic chyme that enters the small intestine from the stomach.

The secretion of pancreatic enzymes and bicarbonate is stimulated by two hormones—cholecystokinin (CCK) and secretin. These are secreted into the bloodstream by endocrine cells found in the wall of the small intestine. All of this is in response to the inflow of chyme from the stomach. How does your body deal with fats. They aren't so soluble in water. Consider how washing dishes or greasy pans works. You add some soap and this promotes the solution of fats into the water.

Now suppose you eat a nice meal of raclette cheese (this stuff is perhaps 120% fat and is wonderful). How does your body handle it. In order for the large fat molecules to be efficiently broken down by the lipases (enzymes) the large clusters of this stuff must be broken down into much smaller clusters. This is done by the addition of bile salts produced by the liver and stored in the gall bladder. The gall bladder empties into the small intestine.

Bile salts are similar to detergents. In the presence of bile salts, fat globules are reduced to stable, microscopic droplets that can be attacked efficiently by the fat digesting enzymes, the lipases. The first step in the absorption of food molecules is their movement from the lumen of the small intestine into the epithelial cells that make up its lining. The inner surface of the small intestine is a highly textured surface because of the presence of microscopic finger-like projections called villi. These villi greatly increase the surface area available for absorption. Each villus is, in turn, covered with small projections called microvilli which further increase surface area. Together, the villi and microvilli yield a surface area more than 150 times that of your skin.

Each villus is laced with a rich capillary network surrounding a centrally located lymphatic vessel called a lacteal. The lacteal absorbs the products of lipid digestion such as fatty acids. From the lacteal, microscopic fat droplets (of fatty acid) are transported through a series of lymphatic vessels that eventually drain into a large vein in the neck.Most non-fatty nutrients diffuse directly into the blood capillaries of the intestinal villi where they are carried to the liver and removed from the bloodstream. The liver is the major metabolic regulatory center, controlling blood-glucose levels and releasing glucose into the blood as needed. By the time digested food gets to the end of the small intestine, virtually all of its nutrients have been removed along with most of its water. The nutrient depleted chyme is pushed into the large intestine or colon.

What happens here? Water is returned and the remaining contents are converted into a material known as feces. You may know this material by other names.

Pressure sensitive neurons detect when solids have accumulated at the terminal end of the large intestine which is called the rectum and respond by initiating a defecation reflex. Because one of the two anal sphincters is under voluntary control, we humans can consciously delay expulsion of feces until an appropriate time.

Projecting from the large intestine is a short, blind tube, the appendix. Inflammation of the appendix, literally appendicitis, if unattended, can lead to a rupturing of the structure which releases intestinal contents into the abdominal cavity. There are numerous bacteria present in feces and their presence in the abdominal cavity can lead to a fatal infection.

The huge numbers of bacteria that live in the large intestine are not freeloaders. Bacteria constitute almost half the dry weight of large intestine contents. These bacteria metabolically attack organic substances remaining in chyme and often produce unpleasant smelling by-products. If gaseous, we refer to these by-products as flatus. You may know this gaseous material by other names.

These same bacteria also produce vitamin K, biotin, folic acid and other nutrients that we humans absorb. Bacteria also contribute by competing for space and nutrients with less welcome bacteria. Consider how upset things become after taking substantial amounts of antibiotics that diminish intestinal bacteria. The result is often a bout of diarrhea and/or an opportunistic yeast infection.

There is still a lot controversy over such topics as the impact of dietary sugar, cholesterol and saturated fats on our health.Nutritionists, however, generally agree that our diet should balance carbohydrates, triglyceride lipids (fats and oils) and proteins. Foods that balance these groups of foodstuff should also easily provide us with enough energy, organic building blocks, vitamins and minerals for sustenance.

An average person engaged in a relatively sedentary life style requires about 2,500 Calories per day to maintain their bodies in a stable state. A person who engages in strenous activity, maybe a professional athlete, may require over 4,000 Calories. Metabolic rates also change with age.

Carbohydrates provide the most readily available form of glucose and therefore, the most readily available form of usable energy. Glucose is an all purpose energy source. It is also the one sugar usable by all brain and nerve cells. Not all carbohydrates are easily digestible. Cellulose is a carbohydrate but we humans cannot digest it. Yet cellulose is useful to humans. Consider that when we eat celery, we consume considerable cellulose which ends up contributing to the bulk in our digestive tract and provides something for peristalsis to act upon. Low-cellulose diets can lead to constipation and have been linked to cancer. The only reason one can have for suggesting foods rich in both polysaccharides and undigestible fibers over that of polysacchride rich candy is the presence of the fibers is useful for healthy digestion.

Fats are chemical, highly reduced and are rich sources of energy. Two fatty acids, linolenic and linoleic, are *essential* fatty acids necessary for biological membrane construction and cannot be manufactured by the human body. Diets rich in saturated fats and cholesterol MAY predispose susceptible individuals to cardiovascular disease. Nutritionists tend to agree that a healthy diet should be low in fat, providing less than 30% of our caloric intake.

This is not so easy to do because fats tend to provide attractive taste and texture to foods. We humans LIKE fats. One of the large food processing companies has a product called Olestra, which is a synthetic material that provides the taste and texture of fat but is not digestible. It simply passes through like cellulose. It has just won FDA approval for human consumption. Seems it gives some individuals diarrhea though.Dietary proteins are needed to provide amino acids from which we assemble our enzymes, antibodies and other kinds of proteins. We can obtain protein from virtually any food. We can manufacture all but eight amino acids. These eight essential amino acids must be ingested. The absence of even one of these essential amino acids can halt protein synthesis.These required amino acids are synthesized by plants and bacteria.

We also need vitamins. A vitamin is an organic compound needed in trace amounts for normal health but is one that we cannot synthesize. There are 13 that we humans must acquire through our diet or risk suffering vitamin deficiency diseases, some of which can be fatal.

Do you need to take vitamin pills? If you eat a well-balanced diet, probably not.

A dietary supply of certain inorganic minerals is also necessary for proper nutrition. Without calcium or magnesium, a large number of enzyme mediated reactions in your body would simply not be able to occur. These elements act as factors necessary for the enzymes to operate.

Calcium and phosphorous are needed for bone growth. Calcium is also needed for muscle function. Iron forms the core of electron transport pigment proteins and hemoglobin. Phosphorous is also needed for ATP synthesis.

A number of other elements such as iodine, copper, molybdenum, manganese and chromium are needed in tiny amounts for certain enzyme reactions or as co-factors and are referred to as trace elements required in our diets.

Under these subheads we are going to discuss the different juices and secretions involved in the digestion.

THE SALIVA

The salivary glands in mammals are exocrine glands, glands with ducts, that produce saliva. They also secrete amylase, an enzyme that breaks down starch into maltose. In other organisms such as insects, salivary glands are often used to produce biologically important proteins like silk or glues, and fly salivary glands contain polytene chromosomes that have been useful in genetic research. The glands are enclosed in a capsule of connective tissue and internally divided into lobules. Blood vessels and nerves enter the glands at the hilum and gradually branch out into the lobules.

In the duct system, the lumens formed by intercalated ducts, which in turn join to form striated ducts. These drain into ducts situated between the lobes of the gland (called interlobar ducts or secretory ducts). All of the human salivary glands terminate in the mouth, where the saliva proceeds to aid in digestion. The saliva that salivary glands release is quickly inactivated in the stomach by the acid that is present there. The salivary glands are situated at the entrance to the gastrointestinal system to help begin the process of digestion.

The parotid glands are a pair of glands located in the subcutaneous tissues of the face overlying the mandibular ramus and anterior and inferior to the external ear. The secretion produced by the parotid glands is serous in nature, and enters the oral cavity through the Stensen's duct after passing through the intercalated ducts which are prominent in the gland. Despite being the largest pair of glands, only approximately 25% of saliva is produced by the glands.Saliva contains a mixture of enzymes like salivary amylase (ptyalin), maltase(trace amounts), lysozyme (which disinfect and kills bacteria and germs which enter the mouth), salts and water. Saliva helps converting starch into maltose which is then converted patially to glucose by the maltase.

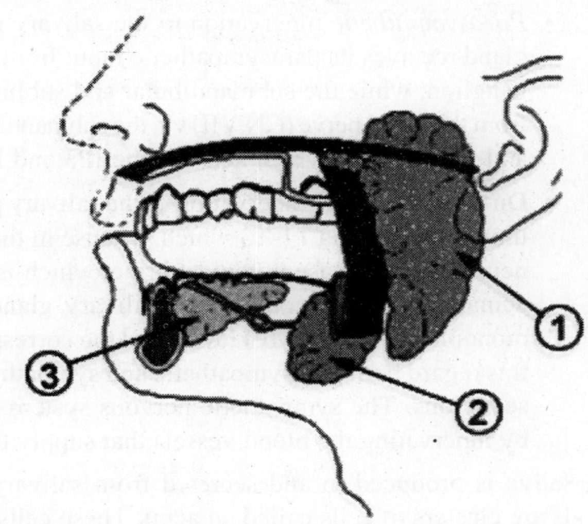

Fig. 9.22 : Salivary Glands: 1 is Parotid gland, 2 is Submandibular gland and 3 is Sublingual gland

The submandibular glands are a pair of glands located beneath lower jaws, superior to the digastric muscles. The secretion produced is a mixture of both serous and mucous and enters the oral cavity via Wharton's ducts. Approximately 70% of saliva in the oral cavity is produced by the submandibular glands, even though they are much smaller than the parotid glands.

The sublingual glands are a pair of glands located beneath the tongue to the submandibular glands. The secretion produced is mainly mucous in nature, however it is categorized as a mixed gland. Unlike the other two major glands, the ductal system of the sublingual glands do not have striated ducts, and exit from 8-20 excretory ducts. Approximately 5% of saliva entering the oral cavity come from these glands.

There are over 600 minor salivary glands located throughout the oral cavity within the lamina propria of the oral mucosa. They are 1-2mm in diameter and unlike the other glands, they are not encapsulated by connective tissue only surrounded by it. The gland is usually a number of acini connected in a tiny lobule. A minor salivary gland may have a common excretory duct with another gland, or may have its own excretory duct. Their secretion is mainly mucous in nature (except for Von Ebner's glands) and have many functions such as coating the oral cavity with saliva. Problems with dentures are usually associated with minor salivary glands. Von Ebner's glands are glands found in circumvallate papillae of the tongue. They secrete a serous fluid that begin lipid hydrolysis. They are an essential component of taste. Salivary glands are innervated, either directly or indirectly, by the parasympathetic and sympathetic arms of the autonomic nervous system. Both result in increased amylase output and volume flow.

- *Parasympathetic* innervation to the salivary glands is carried via cranial nerves. The parotid gland receives its parasympathetic input from the glossopharyngeal nerve (CN IX) via the otic ganglion, while the submandibular and sublingual glands receive their parasympathetic input from the facial nerve (CN VII) via the submandibular ganglion. These nerves release acetylcholine and substance P, which activate the IP3 and DAG pathways respectively.
- Direct *sympathetic* innervation of the salivary glands takes place via preganglionic nerves in the thoracic segments T1-T3 which synapse in the superior cervical ganglion with postganglionic neurons that release norepinephrine, which is then received by â-adrenergic receptors on the acinar and ductal cells of the salivary glands, leading to an increase in cyclic adenosine monophosphate (cAMP) levels and the corresponding increase of saliva secretion. Note that in this regard both parasympathetic and sympathetic stimuli result in an increase in salivary gland secretions. The sympathetic nervous system also affects salivary gland secretions indirectly by innervating the blood vessels that supply the glands.

Saliva is produced in and secreted from salivary glands. The basic secretory units of salivary glands are clusters of cells called an acini. These cells secrete a fluid that contains water, electrolytes, mucus and enzymes, all of which flow out of the acinus into collecting ducts. Within the ducts, the composition of the secretion is altered. Much of the sodium is actively reabsorbed, potassium is secreted, and large quantities of bicarbonate ion are secreted. Bicarbonate secretion is of tremendous importance to ruminants because it, along with phosphate, provides a critical buffer that neutralizes the massive quantities of acid produced in the forestomachs. Small collecting ducts within salivary glands lead into larger ducts, eventually forming a single large duct that empties into the oral cavity.

Most animals have three major pairs of salivary glands that differ in the type of secretion they produce:

- *parotid glands* produce a serous, watery secretion
- *submaxillary (mandibular) glands* produce a mixed serous and mucous secretion
- *sublingual glands* secrete a saliva that is predominantly mucous in character

The basis for different glands secreting saliva of differing composition can be seen by examining salivary glands histologically. Two basic types of acinar epithelial cells exist:

- *serous cells*, which secrete a watery fluid, essentially devoid of mucus
- *mucous cells*, which produce a very mucus-rich secretion

Acini in the parotid glands are almost exclusively of the serous type, while those in the sublingual glands are predominantly mucus cells. In the submaxillary glands, it is common to observe acini composed of both serous and mucus epithelial cells.

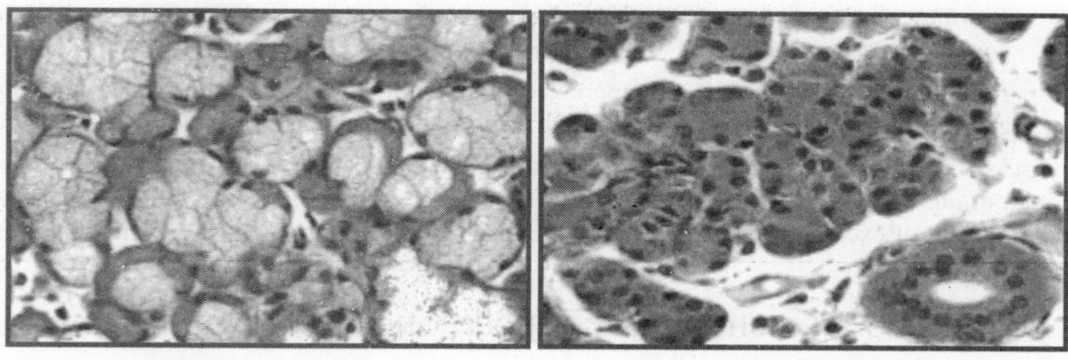

Mandibular gland (mixed) Parotid gland (serous)

Fig. 9.23 : The Microscopic Image of Mandibular Glands and Parotid Glands

In the histologic sections of canine salivary gland shown above, the cells stained pink are serous cells, while the white, foamy cells are mucus-secreting cells.

Secretion of saliva is under control of the autonomic nervous system, which controls both the volume and type of saliva secreted. This is actually fairly interesting: a dog fed dry dog food produces saliva that is predominantly serous, while dogs on a meat diet secrete saliva with much more mucus. Parasympathetic stimulation from the brain, results in greatly enhanced secretion, as well as increased blood flow to the salivary glands.

Potent stimuli for increased salivation include the presence of food or irritating substances in the mouth, and thoughts of or the smell of food. Knowing that salivation is controlled by the brain will also help explain why many psychic stimuli also induce excessive salivation - for example, why some dogs salivate all over the house when it's thundering

What then are the important functions of saliva? Saliva serves many roles, some of which are important to all species, and others to only a few:

- *Lubrication and binding:* the mucus in saliva is extremely effective in binding masticated food into a slippery bolus that (usually) slides easily through the esophagus without inflicting damage to the mucosa. Saliva also coats the oral cavity and esophagus, and food basically never directly touches the epithelial cells of those tissues.
- *Solubilizes dry food:* in order to be tasted, the molecules in food must be solubilized.
- *Oral hygiene:* The oral cavity is almost constantly flushed with saliva, which floats away food debris and keeps the mouth relatively clean. Flow of saliva diminishes considerably during sleep, allow populations of bacteria to build up in the mouth — the result is *dragon breath* in the morning. Saliva also contains lysozyme, an enzyme that lyses many bacteria and prevents overgrowth of oral microbial populations.
- *Initiates starch digestion:* in most species, the serous acinar cells secrete an alpha-amylase which can begin to digest dietary starch into maltose. Amylase is not present, or present only in very small quantities, in the saliva of carnivores or cattle.

- *Provides alkaline buffering and fluid:* this is of great importance in ruminants, which have non-secretory forestomachs.
- *Evaporative cooling:* clearly of importance in dogs, which have very poorly developed sweat glands - look at a dog panting after a long run and this function will be clear.

Diseases of the salivary glands and ducts are not uncommon in man, and excessive salivation is a symptom of almost any lesion in the oral cavity. The dripping of saliva seen in rabid animals is not actually a result of excessive salivation, but due to pharyngeal paralysis, which prevents saliva from being swallowed.

THE CHEMICAL COMPOSITION OF SALIVA

It is a fluid containing:

- Water
- Electrolytes:
  - 2-21 mmol/L sodium (lower than blood plasma)
  - 10-36 mmol/L potassium (higher than plasma)
  - 1.2-2.8 mmol/L calcium (similar to plasma)
  - 0.08-0.5 mmol/L magnesium
  - 5-40 mmol/L chloride (lower than plasma)
  - 25 mmol/L bicarbonate (higher than plasma)
  - 1.4-39 mmol/L phosphate
  - Iodine (mmol/L usually higher than plasma, but dependent variable according to dietary iodine intake)
- Mucus. Mucus in saliva mainly consists of mucopolysaccharides and glycoproteins;
- Antibacterial compounds (thiocyanate, hydrogen peroxide, and secretory immunoglobulin A
- Epidermal growth factor or EGF
- Various enzymes. There are three major enzymes found in saliva.
  - α-amylase (EC3.2.1.1). Amylase starts the digestion of starch and lipase fat before the food is even swallowed. It has a pH optima of 7.4.
  - lingual lipase. Lingual lipase has a pH optimum ~4.0 so it is not activated until entering the acidic environment of the stomach.
  - Antimicrobial enzymes that kill bacteria.
- Lysozyme
- Salivary lactoperoxidase
- Lactoferrin
- Immunoglobulin A
  - Proline-rich proteins (function in enamel formation, Ca^{2+}-binding, microbe killing and lubrication)

– Minor enzymes include salivary acid phosphatases A+B, N-acetylmuramoyl-L-alanine amidase, NAD(P)H dehydrogenase (quinone), superoxide dismutase, glutathione transferase, class 3 aldehyde dehydrogenase, glucose-6-phosphate isomerase, and tissue kallikrein (function unknown).

- **Cells:** Possibly as much as 8 million human and 500 million bacterial cells per mL. The presence of bacterial products (small organic acids, amines, and thiols) causes saliva to sometimes exhibit foul odor.
- **Opiorphin**, a newly researched pain-killing substance found in human saliva.

Different reagents used to determine the content of saliva \1. Molisch test gives a positive result of purple color that is costituent to the presence of carbohydrates

THE DIGESTION IN STOMACH

The stomach is a small,'C'-shaped pouch with walls made of thick, elastic muscles, which stores and helps break down food. Food enters the stomach through the cardiac orifice where it is further broken apart and thoroughly mixed with gastric acid, pepsin and other digestive enzymes to break down proteins. The acid itself does not break down food molecules, rather it provides an optimum pH for the reaction of the enzyme pepsin and kills many microorganisms that are ingested with the food. The parietal cells of the stomach also secrete a glycoprotein called intrinsic factor which enables the absorption of vitamin B-12. Other small molecules such as alcohol are absorbed in the stomach, passing through the membrane of the stomach and entering the circulatory system directly. Food in the stomach is in semi-liquid form, which upon completion is known as chyme.

The transverse section of the alimentary canal reveals four distinct and well developed layers within the stomach:

- **Serous membrane,** a thin layer of mesothelial cells that is the outermost wall of the stomach.
- **Muscular coat,** a well-developed layer of muscles used to mix ingested food, composed of three sets running in three different alignments. The outermost layer runs parallel to the vertical axis of the stomach (from top to bottom), the middle is concentric to the axis (horizontally circling the stomach cavity) and the innermost oblique layer, which is responsible for mixing and breaking down ingested food, runs diagonal to the longitudinal axis. The inner layer is unique to the stomach; all other parts of the digestive tract have only the first two layers.
- **Sub mucosa,** composed of connective tissue that links the inner muscular layer to the mucosa and contains the nerves, blood and lymph vessels.
- **Mucosa** is the extensively folded innermost layer filled with connective tissue and covered in gastric glands that may be simple or branched tubular, and secret mucus, hydrochloric acid, pepsinogen and renin. The mucus lubricates the food and also prevents hydrochloric acid from acting on the walls of the stomach.

Phases of Gastric Secretion

- *Cephalic phase:* This phase occurs before food enters the stomach and involves preparation of the body for eating and digestion. Sight and thought stimulate the cerebral cortex. Taste and smell stimulus is sent to the hypothalamus and medulla oblongata. After this it is routed through the vagus nerve and release of acetylcholine. Gastric secretion at this phase rises to 40% of maximum rate. Acidity in the stomach is not buffered by food at this point and thus acts to inhibit parietal (secretes acid) and G cell (secretes gastrin) activity via D cell secretion of somatostatin.

- *Gastric phase:* This phase refers to the stimulatory effect of food in the stomach, although the purely mechanical effect is secondary to the chemical effect. This chemical effect, especially pronounced in the case of water-soluble products of meat and various peptones and peptides, accounts for most of the stimulation in this phase. The mechanism of the chemical secretion is not yet clear, but the leading theory holds that these stimulants, called "secretogogues", bring about the production of a hormone by the pyloric mucosa. This substance then is absorbed into the blood stream and activates the secreting glands of the stomach on reaching them. Many of these stimulants cause a copious secretion when introduced into the stomach, but they have little effect on intravenous injection. Edkins found that certain extracts of ground pyloric mucous membrane evoked a copious flow of gastric juice when administered intravenously. He coined the name "gastrin" for the active material in these extracts and felt that this substance, prepared by him *in vitro,* was normally formed *in vivo* from peptones, meat extractives, and other material contacting the pyloric mucosa. More recently other tissue extracts were found to contain a substance or substances similarly active. Histamine was found to be a powerful stimulant for gastric secretion; it was isolated from gastric mucosa and other tissues, and consequently gastrin and histamine were supposed to be one and the same substance. This view is now held by many, although considerable experimental work tends to invalidate such a theory. Some reports indicate that the active principle is protein in nature, although Edkins demonstrated an increased, rather than a decreased, activity on boiling his preparations. It is impossible at present to be certain whether gastrin is identical with histamine.
- *Intestinal phase:* This third phase of gastric secretion is not well understood. Certain foods placed directly in the duodenum of experimental animals cause gastric secretion, and drainage of the duodenal contents decreases gastric secretion. It is thought that the mechanism of stimulation is similar to that concerned in the gastric phase; i.e., hormonal. The composition of gastric juice. Gastric juice is a watery solution produced by three types of cells of the gastric mucosa; pepsin by the chief cells, HCl by the parietal cells, and mucin by the columnar epithelial cell&. Gastric juice, as such, is a mixture containing about 99.4 per cent water. The solids are composed of the organic substances mucin (glycoprotein), pepsin, and possibly small amounts of lipase and other enzymes of little or no importance in digestion. Some organic acids such as lactic are always found in small quantity. The inorganic constituents are H^+, Na^+, K^+, Cl^-, and small amounts of phosphates. Pepsin is secreted by the chief cells of the gastric glands in an inactive state called "pepsinogen." It is activated by hydrogen ion yielding the enzyme pepsin. At pH 4.6 or below the activation becomes autocatalytic; that is, pepsin activates pepsinogen. This enzyme starts the digestion of many native proteins, splitting them into smaller fragments. Under the proper conditions small peptides and even amino acids may be produced, The extent of pepsin action depends upon the length of time of contact with food in the stomach, the pH,

and other factors. The optimum pH of the enzyme is low and varies according to the nature of the substrate molecules from around 1.5 to 2.5; hence it is inactivated shortly after entering the intestine. It is questionable if rennin is a constituent of human gastric juice. According to Dotti and Kleiner, none is present in adult gastric juice. In ruminants rennin is found in the fourth stomach. It has recently been crystallized and is distinct from pepsin. Pepsinogen and pepsin have also been crystallized. Casein of milk is acted on in some way by rennin to form paracasein, which in the presence of calcium ions forms insoluble calcium paracaseinate. Many proteolytic enzymes including pepsin have this milk-clotting power, but to a lesser degree."Phosphoamidase" is the name applied to certain proteolytic enzymes capable of clotting milk and also able to hydrolyze such a compound as N-(p-chlorophenyl)amidophosphoric acid. It was found that rennin, chymotrypsin, and pepsin have a fairly constant ratio of milk-clotting and phosphoamidase activity (breaking a P-N bond), and the suggestion was made (17) that the first step in the coagulation of casein is not proteolysis, but the hydrolysis of P-N bonds. It has been reported that pure crystalline rennin shows such activity only upon the addition of a thermostable activator from crude rennin or milk. The purpose of clotting milk and other liquid proteins, such as egg white, in the stomach is to prolong the stay of these substances to allow more time for digestive action. Gastric lipase has been reported as a constituent of gastric juice, but its presence as a secretory product is doubtful. Its action in the stomach could be of little or no importance in digestion. Buchs reports that pig gastric mucosa contains" enormous" quan-tities of cathepsin which can be extracted by highly acid solution. Extracts prepared with weakly acid solutions or with glycerin contain only traces of the enzyme. Lysozyme is apparently a normal constituent of gastric juice.The important small mucoproteinintrinsic factor is produced in the stom-ach.

Hydrochloric Acid of Gastric Juice

Gastric juice, as stated before, is a mixture of various secretions. Its composition varies considerably, depend-ing on the type of stimuli acting to bring about secretion. Mixed gastric juice has been analyzed by many workers, but pure parietal cell secretion has not been collected. Its composition has been arrived at indirectly, especially by Hóllander. He plotted total acidity of gastric juice against neutral chloride content and found that the curve was a straight line and that an inverse ratio exists between these two factors. On extrapolating the curve to a point representing zero concentration of neutral chloride, the correspond-ing value for the acidity was found to be 165 meq per liter. This was taken to be the HCl concentration; this concentration is isotonic with blood. At this point phosphates also showed zero concentration, and thus it was Hollander's belief that pure parietal cell secretion is a solution of isotónic HCl only. Other workers confirmed important parts, but not all, of his work. Gray confirmed the total chloride content, but arrived at a lower value for HCl concentration. His calculations show values of 166 meq of chloride ion, 159 meq of hydrogen ion (HCl), and 7 meq of potassium ion per liter as the com-position of parietal cell secretion. Such a solution is also isotonic. It would be about 0.16 *N* HCl with a pH close to 0.9. According to Hollander, the composition is practically independent of the rate of formation, the strength of the stimulus, and probably the nature of the stimulus.

All evidence points to the blood as the immediate source of the chloride ion used to form HCl. In the gastric circulation the arterial blood has a higher chloride content than venous blood while active secretion is in progress. Even the general circulation shows this effect. As quickly as one to two minutes after the injection of radioactive chloride into the general circula-tion, it appears in the gastric juice. There is little doubt that the blood sup-plies the chloride ion and that the lymph and tissues are drawn upon to make up the deficit.

There remains to be established the mechanism of HCl formation. For many years this problem has received philosophical attention, but only re-cently have theories been postulated on experimental evidence.

Theories of HCl production. The mechanism of HCl production is not known. Many facts are established and various theories have substantial experimental backing. The earlier theory of Davenport though well founded in many respects, is now thought to be incomplete as an explana-tion for production of all the HCl of gastric juice. Davenport demonstrated a high concentration of carbonic anhydrase in the parietal cells of the gastric mucosa. This enzyme brings about the reversible hydration of

$$CO_2: H_2O + CO_2 \rightleftarrows H_2CO_3.$$

The theory holds that CO_2 is converted into H_2CO_3 in the parietal cells and that the carbonic acid ionizes to form $H^+ + HCO_3^-$. The hydrogen ions in some way are secreted, and since electrical neutrality must be maintained, negative ions accompany them. Chloride ions are taken from the blood to be secreted with the H^+. The bicarbonate ions remaining are used to replace the chloride ions taken from the blood. This theory cannot account for total H^+ production, since Davies and coworkers and others have established that more H^+ is produced than molecules of oxygen used by a secreting gastric mucosa. This indicates that more CO_2 is required than that available from oxidative processes within the parietal cells. The prob-lem is not as simple as might be depicted by the equation:

$$C_6H1_2O_6 \rightarrow \text{-}6CO_2 + 6H_2O \rightarrow 6H_2CO_3 \rightarrow 6H^+ + 6HCO_3^-.$$

Davies and coworkers proposed that although glucose may be the source of some of the H^+ (according to the equation above) water is one substance which is capable of supplying equivalent amounts of acid and base in sufficient quantity to account for the rapid production of gastric acid. Theories were proposed to account for breaking water into H^+ and OH^- ions, and mechanisms put forth to account for the required energy for this action and for the transport of the H^+ produced. The equations

$HOH \rightarrow H^+ + OH^-$ and $OH^- + CO_2 \rightarrow HCO_3^-$ depict in general terms part of what may proceed according to this theory.

It is discouraging that as late as 1959 Davenport in a review stated that all that is known of the metabolic aspects of gastric HCI secretion was covered in a symposium held in 1955. Thus, progress in our understand-ing of this significant field is slow.

The Secretory Mechanism: The secretory mechanism consists of at least four parts: (1) Aerobic and anaerobic reactions in the parietal cells produce high energy phosphate bonds (–P). These provide energy to drive the secretory reactions. (2) The secretory reactions may be indicated in three steps; *(a)* One (–P) combines with a reduced low-energy precursor to form a reduced high-energy precursor. *(b)* This high-energy compound is oxidized yielding a H^+, a low-energy precursor, and an electron. The energy of the parent compound is utilized in transporting the H^+ against a concen-tration gradient.

(c) The oxidized low-energy precursor becomes reduced by substrate electrons and is available to undergo reaction (2(a))-that is, combine with (–P). The electron liberated in reaction *(2(b))* is finally accepted by oxygen. One oxygen molecule is reduced for each four H^+ secreted producing four hydroxyl ions (or their equivalent). Four electrons are required to produce $2OH^-$ from O_2. (4) The OH- produced are neutralized by various buffers, including the system

$$CO_2 + H_2O \rightarrow H_2CO_3 \text{and } H_2CO_3 \rightarrow H^+ + HCO_3^-,$$

in which case the H^+ reacts with the OH- to form water.

The high energy phosphate bond may reside in ATP or creatine phosphate. The compounds yielding H^+ for gastric juice are substrate molecules.

Stimulation and Inhibition of Gastric Secretion

Fat in the intestine, causes the production of a substance, apparently hormonal in nature, which inhibits gastric secretion and gastric motility. Active preparations have been obtained from intestinal mucosa by Ivy and associates. On injecting such material, secretion is inhibited over a period of hours. The active principle has been named enterogastrone. Interesting applications of such preparations in allaying ulcer development and in treating certain ulcers have been reported. The motility inhibition may be due to a separate factor in the preparations. Urogastrone from urine also inhibits gastric secretion.

Small repeated doses of alkali or of dilute alcohol or of bile have a secretogogue effect on the stomach mucosa. Dilute HCl also stimulates secretion, but higher concentrations are inhibitory. Repeated injections of acetyl 3-methyl choline chloride (Mecholyl) cause a copious secretion of gastric juice. Pilocarpine also stimulates secretion. According to Hollander, and contrary to the previously accepted view, this drug stimulates the production of HCl, pepsin, and water, and not primarily of mucin.

Histamine is one of the most potent gastric secretogogues. It is used in clinical studies to test the response of the gastric secretory mechanism. This will be brought out further under the discussiop of gastric analysis. A number of factors tend to neutralize or dilute the acid of the stomach. The glycoprotein mucin an ingested food proteins buffer the acid. Saliva dilutes the acid, and the alkaline duodenal regurgitations neutralize some of it. The significance of gastric urease in relation to ammonia formation from urea and the part these play in neutralizing gastric juice in normal human beings and in certain cases of peptic ulcer have been studied in some detail

Functions of HCl in Gastric Juice

Hydrochloric acid is essential for the activation of pepsinogen in the stomach. It also has an important antiseptic action resulting in less gastric fermentation and also aids in the prevention of body infections. Its direct action on foodstuffs is probably of minor importance. On entering the duodenum with the chyme, it has the important function of activating prose cretin (see intestinal digestion).

In man stomach emptying after a mixed meal is completed in three to five hours. Fatty foods leave the stomach slowly, proteins more rapidly, and carbohydrate foods most rapidly. Liquids begin to leave almost as soon as they enter the stomach. Many other factors are involved in regulating the rate of emptying of the stomach. The mechanism responsible for opening the pyloric sphincter and closing it after a small amount of chyme has entered the duodenum is not too clear. It is established that the acidity of the chyme is not a specific stimulus for this, since the stomach is capable of emptying an alkaline content. A number of factors, of which the acidity of the chyme is but one, are known to be capable of causing closure of the sphincter. The rhythmic opening of the pyloric sphincter is related to gastric motility.

Gastric Analysis

The clinical value of an examination of gastric contents is somewhat restricted. The material is difficult to obtain compared to blood or urine; the methods of analysis are far from perfected, and often the results of analysis are difficult to interpret. Most frequently the free and the combined hydrochloric acid are determined. Since lactic acid may be produced in larger amounts by fermentation when the HCl content is low, the organic acid fraction is sometimes determined. Pepsin activity, blood, and bile are estimated less frequently. Microscopic examination may be used to establish the nature and number of various cells present. The emptying time of the stomach is also often of interest.

A test meal-such as the Ewald meal, consisting of dry toast and tea is given the subject in the postabsorptive state. Many test meals have been proposed and used by various workers.

Histamine as previously pointed out stimulates gastric secretion. It is sometimes used in place of a test meal. It has the advantage that by the use of enough histamine or by a second injection maximal stimulus may be obtained and sustained secretory function studied. It also eliminates many factors concerning the test meal itself, such as dislike for tea and variations in the composition of the toast or the tea. Histamine stimulation has been criticized, however, because it provides no food in the stomach, and since the gastric stimulating hormone and histamine have not been positively identified as the same substance, it is felt by many that it may not be a physiologic stimulus.

One of the above procedures is generally used to evoke a flow of gastric juice, and samples are then aspirated through a stomach tube for study. It was customary some years ago to remove a sample one hour after the test meal, but more adequate information are obtained by the fractional technique of Rehfuss. The stomach is first emptied of the residual gastric content and then the test meal is given. Successive samples are removed at IS-minute intervals for an hour, or longer if it seems advisable.

The samples are strained or filtered, and certain qualitative and quanti-tative estimations conducted.

Hydrochloric Acid

An aliquot of the filtered gastric contents is titrated with standard alkali to a pH of about 3.5. Topfers reagent (dimethyl amino-azobenzene), or other indicators which may yield a sharper end point are used. Standard buffered mixtures may be employed for color reference in the titration. This gives the free hydrochloric acid. Reference to the titration curves of strong acids indicates that HCl is almost completely converted to salt at pH 3.5.

Total acidity may be determined in the same sample by adding phenol-phthalein and continuing the titration to the end point of this indicator. Phenol red is preferred by some because it gives a color change around pH 7 (phenolphthalein changes at about pH 8.5), and titrating to this pH excludes some of the buffer material which is titrated when phenolphthalein is used.

By subtracting the free from the total acidity, the combined acidity is obtained. This value includes the HCl present as protein salts, the organic acids, and the titratable acidity of certain buffers such as phosphates.

It is common practice to express the results in milliliters of 0.1 *N* NaOH required to titrate 100 ml of gastric juice. These values are called "units." Total acidity of gastric juice removed from the stomach one hour after the test meal varies in normal individuals over the wide range of from 35 to 70 units. It is normally highest at this time, and this level is maintained for about one-half hour and declines to the resting level after another one or two hours. Free acidity normally runs parallel to total acidity and varies from 20 to 50 units. Like many clinical determinations, the titration of free HCl in gastric juice is not an exact method. In fact, it need not be, since the normal variations are great. Only values outside the normal range are clinically significant, and they can be found with the methods employed.

Achlorhydria refers to the absence of HCl in the gastric juice. It is common in pernicious anemia, but a high percentage of older persons also show this condition.

Pepsin

In one method for the determination of peptic activity, capillary glass tubes are filled with egg albumin. After heat coagulation of the protein, the tubes are placed in containers with known amounts of gastric juice adjusted to a definite pH. The peptic activity is calculated after measuring the length of the albumin column liquefied by the enzyme at the ends of the tubes in a definite time interval. Hemoglobin, gelatin, and other proteins are used as substrates in the various techniques described.

Blood

Blood may be tested for by the very sensitive benzidine reaction. A green or blue color develops in the presence of minute quantities of blood on the addition of an acetic acid solution of benzidine and then the addition of a little hydrogen peroxide.

The significance of abnormal findings in gastric analysis cannot be discussed here. Textbooks on laboratory diagnosis should be consulted for such information.

THE DIGESTION PROCESS IN INTESTINE

After the chyme enters the intestine, three distinct secretions are involved in the completion of the digestive processes: the pancreatic juice, the intestinal juice or succus entericus, and the bile.

Pancreatic Juice

The digestive juice secreted by the pancreas is a clear watery solution containing about 1.5 per cent solids. Approximately two-thirds of the solids are inorganic and one-third organic. The latter consists of proteins, including several enzymes and smaller molecules. The concentration of inorganic salts is about isomolar with that of blood plasma, close to 160 millimolar, and the positive and negative ions are equivalent.

The pH of pancreatic juice varies from 7 to 8 roughly. The alkalinity is due primarily to HCO_3^-. It has been estimated that an adult secretes about 650 ml of pancreatic juice in 24 hours. The specific gravity averages 1.008.

The blood plasma is the source of the cations, the bicarbonate, and perhaps the other inorganic ions. In cats the alkali reserve of the resting pancreas and of this gland after copious secretion stimulated by secretin is the same. Also, the concentrations of sodium, potassium, and calcium in the cat's pancreas are not changed after profuse secretion.

It was administered that bicarbonate containing radioactive carbon to dogs by intravenous injection. They found the radioactive bicarbonate in the pancreatic juice in a short time and at levels four to five times that of blood serum. Since the total CO_2 content of the pancreatic juice was found to be four or five times that of serum, these workers concluded that the blood is the primary source of CO_2 in pancreatic juice.

The enzymes of this digestive juice will be considered presently.

Stimulation of pancreatic secretion. When chyme enters the duodenum, secretin in some way is made available to enter the blood stream and travel to the pancreas, where it exerts its hormonal effect of stimulating this gland to. secrete. The above, statement is based on good evidence which very briefly is as follows: In 1902 researchers found that a hydrochloric acid extract of mucosa of the duodenum contained something that on injection into an animal's blood brought about pancreatic secretion. They also demonstrated that introduction of HCl into the intestine caused the pancreas to secrete even after denervation of this organ. They concluded that the mechanism of stimulation is a chemical or hormonal one, and they proposed the name "secretin" for the hormone involved.

Later work has established that their general assumptions are correct. For example, transplanted a loop of intestine and also the tail of the pancreas into subcutaneous tissue in dogs. This procedure destroyed the nerve supply of these organs. When acid was placed in the isolated loop, the transplanted pancreas secreted. A number of similar experiments afford proof for the hormonal nature of pancreatic stimulation.

Various workers have proposed that secretin exists in a prosecretin form in the intestinal mucosa and that acid brings about the required change in some way. Potent extracts can be prepared by extracting intestinal mucosa with dilute HCl. But other substances in contact with the mucosa also cause the production or liberation of the hormone. Fatty acids, alcohol, or even water are effective.

The purification of secretin has been reported by various workers. Agren made careful analyses ohighly purified crystalline secretin.

He reported that one molecule contains two molecules of arginine and proline, one molecule of histidine, aspartic, and glutamic acids, and three of lysine. A molecular weight of 5000 was found by the ultracentrifuge method of Svedberg. Agren suggested that the polypeptide contains 50 amino acid residues. Niemann calculated from Agren's data that the secretin molecule probably contains 36 amino acids and proposed the following partial empirical formula:

$$L_3Ar_2P_2H_1Gl_1As_1M_1X_{25}$$

where L = lysyl, Ar = arginyl, P = prolyl, H = histidyl, GI = glutaminyl, As = asparaginyl, M = methionyl, and X = unknown amino acid residues.

Blood serum apparently contains an enzyme termed" secretinase" by Ivy and coworkers, since incubation of secretin with serum results in destruction of the hormone. Harper and Raper have extracted a second hormone from the small intestine which stimulates the secretion of pancreatic enzymes. They named the substance" pancreozymin." Enzymes of pancreatic juice. Trypsin and chymotrypsin are elaborated in their zymogen or inactive forms called" trypsinogen" and" chymo-trypsinogen." These have been crystallized. Enterokinase is an enzyme apparently produced in the small intestine and it is responsible for converting trypsinogen into active trypsin. A small amount of trypsin can then act autocatalytically to convert more trypsinogen and also chymotrypsinogen into active enzymes.

The mechanism of this reaction has been studied by various investigators. It is now clear that activation of trypsinogen and chymotrypsinogen involves cleavage of a single peptide bond between a basic amino acid and an isoleu-cyl–valyl sequence. Upon activation of trypsinogen by trypsin, there is released a peptide with the probable structure valyl-(aspartyl)4-lysine, and active trypsin. It appears therefore that the single hydrolytic event leading to active enzyme involves hydrolysis of a lysyl-isoleucine bond.

Researchers studied various physical and chemical properties of the precursor proteins during the course of activation. Their data indicate that the decrease in specific rotation of chymotrypsinogen and trypsinogen is directly correlated with the appearance of enzyme activity. It was suggested that the observed changes in optical rotation arise from structural rearrangements and that this change is intra- rather than intermolecular. The activation of zymogens is reviewed by Neurath

Trypsin and chymotrypsin are generally classed as proteinases (endopeptidases), since they attack peptide chains not at terminal bonds, but at more centrally located bonds. Thus small peptide chains result from the action of these enzymes. This is not to be construed as meaning that these enzymes cannot hydrolyze terminal groups under the proper conditions. Certainly work with specific polypeptide substrates indicates that the amino acid arrangement on the peptide chain is of great importance in establishing what bonds will be broken by any proteolytic enzyme. It is, however, safe to assume that these two enzymes (proteinases) do produce large protein break-down products during intestinal digestion. These are called "proteoses", "Hpeptones" and "polypeptides". Some amino acids are undoubtedly produced also. The enzyme specificity determines what type of protein is readily attacked by the enzymes. Thus, trypsin is highly active toward the basic proteins, histones, and protamins, while pepsin of the gastric juice does not attack them appreciably.

Carboxypeptidase is another proteolytic enzyme in pancreatic juice. It splits the peptide bond at the end of the peptide chain having an amino acid with a free carboxyl group. The action can be repeated until an entire peptide chain is reduced to amino acids if the proper arrangement exists. If the amino acid configuration is not one that this enzyme can handle, other enzymes in intestinal juice may finish the hydrolysis.

Steapsin or pancreatic lipase is a potent lipolytic enzyme of pancreatic juice. It appears that the enzyme as elaborated is not very active, but a number of substances, such as soaps, bile salts, certain proteins, and amino acids, can activate it.

It is well established that triglyceride hydrolysis is far from complete in the intestine. Mattson has reviewed this field and reported much of his own work. This investigator studied digestion products of triglycerides with fatty acids in known positions. After feeding 2-oleoyl dipalmitin to rats for instance, lipids were isolated from the intestinal lumen. The monciglycerides were found to be predominantlymonoolein, and the free fatty acids were palmitic acid. From such *in vivo* work and various *in vitro* studies with pancreatic lipase, Mattson concludes that the hydrolysis of triglycerides is a series of directed step-by-step reactions from triglyceride to 1,2-diglyceride to 2-monoglyceride and that the enzyme is specific for ester groups on the primary hydroxyl groups. Free glycerol could result from enzymatio isomerization of the 2-monoglyceride to the 1-monoglyceride followed by lipolytic hydrolysis.

Amylopsin or pancreatic amylase (an α-amylase) hydrolyzes starches, dextrins and glycogens. The enzyme attacks more interior linkages in the branched oligosaccharides, such as starch and glycogen, producing units larger than maltose and then producing maltose. Amylose (linear molecule) following exhaustive hydrolysis with amylase is oonverted completely to maltose (87%) and glucose (13%). The 1,4-α glucoside linkage is attacked in all cases.

Intestinal Juice

The intestinal juice or succus entericus is a mixture of the secretions of various glands of the intestinal mucosa. It is concerned with digestion by virtue of its mucin and fluid content which help lubricate and aid in keeping the partly digested food particles in motion along the intestinal tract. The enzymes of the small intestinal mucosa probably operate intracellulary, and the enzymes present in the gut contents may arise from desquamation rather than from secretion. The composition of intestinal juice varies from time to time and from one level of the intestine to another. In general, however, the mixed juice may contain from 1 to 2 per cent solids, primarily mucin, enzymes, and inorganic salts. The organic and inorganic matter is distributed about equally. The pH of the juice is generally 7 to 8. Various types of cells are always present. The secretion of intestinal juice is stimulated by food in the intestine. This acts mechanically through reflex nervous stimulation.

Enzymes of Intestinal Cells

Aminopolypeptidase hydrolyzes small peptides at the bond involving an amino acid with a free amino group. It can continue to remove amino acids from this end of the chain, provided the proper arrangement of amino acids exists. A dipeptidase breaks dipeptides into two amino acids. A prolinase is apparently a specific enzyme and is responsible for the hydrolysis of small peptides containing proline probably as a terminal group. The enzymes noted above, along with carboxypeptidase from pancreatic juice and probably others not yet well defined, constitute what previously was referred to as "erepsin." They complete protein digestion to the amino acid stage. In general, they are inactive on native protein molecules; other enzymes, such as pepsin and trypsin, must initiate the hydrolytic processes. Nucleoproteins are hydrolyzed by the digestive enzymes. It is not clear where the various enzymes involved are produced.

It is known that intestinal cells contain a polynucleotidase which splits nucleic acids into individual nucleotides. A nucleotide contains ribose or deoxyribose, phosphoric acid, and a purine or pyrimidine base. The nucleotides are further hydrolyzed by an intestinal phosphatase called "nucleotidase," yielding phosphate and a nucleoside. Finally, a part of the nucleoside, in the presence of phosphate, is converted by nucleosidase (a phosphorylase) to base and a sugar phosphate. Intestinal contents also contains enterokinase, the enzyme responsible for the activation of trypsinogen elaborated in the pancreatic juice. The disaccharid hydrolyzing enzymes lactase, sucrase, (invertase), and maltase are found in mucosal cells. Although little progress has been made in the purification of intestinal enzymes hydrolyzing the phospholipids, it is probable that lecithins, cephalins, etc., are hydrolyzed by the same enzymes and the name "phospholipase" is better suited than the older term "lecithinase."

Different phospho-lipases capable of splitting different linkages in these molecules occur in the intestine. Hydrolysis at the linkage between the primary alcohol of glycerol and the phosphate (phospholipase C action) yields a diglyceride and phosphorylcholine. The diglyceride would be handled as fat and the phosphoryl-choline absorbed or further split by a phosphatase. Original removal of the fatty acid on the primary alcohol group of glycerol yields a lysolecithin (or cephalin), a strong hemolytic substance. Phospholipase A brings about this action; the enzyme is found in animal and plant tissues and in snake venom. Phospholipase B, also found in animal tissues, abolishes the hemolytic action of the lysolecithins by splitting off the secondary hydroxyl fatty acid, forming glycerylphosphorylcholine.

THE BILE JUICE

The third intestinal secretion important in digestion is bile. It is secreted continuously by the liver cells and is stored in the gall bladder. Though small amounts may enter the intestine regularly, the ingestion of food and other stimulants bring about emptying of the gall bladder, and thus larger quantities of bile enter the intestine at this time. Bile contains, in addition to the liver secretion, constituents added by the mucosa of the gall bladder and the biliary passage. During its stay in the gall bladder, water and certain inorganic salts are absorbed. This important concentration results in bile which may contain ten times the solids found in liver bile. A mucin-containing fluid is secreted by the gall bladder, and this, along with the concentration, results in more viscous bile as it leaves this organ. Certain excretory products may be added to the bile by the gall bladder.

Composition of Bile

The concentrating effect of the gall bladder is brought out well by the data in Table below, which shows averages and ranges for various constituents of bladder bile compared with those of liver bile. Bladder bile is fairly viscous; it is yellow, green, or other colors, depending on the species and on how long after death it is obtained. The amount of bile secreted by man has been estimated at from 500 ml to 1100 ml per 24 hours.

Hepatic duct bile is generally alkaline because of the bicarbonate present. It may have a pH as high as 8.6 or as low as 6.9. An average of a number of reports gives a pH 7.7. In the gall bladder part of the bicarbonate is absorbed; as a result of this and also on account of the Donnan membrane equilibrium, bladder bile may be nearly neutral or in some instances as acid as pH 5.5.

Hepatic bile, intestinal juice, gastric juice, and saliva all have approximately the same total electrolyte content. They are also about isomalar with blood plasma. Bladder bile is the exception and averages around one-third more in electrolyte concentration. Concentration of hepatic duct bile by the gall bladder can increase the specific gravity from 1.010 to as much as 1.040. Three important constituents of bile, cholesterol, bile salts, and bile pigments, as well as the chemically related porphyrins, will be discussed individually.

Cholesterol

Bile contains considerable cholesterol (Table 9.1). The sterol is synthesized and excreted by the liver. The concentration of cholesterol in bile is usually not related to diet, although in a patient with an excessively high serum cholesterol level it was shown by Hellman and co-workers that the feeding of corn oil lowered serum cholesterol to about the same extent as the increase in fecal sterols. After butter feeding the increased serum level was about the same as the decreased fecal sterols.

Table 9.1 : The Composition of Bile Juice

Constituent	Bladder Bile	Fistula Bile
Water	888.6	973.4
Total solids	111.4 (47-165)	26.6 (10-40)
Bile acids	51.8 (14-92)	10.9 (4.2-18.3)
Mucin and pigment	34.2 (18-43)	6.1 (4.3-9.3)
Total lipid	22.5 (19-26)	3.4 (2.9-4.2)
Neutral fat	3.7 (1.5-5.6)	1.1 (0.4-3.0)
Fatty acids	9.7 (9-10.9)	1.1 (0.8-1.4)
Phospholipids	2.0 (1.8-2.2)	0.6 (0.50.6)
Cholesterol	6.3 (3.5-9.3)	1.2 (0.8-1.7)
Inorganic ions	8.5	7.5 (5.8-9.2)

Cholesterol is converted into bile acids by the liver (see further under Bile Acids). It was reported that lowering of serum cholesterol by dietary means resulted in an increase in fecal bile acids.

Some of the cholesterol excreted into the intestine is converted by bacterial enzymes into various products, including coprostanol in humans.

Bile Acids and Bile Salts

The parent bile acids found in human bile are hydroxylated cholanic acids. In all cases the hydroxyl groups have the orientation. The ring system is saturated. Quantitatively, the principal acids are cholic acid *(3a, 7 a,* 12a-trihydroxycholanic acid), chenodeoxycholic acid *(3a,* 7a-dihydroxycholanic acid), and deoxycholic acid *(3a,* 12a-dihydroxycholanic acid). A variety of other bile acids are found in various animal species. In man and in several animal species it is es-tablished that the bile acids are derived from cholesterol. In human and other mammalian bile the acids are conjugated with glycine and taurine in variable ratios. These conjugates are present as sodium salts -the bile salts-due to nearly complete ionization of the acids at physiologic pH. The conjugation involves loss of water from the carboxyl of the bile acid and the -NH_2 of the amino acid as in peptide bond formation. Glyco-cholic acid and taurocholic acid are the conjugation products of cholic acid. Elliot proposed that cholyl CoA is an intermediate in the conjugation and that its formation by guinea pig liver microsomes may be represented thus:

Cholic Acid + ATP + CoA $\rightleftarrows$ Cholyl CoA + AMP + PP

It is likely that conjugation in all cases involves activated bile acids. Taurine (NH_2-CH_2-CH_2-SOaH) is derived from cysteine through cysteinsulphiinic acid and 2-aminoethanesulfinic acid. The concentration of bile salts varies considerably. An average value for human bladder bile is 3 or 4 per cent.

Chenodeoxylic acid is the same as cholic acid except for Hydrogen replacing the OH at position 7 is replaced by H.

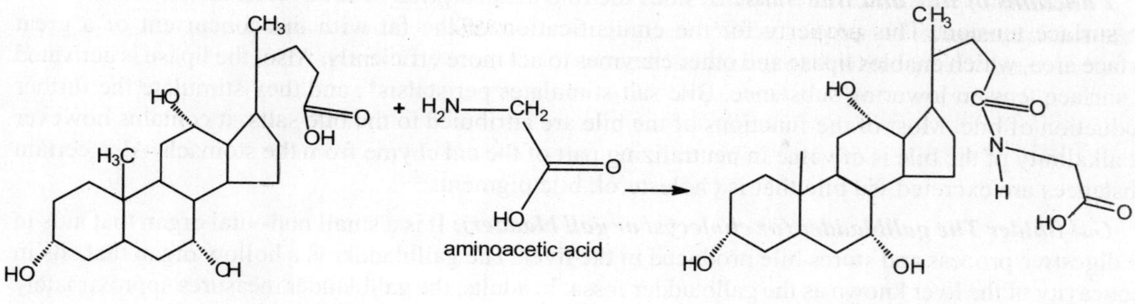

aminoacetic acid

Glycocholic acid

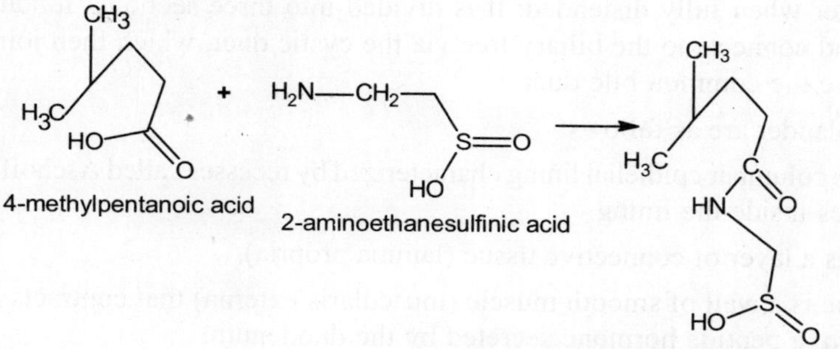

4-methylpentanoic acid

2-aminoethanesulfinic acid

[(4-methylpentanoyl) amino] methane sulfinic acid

Taurocholic acid

The steps in conversion of cholesterol to bile acids are not yet clearly defined. It has been seen that 3α, 7α dihydroxycoprostane and 7 α hydroxycholesterol are transferred into taurocholic acid and taurochenodeoxycholic acids, these investigations suggested that hydroxylation of cholesterol at carbon 7 is primary step in bile acid synthesis. Other steps have been proposed. Deoxy cholic acid is formed from cholesterol and is readily converted into cholic acid

Circulation of Bile Salts

Normally small amounts of bile salts occur in the feces. Synthesis of these compounds keeps the total body supply about constant. During normal digestion, when the bulk of the bile is delivered to the intestine, the bile salts are largely resorbed by the portal blood and are removed by the liver and reexcreted by the bile. This circulation of bile salts has very important functions. It appears that fatty acids as well as certain fat soluble vitamins are absorbed from the intestine through mechanisms involving the bile salts. Combinations of fatty acids and bile salts known as "cholic acids" are water soluble and as such they are more readily absorbed. A discussion of this problem can be found under fat absorption in the subhead Vitamin K and cholesterol are poorly absorbed formed the intestine in the absence of bile salts.

In this review on bile salt circulation it is indicated that the bile salts take part in the circulation about three times. So the absorption of digestion products mediated by a given quantity of bile salts is greatly increased through the circulation mechanism.

Functions of Bile and Bile Salts: Besides the role in absorption of food stuff, the bile also lowers the surface tension. This property for the emulsification of the fat with the concurrent of a great surface area, which enables lipase and other enzymes to act more efficiently. Also, the lipase is activated by surface tension lowering substance. Bile salt stimulates peristalsis*, and they stimulate the further production of bile. Most of the functions of the bile are attributed to the bile salts, it contains however the alkalinity pf the bile is of value in neutralizing part of the cid chyme from the stomach. Also certain substances are excreted via bile that is cholesterol, bile pigments.

Gal Balder The gallbladder (or cholecyst or gall bladder): It is a small non-vital organ that aids in the digestive process and stores bile produced in the liver. The gallbladder is a hollow organ that sits in a concavity of the liver known as the gallbladder fossa. In adults, the gallbladder measures approximately 8 cm in length and 4 cm in diameter when fully distended. It is divided into three sections: fundus, body, and neck. The neck tapers and connects to the biliary tree via the cystic duct, which then joins the common hepatic duct to become the common bile duct.

The different layers of the gallbladder are as follows:

- The gallbladder has a simple columnar epithelial lining characterized by recesses called Aschoff's recesses, which are pouches inside the lining.
- Over the epithelium there is a layer of connective tissue (lamina propria).
- Above the connective tissue is a wall of smooth muscle (muscularis externa) that contracts in response to cholecystokinin, a peptide hormone secreted by the duodenum.
- There is, in essence, no submucosa separating the connective tissue from serosa and adventitia, but there is a thin lining of muscular tissue to prevent infection.

The adult human gallbladder stores about 50 millilitres (1.8 imp fl oz; 1.7 US fl oz) of bile, which is released when food containing fat enters the digestive tract, stimulating the secretion of cholecystokinin (CCK). The bile, produced in the liver, emulsifies fats in partly digested food.

After being stored in the gallbladder, the bile becomes more concentrated than when it left the liver, increasing its potency and intensifying its effect on fats.

Gallstones

Gall Bladder Stones, also known as Gallstones, are a few small, pebble-like substances that develop in the gallbladder of a person. The Gallbladder is a pear-shaped organ of the body, located below the liver and is attached to the under-surface of the liver on its right side. It stores the bile secreted by the liver. Bile is a waste product that contains water, cholesterol, fats, bile salts, proteins, and bilirubin. The gall bladder stones usually form when the bile stored in the Gallbladder hardens into pieces of stone-like material. Bile hardens mainly when it contains too much cholesterol, bile salts, or bilirubin. The gallstones are actually an inflammatory condition and are also known as cholecystitis. Though most gall bladder stones develop due to an accumulation of cholesterol, some of them may contain calcium salts and bile pigments. The women are more affected by this disease in comparison to men.

Gall bladder stones are of two types like cholesterol stones and pigment stones. The cholesterol stones are usually of yellow-green colour and are made primarily of hardened cholesterol. This type of

---

Peristalisis: the process of wavelike muscle contractions of the alimentary tract that moves food along.

gall bladder stones account for about 80 per cent of all gallstones. On the other hand, the pigment stones are actually small, dark stones made of bilirubin. The size of gallstones may range from that of a grain of sand or to that of a golf ball. One large stone, hundreds of tiny stones, or a combination of the two, can be developed in the gall bladder. There are many causes behind the formation of gall bladder stones. The main cause of gall bladder stones is disturbances in the composition of the bile. A change in the ratio of cholesterol and bile salts may result in the formation of deposits. The deposits may be in the form of fine gravel in the initial period. They constitute the nucleus for further deposits and eventually lead to the formation of larger stones. An irritation of the lining of the gall bladder, caused by inflammation is also considered as a cause of gall bladder stones. Having excessive cholesterol or calcium in the diet can cause gallstones, and the way body processes these substances also plays an important role in the formation of gallstones. Some people may also inherit narrow bile ducts that later increases their chances of developing gallstones. Obesity is considered one of the major causes of gall bladder stones, as well.

There are several symptoms of gall bladder stones that can help in identifying them at an early stage. The most common symptoms of gall bladder stones include heart attacks, ulcers, appendicitis, pancreatitis, hiatal hernia, hepatitis, etc. The symptoms of blocked bile ducts are often called as gallbladder "attack", as they occur suddenly. A typical gall bladder attack may cause various difficulties like, steady pain in the right upper abdomen that increases rapidly and lasts from 30 minutes to several hours; pain in the back between the shoulder blades; pain under the right shoulder; etc. All these difficulties are also considered as symptoms of gall bladder stones. The other common symptoms of gall bladder stones include nausea and vomiting; fever of low-grade or chills; yellowish colour of the skin or whites of the eyes; clay-coloured stools; etc. Symptoms like indigestion, gas, constipation, disturbed vision, intolerance to fats, dizziness, jaundice, anæmia, acne, etc. are also considered as common symptoms of gall bladder stones. However, some people with gall bladder stones may not show any of the symptoms and these gallstones are termed as "silent stones". These stones do not interfere with gallbladder, liver, or pancreas function and do not need treatment, as well. The diagnosis of gall bladder stones is done in different ways. Most of the times, diagnosis is done based on a person`s symptoms. A simple physical examination like gently pressing just below the ribs on the right side of the chest with fingers may help in diagnosing gall bladder stones. Sometimes, blood tests or a simple x-ray can show signs of obstructed ducts. Apart from these, there are two tests that are used most commonly for diagnosis of gall bladder stones. These tests are named as abdominal ultrasound and the oral cholecystogram (OCG) or cholescintigraphy and both the tests can detect gallstones with great accuracy. Apart from the conventional treatment procedures, many alternative medications are often used for the treatment of gall bladder stones. The alternative treatment methods like Ayurveda, Homoeopathy, Naturopathy or nature care, water treatments, Yogic Asanas, dietary restrictions, Magnetic Therapy, etc. are used quite extensively for treatment of gall bladder stones. The Oral bile acid dissolution therapy; contact solvent dissolution; mechanical extraction through a catheter placed into the gallbladder either through the skin or through and endoscope; fragmentation through shock-wave lithotripsy combined with bile acid dissolution; etc. are also used as altérnative medications for treatment of gall bladder stones. The gall bladder stones are considered as one of the most common types of diseases at present. Several people get affected by this disease every year. Though, the diagnosis and treatment of the gall bladder stones is quite easy to be done, delay in detection of the disease may cause greater harms to the affected person.

The Composition of the Bile Pigments

The main bile pigments are the bilirubin and biliverdin. The colour of the bile is due to these and also because of the derivatives of these. There are considered to be extraordinary products. Oxidations and reductions of these form a great variety of other pigments. Some of the pigment formations are found in the urine, in feces and also in the bile, and in certain instances in the skin also in other parts of the body.

It is an established fact the bile pigment is produced reticuloendothelial systems in various parts of the body. The bone marrow appears to be best active cite of formation.

The ultimate sources of bile pigment are the hæmoglobin of the broken down cells and to a lesser degree myoglobin. See the breakdown process in the reactions at next page.

The First step in the bilirubin formation involves the hydrolysis of hæme to the protein globin and the iron containing porphyrin hæme, by oxidative action the iron containing porphyrin ring is opened and the compound called α-hydrooxyme results on further oxidation to verdohæmin formation. The opening of the porphyrin ring enables the removal of the iron and the results in the formation of biliverdin. The investigation of the formula of the hæme and biliverdin clearly indicates the place of opening of the methyl group ($-CH_3$), i.e. between two different pyrrole carbon atoms, one holding a vinyl and the other a methyl group. The iron released is utilized for new hæmoglobin formation or is converted into ferritin for storage. Some may be excreted.

It also appears that other precursors of biliverdin exist in which the ring opening occurs before the removal of the protein. The terms verdoglobin and choleglobin have been applied to these substances. Choleglobin may be defined as bile pigment hæmoglobin. After the excretion of bile into the intestine, further changes occur with the formation of a number of new compounds. Several of these are established as to identify and chemical structure.

It is generally agreed that bilirubin on entering the intestine is reduced by addition of hydrogen ion, by bacterial enzymes to urobilinogen. Some of this material is oxidized to urobilin. The addition and removal of hydrogen can occur at various atoms in the rings, so that a number of different compounds are possible. The urobilinogen not oxidized in part absorbed by the portal system, removed by the liver, oxidized there to bilirubin and re-excreted in the bile. Small amounts of urobilinogen may escape removal from the blood by the liver and so be excreted by the kidney. In voided urine this material is oxidized to urobilin. Normally there is very little of this in urine, but in instances of impaired liver function and in obstructive jaundice larger amounts are found.

It is interesting that the small amount of urobilinogen normally present in bile-from incomplete oxidation to bilirubin by the liver-disappears in experimental bile fistula. This is explained by the fact that urobilinogen is produced in the intestine, absorbed, and re-excreted. If no bilirubin is excreted into the intestine, no urobilinogen is formed, and that already present is soon removed with the feces. Consequently, the fistula bile becomes urobilinogen-free.

Figure 9.23 is a schematic representation of the circulation and excretion of bile pigments and derivatives.

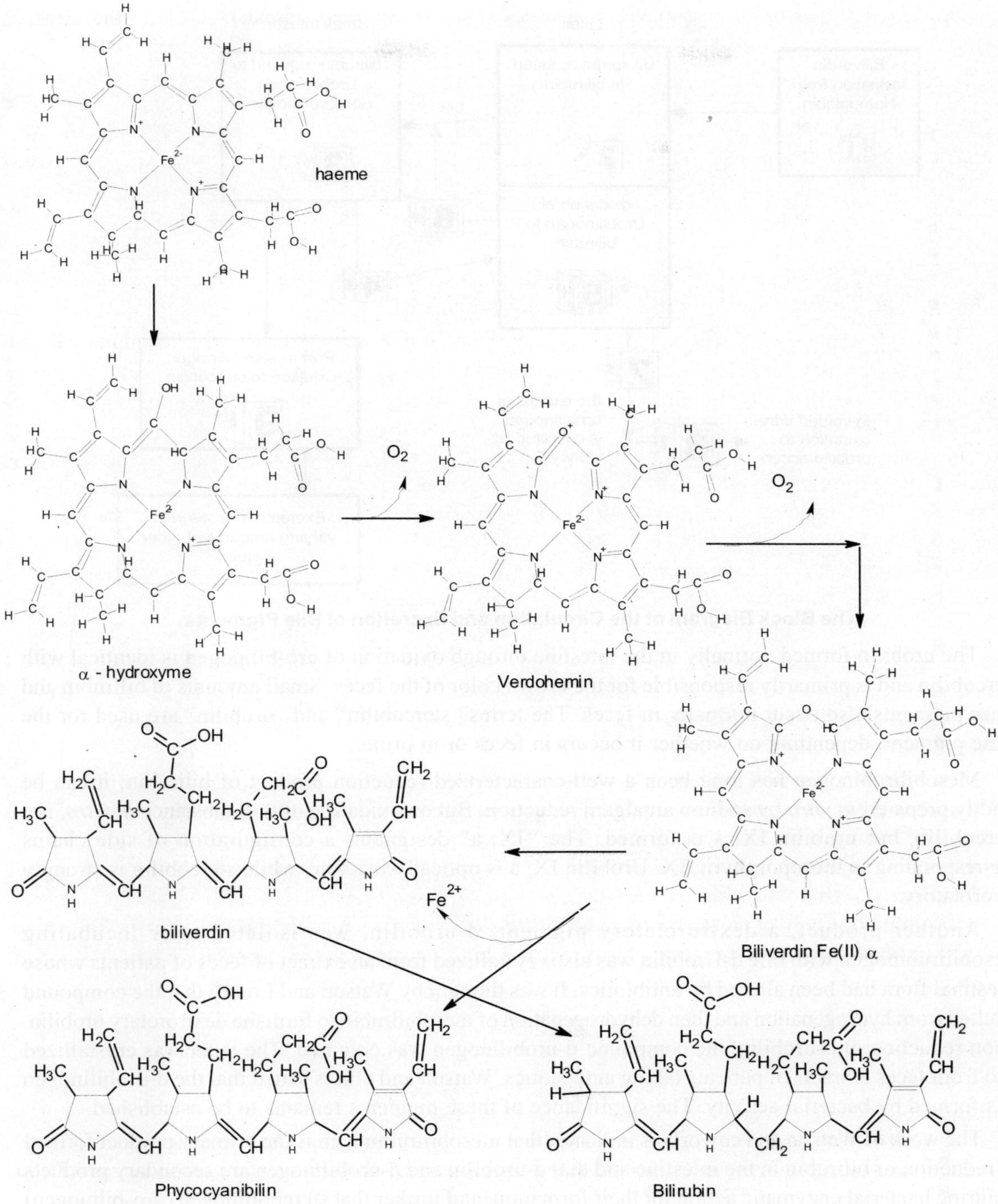

haeme

α - hydroxyme

O_2

Verdohemin

O_2

Biliverdin Fe(II) α

Fe^{2+}

biliverdin

Phycocyanibilin

Bilirubin

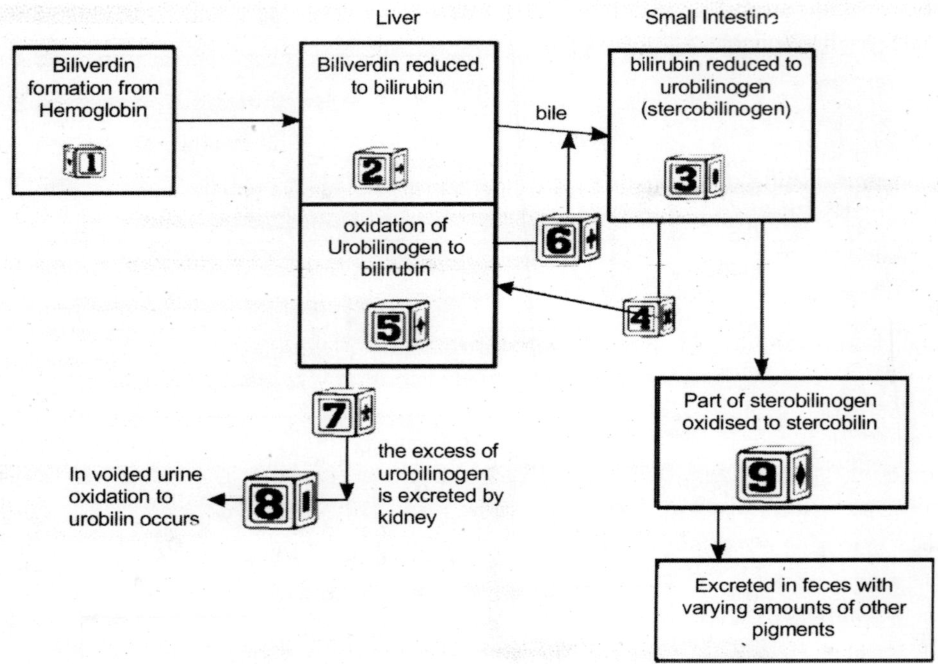

The Block Diagram of the Circulation and Excretion of Bile Pigments

The urobilin formed normally in the intestine through oxidation of uro-bilinogen is identical with stercobilin and is primarily responsible for the brown color of the feces. Small amounts of bilirubin and other pigments also occur normally in feces. The terms" stercobilin" and" urobilin" are used for the same pigment, depending on whether it occurs in feces or in urine.

Mesobilirubinogen has long been a well-characterized reduction product of bilirubin; it can be readily prepared *in vitro* by sodium amalgam reduction. But on oxidation of this substance *in vitro,* not stercobilin, but urobilin IX, a is formed. The "IX, a" designates a configuration of side chains cor-responding to mesoporphyrin IX. Urobilin IX, a is optically inactive, while stercobilin is strongly levorotatory.

Another product, a dextrorotatory pigment, d-urobilin, was isolated after incubating mesobilirubinogen with bile d-Urobilin was also crystallized from an extract of feces of patients whose intestinal flora had been altered by antibiotics. It was thought by Watson and Lowry that the compound resulted from hydrogenation and then dehydrogenation of mesobilirubin to form the dextrorotary urobilin. Upon reduction of d-urobilin, the compound d-urobilinogen was obtained. The latter was crystallized also from feces extracts of patients taking anti-biotics. Watson and others stated that the d-urobilinogen was formed by bacterial activity. The significance of these pigments remains to be established.

The work of Watson and coworkers indicates that mesobilirubinogen is the primary product formed by reduction of bilirubin in the intestine and that d-urobilin and d-urobilinogen are secondary products requiring bacterial enzymatic action for their formation, and further that stercobilinogen (uro-bilinogen) is formed from mesobilirubinogen in the intestine and can be oxi-dized to stercobilin (urobilin). Fig. 9.24 summarizes this concept.

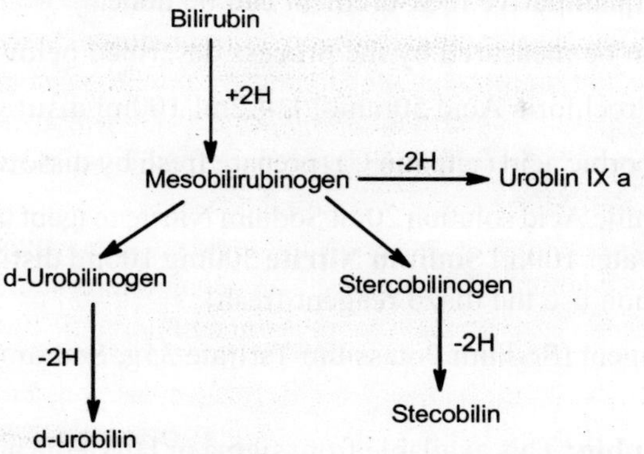

Bilirubin
↓ +2H
Mesobilirubinogen —(-2H)→ Uroblin IX a
Mesobilirubinogen → d-Urobilinogen —(-2H)→ d-urobilin
Mesobilirubinogen → Stercobilinogen —(-2H)→ Stecobilin

A normal adult may excrete 1 or 2 mg of bile pigment in the urine and as much as 250 mg in the feces in one day.

The accompanying structures show a number of the compounds discussed:

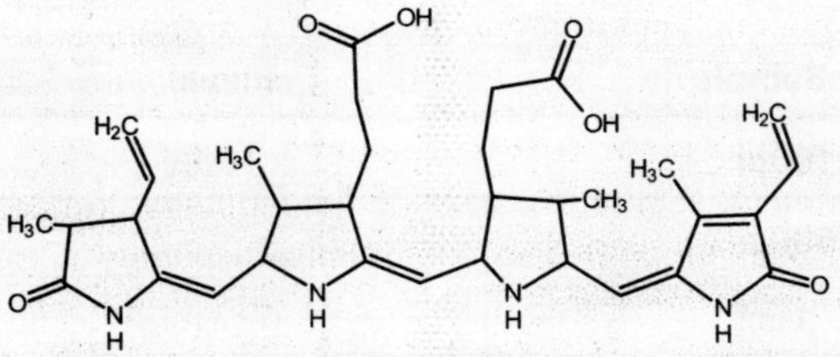

BILIVERDIN

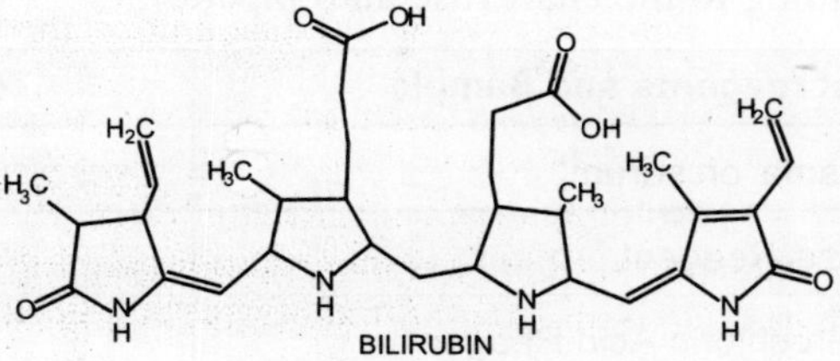

BILIRUBIN

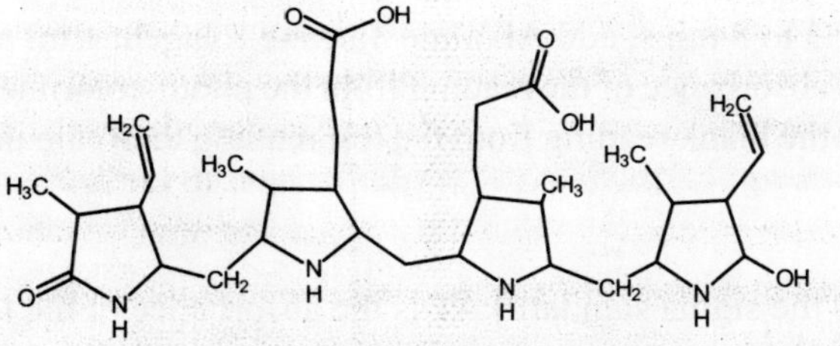

MESOBILIRUBINOGEN

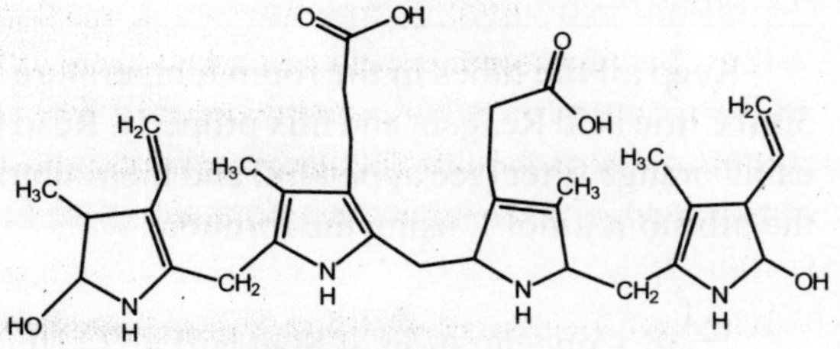

STERCOBILINOGEN

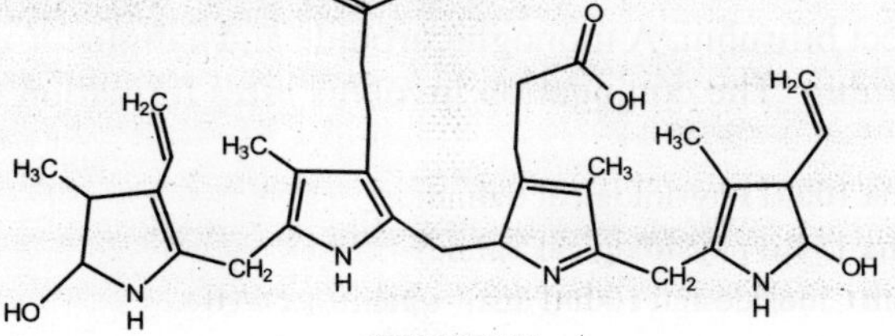

STERCOBILIN

The Van den Berg reaction has been used for many years in the determination of bilirubin in blood plasma. In practice plasma is treated with diazotized sulphinic acid in acid solution, producing a coloured complex from which the quantitative measurement can be done.

The total bilirubin can be measured by the process described below: -

Reagents reqd: - Hydrochloric Acid 50mmol/L 4.2ml/100ml distil water (mix together)

Sulphanilic Acid, Ascorbic acid (vitamin C) [prepare fresh by dissolving 0.2 gm in 5ml distil water]

Diazo reagent Sulphanilic Acid solution 20ml Sodium Nitrite reagent 0.5ml mix together {**sulpahnilic acid** 500mg 1.5ml distil water 100ml **Sodium Nitrite** 500mg 100ml distil water prepare both separately and keep under refrigeration use the diazo reagent fresh}.

Alkaline Tartrate Reagent (Sodium Potassium Tartrate 35g, Sodium Hydroxide 7.5g, 100ml Distil Water)

Control Serum Bilirubin: This available from sigma or Hi Media as Seronorm Bilirubin Pædiatric or Bilirubin reagent also from Loba (.20196g in 1L Distil water will give a conc. of 340µmol/L from Loba 500mg pack).

Procedure: Take two test tubes one marked "test" and the other as "blank" start the procedure according to the chart (use auto pipette)

Test reagents and Sample	Test Sample	Blank
Plasma or serum*	100µl	100µl
Diazo Reagent	500µl	—
Hydrochloric Acid Reagent	1 ml	1 ml
Sulphanilic Acid Reagent	—	500 µl

Keep all the tubes in the room temperature for 5 to 8 mins, add Alkaline Tartrate Reagent 1 ml and 50µl Citric acid Reagent and mix properly. Read the absorbance of the reading from the photo-colorimeter using orange filter (see appendix) and then subtract the blank reading from the test reading then calculate the bilirubin µmol/L using the formula.

$\frac{AT}{AS} \times$ Concentration of Standard or 34 [this is the sigma standard] AT is the absorbance of the test sample and AS is the absorbance of the Control Serum Bilirubin.

Bilirubin glucuronide is now established as the direct acting form, and unconjugated bilirubin, due to its insolubility in aqueous solution, is the indirect bilirubin. A monoglucuronide and a diglucuronide are now established as constituents of direct bilirubin. The conjugation involves ester formation with

* For Infants dilute the Plasma into 1:2 (i.e. 100µl plasma 100µl Physiological Saline)
For neonatal dilute the Plasma into 1.4 (i.e. 50µl plasma 150µl physiological Saline)
This physiological Saline is a mixture of 850 mg Sodium Chloride and 100ml distil water.

the carboxyl group of the propionic acid residues of bilirubin and the C-l-hydroxyl of glycuronic acid. In bile the diglucuronide is the main pigment with smaller amounts of monoglucuronide. Unconjugated bilirubin appears not to occur normally in bile. The glucuronides are bound to lipoprotein along with cholesterol and bile acids in gall bladder bile, while in hepatic bile the pigments are bound to proteins.

The following mechanism has been proposed for the formation of bilirubin glucuronide: glucose → glucose-6-PO_4 → glucose- → 1–PO_4 → uridine diphosphate glucose (UDPG) → uridine diphosphate glucuronic acid; this reacts with bilirubin in the presence of glucuronyl transferase to give bilirubin glucuronide.

The Porphyrins

In a variety of disease states abnormal amounts of other pigments may be excreted in the urine and feces. These pigments are closely related to the protoporphyrin in the hemoglobin molecule, but they are chemically distinct from it.

It is now clear that the porphyrins are intermediates in the synthesis of protoporphyrin of heme and are not, as was suspected some years ago, break-down products of hæme.

Researcher synthesized the four possible isomers containing only methyl and ethyl groups on the porphyrin ring. These were named" aetioporphyrin I," "II," "III," and" IV." They do not occur in nature, but naturally occurring porphyrins are classified according to the correspondence of the positions of the side groups with the ethyl and methyl groups of the synthetic aetiopor-phyrins. Fifteen isomers are possible by substituting four methyl, two propionic acid, and two vinyl groups on the basic porphyrin ring. These are called the "protoporphyrins", and a number of natural porphyrins contain these groups. By mild reduction of the protoporphyrins, hydrogen adds to the double bond of each vinyl group, and the resulting 15 possible isomers con-taining ethyl in place of vinyl groups are known as the" "mesoporphyrins." At the present time our knowledge of the biosynthesis of the porphyrins and thus of protoporphyrin, heme, chlorophyll, etc. is fairly complete. Each carbon atom and each nitrogen atom of the porphyrin structure is identified as to source, largely through the work of Shemin and his coworkers.

It was found as early as 1948 that red blood cells of birds can synthesize heme *in vitro*, and soon after cell free extracts of duck erythrocytes were obtained which also accomplished heme synthesis from the proper substrates. Through the use of labeled compounds it was shown that the N atom and the a-carbon atom of glycine were incorporated into the pyrrol structure, but that the carboxyl carbon was not used. Acetate, or succinate arising from acetate, is also incorporated into the pyrrol precursor (porpho-bilinogen). When four porphobilinogen molecules unite in the proper manner, a porphyrin molecule results. In protoporphyrin (see structures) eight of the 34 carbon atoms and the four nitrogen atoms arise from glycine; the other 26 carbon atoms are derived from succinate. Although eight glycine molecules are involved in the synthesis of one porphyrin molecule, four of the N atoms and eight carbon atoms (each carboxyl) are lost in the synthetic process. Upon degradation of protoporphyrin synthesized from glycine-2-C14 it was found that four of the carbon atoms were in the four methene bridges and the other four were distributed one in each pyrrol ring (next to the nitrogen). The remaining ring carbons and side-chain carbons are thus derived from succinate. Studies with carboxyl labeled and methylene labeled succinate have amply confirmed this.

Succinyl CoA + Glycine → 2-amino-3-oxohexanedioic acid

2-amino-3-oxohexanedioic acid

α- amino - β - Ketoadephic acid

CO_2

2

5-amino-4-oxopentanoic acid

Aminoelvulinic Acid

4

Prophobilinogen

3-[5-(aminomethyl)-4-(carboxymethyl)-1*H*-pyrrol-3-yl]propanoic acid

Uroporphyrinogen

co - proporphyrinogen

Oxidation

Decarboxylation

protoporphyrin iX

$+Fe^{++}$

HAEME

Succinyl CoA (active succinate) reacts with glycine to form α-amino-, β-keto-adepic acid. This compound is decarboxylated (the glycine carboxyl) to form o-aminolevulinic acid, and two molecules of this compound condense with the loss of water to form porphobilinogen. Now four molecules of por-phobilinogen condense with the loss of $4NH_3$ and the probable first intermediate formed is uroporphyrinogen, which is decarboxylated (four acetic acid side chains $\xrightarrow{-CO_2}$ four methyl side chains) yielding the intermediate co-proporphyrinogen. Oxidative decarboxylation (two propionic acid side chains $\xrightarrow{-CO_2-H}$ two vinyl side chains) yields protoporphyrin. Falk and coworkers demonstrated the enzymatic formation of uroporphyrin from porphobilinogen. Uroporphyrin and coproporphyrin are oxidative products (loss of H) of uroporphyrinogen and coproporphyrinogen, by conversion of the four saturated carbon bridges ($-CH_2-$) to the methene (–CH=) structures. The enzyme system which converts coproporphyrinogen **III to** protoporphyrin has been studied by the researchers. Wintrobe and others demonstrated the presence of an enzyme in erythrocytes capable of incorporating iron (Fe^{++}) into protoporphyrin to form heme *in vitro.* An enzyme system in rat liver mitochondria which brings about heme synthesis from Fe^{++} and protoporphyrin.

In the structures herewith the α-carbon atom of glycine is marked with an asterisk (*), so that the eight positions the α-carbon atoms assume in the protoporphyrin molecule can readily be visualized. Variations in the manner in which the porphobilinogen molecules unite lead to either an I series or a **III** series porphyrin. The mechanism here is incompletely known, although an enzyme bringing about formation of porphyrin from porphobilinogen has been purified. In the structure of uroporphyrin **III** the side chains on ring D are acetic acid and propionic acid from left to right. Uroporphyrin **I** is identical to the **III** series compound save for a reversal of the positions of these two groups on this D ring. Coproporphyrin **III** has methyl and propionic acids groups on the **D**-ring and coproporphyrin **I** is the same but has the positions of these two groups reversed. Shemin has reviewed in detail the earlier work covering the biosynthesis of the porphyrins.

The enzymatic formation of o-aminolevulinic acid by particulate free extracts from two microorganisms was shown to involve conversion of succinate to succinyl-CoA and the condensation of this active form with a pyridoxal phosphate derivative of glycine. δ-Amino-levulinic acid and CO_2 were the reaction products:

THE GASTRO ENDOCRINAL ABSORPTIONS AND THEIR TYPES

The process of absorption refers to the normal passage of substances into the blood and lymph. Primarily, we are concerned with this process as it occurs in the small intestine. It has been pointed out that the primary purpose of digestion is to break down the foodstuffs into small water-soluble or otherwise diffusible particles to facilitate the process of absorption. It should be kept in mind also that digestion and absorption processes have the added function of minimizing the entrance of toxic molecules into the blood. Thus, food proteins and peptides, many of which are highly undesirable in the blood, are hydrolyzed to the harmless and, indeed, valuable and essential amino acids preceding their absorption. Fatty acids from fat hydrolysis are resynthesized into fats before their entrance into the blood stream. The fatty acids but not the neutral fats, of course are toxic in high concentration. The polysaccharide carbohydrates and even the disaccharides are of little value in the blood stream, and so they are hydrolyzed to monosaccharides before absorption.

The mechanisms accounting for the absorption of amino acids, sugars, and fatty acids will be discussed in detail in the following pages. The absorption of minerals, vitamins, and other substances are considered in appropriate chapters.

The small intestine is the principal organ of absorption. In the mouth and esophagus absorption is negligible, except in the case of certain drugs. The stomach is also not suited to food absorption, although alcohol and a few other substances are known to pass this mucosa to some degree. In the large intestine water absorption occurs; normally little or no absorbable food remains in the intestinal contents by the time it reaches this organ.

The small intestine in man is about 27 ft long. Because of the many folds and the millions of villi, the area offered for absorption is estimated at ten or so square meters. This is an important factor in absorption. Another important factor is the time that the digested material is in contact with the intestine. Ordinarily five to eight hours are required for the remains of a meal to pass into the large intestine. Under various conditions, such as diarrhea, this may be greatly altered.

The villi, some five million in man, are the primary sites of absorption. Each contains lymphatic vessels with the flow of lymph controlled by valves so that it can pass out of the intestine only. This lymphatic system makes up one of the two important routes that absorbed particles travel to enter the blood stream. The second consists of the arterial system. In man one artery enters each villus, and these are collected into the portal vein and hence the blood goes to the liver.

The passage of particles through the mucosa of the small intestine cannot be predicted from a knowledge of known chemical and physical laws. For instance, the prominent plant sterol, ergosterol, is not absorbed from the intestine. But cholesterol, the main animal sterol, is absorbed readily. These two molecules differ only slightly in chemical structure. Sulphates are not absorbed readily, but chlorides, phosphates, etc., are. The common hexose sugars enter the blood at a faster rate than do the pentoses, although the latter are smaller molecules. Glucose, NaCl, and other small molecules do not pass the mucosal lining from the blood into the intestinal lumen, as might be expected if the processes were purely those of diffusion, osmosis, etc. It is readily seen from these few examples that absorption from the intestine depends to a considerable degree on a specificity of this living tissue. It allows the passage of specific molecules and not of others, and it regulates the rate of passage of substances. This is one part of the absorption process that is not well understood.

Fat Absorption

Not all fats are absorbed at equal rates or to an equal degree. Lard and other common dietary fats may be almost completely absorbed that is, 98 or more per cent. Sperm oil is absorbed poorly only about 15 per cent and some other fats even less completely. Practically all fat absorption occurs in the small intestine. Emulsification and lipolytic hydrolysis take place rapidly as the chyme enters the duodenum. Fat absorption may be thought of in terms of preparatory phase (digestive changes), transport phase (molecules entering the mucosal cells), and transportation phase (removal from cells and transport by blood or lymphsystems).

Actually the absorption process and the various resyntheses that occur go on in a single layer of epithelial cells. The products of absorption and resynthesis are discharged on the other side of the cells, where they may be picked up by the intestinal lymphatics or the portal blood.

The mechanism of fat absorption is not thoroughly understood today. A number of theories have been presented over the past 50 years, none of which has enjoyed universal acceptance.

In 1896 Munk proposed one of the earliest theories that fat is absorbed as a fine emulsion. He felt that the minute fat droplets might pass the epithelial wall, since it appears as neutral fat in the lymphatic system. He did not offer a good explanation for the force which could bring this about.

Pfliiger in 1900 offered his soap theory of absorption. He felt that the fats are hydrolyzed by lipase, and then the fatty acids react with base of the bile and pancreatic juice to form sodium salts of the fatty acids. These are water-soluble and as such easily transported across the mucosal wall. The glycerol formed in the hydrolysis is also water-soluble and readily diffusible. Somewhere before entering the lymphatics in the villi, the reaction must be reversed and the neutral fat reformed.

After the disclosure that sodium soaps are not stable at pH values below 8 to 9 and that the small intestine is practically never alkaline, this theory lost credence.

A more intriguing theory of absorption is that of Verzar. He, also believes that fat in the intestine is hydrolyzed to fatty acids and glycerin. But he feels that the bile salts act as hydrotropic agents to bring the water-insoluble fatty acids into solution. He was able to demonstrate this action *in vitro* and showed that the fatty acids are partly in a diffusible form, even in acid solution. The proportion of bile acids produced to fatty acids generally ingested is small. Verzar presumes that the bile acids combine with fatty acids (these are known as choleic acids) and convey them through the intestinal wall, whereupon the bile salt fatty acid combination breaks up and the bile salt is set free.

Bile salts are supposedly adsorbed to the surfac of the epithelial cells, where they can dissolve more fatty acid molecules and convey them into the cells.

Another theory of fat absorption is that of Frazer. His views are summarized very briefly as follows. A part of the ingested fat is absorbed as highly emulsified, neutral fat particles. These enter the lacteals and are carried by the lymphatic circulation to the systemic blood. The other fraction of food fat is enzymatically hydrolyzed and the products in the case of long chain triglycerides are fatty acids and mono- and diglycerides, rather than fatty acids and glycerol. The fatty acids through the hydrotropic action (solubilizing action) of bile salts and possibly by other means are transported through the intestinal wall, are resynthesized into neutral fat molecules, and enter the portal blood stream. After arrival at the lever, sufficient metabolic alterations supposedly intervene so that little of this fat finds its way into the systematic circulation.

Most of Frazer's data on blood lipid levels depends on chylomicron counts. Microscopic examination of blood serum with dark ground illumination reveals bright particles in Brownian movement. These are the chylomicrons. They range in size from 35 mμ to 1mμ Supposedly they are covered with a protein layer that helps maintain them in the colloidal state. A correlation has been demonstrated between chylomicron counts and the level of blood fat determined chemically.

Reiser and coworkers prepared synthetic triglycerides in which the glycerol was labeled with C^{14} and the fatty acids with conjugated double bonds. The latter are useful in tracing the paths of fatty acids in the body. Rats with thoracic duct cannulas were given the doubly tagged neutral fat and analyses were made for C^{14} and for the tagged fatty acid in various lipid fractions of the thoracic duct lymph. The authors concluded that about one-half to one-third of the ingested glyceride was hydrolyzed to monoglycerides, and the remainder was completely hydrolyzed during digestion. It was further indicated that the fraction of glycerol that was hydrolyzed in the intestine was not the glycerol used in resynthesis of lymph glycerides. Also, about one-half of the phospholipids of lymph synthesized in part from the administered neutral fat utilized the hydrolyzed fatty acids and endogenous glycerol, while the remaining half arose from absorbed glycerides.

If glycerol of ingested fat is not the glycerol of the triglycerides in the lymph, then what is the source of the glycerol? To answer this question Reiser and coworkers fed the monopalmitic acid ester of dihydroxyacetone to rats. This compound was tagged with C^{14} in both the fatty acid and in the alcohol molecules. Analyses of lymph lipids (thoracic duct) demonstrated that the compound fed was hydrolyzed to the extent of a little more than 70 per cent, but no fatty acid esters of dihydroxyacetone appeared in the lymph. The latter compoqnd was reduced to glycerol and then converted to tri-glycerides. It was postulated that normally during fatty acid absorption, esterification of dihydroxyacetone with fatty acid and subsequent reduction and further esterification occur, probably in the intestinal mucosa.

With newer techniques for the separation of lipid fractions additional information on fat absorption has been developed. Countercurrent distribution has been used by various workers to separate monoglycerides, diglycerides, etc. Coniglio and Cate (5) used this technique in a study of intestinal absorption of carboxyl-labelled palmitic acid or tripalmitin in rats. At various times following the oral administration of the labelled fat of fatty acid the animals were sacrificed and various lipid fractions were separated from intestinal contents, intestinal wall, and liver. C^{14} activity was determined in the purified fractions. During absorption C^{14} activity was found in mono-, di-, and tri-glycerides well as in free fatty acids of the intestinal contents, with the major part of the activity in the free acids and the triglycerides. At the same time the intestine showed most of the label in triglycerides and the mono- and diglycerides had the least. After feeding free palmitic acid more of the activity was in the free fatty acid fraction of intestinal contents, although considerable activity also resided in di- and triglycerides. These findings indicate that glycerides synthesis as well as hydrolysis occurs in the intestinal contents. Other workers have observed this also.

During the course of absorption the phospholipids of the intestinal wall increased in C^{14} activity, while the activity of the triglycerides decreased. Such findings indicate phospholipid synthesis within the intestinal mucosa and are in accord with results of other investigators. It is clear also that mono- and diglycerides are converted into triglycerides (and phospholipids), since the lymph is known not to contain mono- or diglycerides in man

The liver lipids three hours following the ingestion of the labeled compounds showed C^{14} activity in phospholipids and triglycerides, with small but significant activity in mono- and diglycerides. The origin of the last two substances in hepatic tissue is not known.

It is interesting from this standpoint that researchers found considerable mono- and diglycerides in human blood plasma after feeding tagged fatty acid methyl esters. They offered various possible explanations to account for the partial glycerides in blood including the possibility of hydrolysis of triglycerides in transit.

An unusual opportunity for absorption studies was afforded by a human with chyluria. This woman had an anomalous connection between the intestinal lymphatics and the left renal pelvis, allowing drainage of the lymph into the urine. C^{13}-containing oleic and palmitic acids to this individual either free or in triglycerides. Determination of C^{13} activity in the various separated lipid fractions of urine allowed the investigators to follow incorporation of the compounds into the different types of lipids. It was found that oleic and palmitic acids were incorporated into triglycerides to a large degree and that the triglycerides constituted the nain transport form of these acids-about 90 per cent of the total isotope :ontent of lymph. From 3 to 9 per cent of the isotope found its way into phospholipid, and 1 to 7 per cent was found in nonesterified fatty acids.) ther measurements showed that about 90 per cent of the triglycerides of Ghe chyle were of dietary origin, while only around 20 per cent of the phospholipids and non-esterified fatty acids represented absorbed fatty acids. This indicated that these two fractions were largely of endogenous origin, but that some phospholipid was synthesized from the fed fats or fatty acids.

About one-fifth of fed phospholipid was absorbed into the lymph in rats without undergoing hydrolysis in the experiments.

It is pretty well established that in the absorption process fatty acids (fed either as neutral fat or free fatty acids) longer than 12 carbons in chain length are transported almost exclusively via the lymph, while those with less than 12 carbons are transported to a greater extent by the portal blood. Resynthesis of absorbed free fatty acids into glycerides occurs in the intestinal wall and to a lesser extent in other tissues.

However, it appears that the short-chain acids absorbed by the portal system may largely be transported to the liver without first undergoing re-synthesis into triglycerides. Thus, Borgstrom found that during absorption of C^{14} decanoic acid in rats 63 to 73 per cent of the total C^{14} activity in portal blood was present as free decanoic acid and amounted to 4 or 5 mg per cent. It is likely that free fatty acid absorption (portal system) is an active process. The sodium salts of acetic, propionic, and butyric acids could be transferred from the mucosal to the serosal side of isolated everted intestinal loops against a concentration gradient and that the transfer was oxygen dependent.

It follows that in general the fraction of fat absorption via either route will depend upon the percentage of the longer-chain fatty acids in the lipids of the food eaten. In foods eaten by the United States population on the average, the large majority of fatty acids are longer than 12 carbons, so most of fat absorption goes via the lacteals and into the intestinal lymph.

Cholesterol is absorbed in the small intestine and unsaturated fats and fatty acids enhance absorption. Saturated fats in the intestine inhibit cholesterol absorption. Bile salts are required for absorption of cholesterol in the endocrinal system, and the majority of that absorbed is found in the lymph, where around 70 per cent was esterified regardless of the amount absorbed.

Turner has employed fats tagged with I^{131} in studies of absorption and transport of fat. He reviewed the field of fat absorption and his work using the iodine isotope. It is difficult to summarize a field with so much information lacking. However, the following are some of the facts which seem to be established or have experimental backing:

1. Some neutral fat is absorbed without hydrolysis.
2. Hydrolysis products, di- and mono-glycerides, as well as fatty acids, can be absorbed.
3. Bile salts aid absorption of the various fractions.
4. Resynthesis occurs within the single layer of epithelial cells lining the mucosa forming triglycerides.
5. Long-chain fatty acids are transported via the lymphatics, predominantly as triglycerides, and shorter-chain acids by the portal system, mostly as free fatty acids.
6. Phospholipids are partially hydrolyzed but are absorbed without hydrolysis. Phospholipid synthesis occurs in the laning cells from mono- and di-glycerides, the phospholipids are transported into the lacteals.
7. Cholesterol is absorbed in the free state. Bile salts are required, and unsaturated fats enhance absorption.

Non-esterified fatty acids (NEFA) in the blood may constitute an important vehicle of fatty acid transport from adipose depots to other tissues. However, these fatty acids are not "free" but are almost entirely bound to serum proteins and principally to the albumin. In normal humans the NEF A concentration is about 0.5 meq per litre and the "free" fatty acid concentration is perhaps less than 0.01 per cent of this value. The half-life of the NEF A is extremely short, allowing for a rapid transfer of fatty acids to the tissue cells.

Normally the feces contain small amounts of unabsorbed lipids. Continued excretion of large amounts of fats characterizes the condition known as "steatorrhea" (coeliac disease in children). Diarrhea, upset electrolyte balance, and vitamin deficiencies lead to malnutrition. Impaired fat absorption in one type of the disease may result from insufficient bile due to obstruction or other reasons. French and coworkers reported considerable success in the dietary treatment of adult idiopathic steatorrhea. In the steatorrhea that develops in a high percentage of patients following gastrectomy the disturbance was considered by researchers to be a deficiency in fat digestion rather than faulty absorption. Vitamin B_{12} absorption may be impaired following gastrectomy (deficient intrinsic factor), and the vitamin along with intrinsic factor has been used in the treatment of steatorrhea and the anemia which sometimes accompanies the disease.

Carbohydrate Absorption

In man absorption of carbohydrates from sites other than the small intestine is negligible. Under normal circumstances only monosaccharides are absorbed. The other carbohydrates, such as the disaccharides and polysaccharides, must first be hydrolyzed to the mono-saccharide stage. Lactose and sucrose in the blood stream are readily eliminated by the kidneys, although some maltose may be hydrolyzed by the blood maltase and the glucose utilized. The absorption of the monosaccharides is normally quite complete, and most of it enters the portal blood.

The rates of absorption are markedly different for the various monosaccharides, indicating as pointed out previously that the absorption process is more involved than one of simple diffusion. For instance, researcher reported that in the absorption coefficients (milligrams sugar absorbed per 100 g in one hour) are in the following order: galactose> glucose> fructose > mannose > xylose> arabinose. It was also found that glucose from a 25, 50, or 80 per cent solution was absorbed at the same rate. This was considered as evidence that the absorption rate is independent of the concentration in the intestine. It should be noted that the two pentose sugars studied were absorbed more slowly than the hexoses.

Cori felt that each sugar has an independent rate of absorption, but other work indicates that by varying certain experimental conditions, this rate can be altered. Thus, researchers showed that when variable volumes of the same concentration of glucose were given to rats (stomach tube), the absorption coefficient was related to the dose of glucose. Under voluntary feeding conditions they found absorption rates far higher than those previously reported. By maintaining rats in a cold environment, the coefficients were found to be even further elevated, due largely to the increased appetite of the animals.

The mechanism of the absorption of glucose (or other physiologically important sugars) is indeed obscure. Since glucose is absorbed selectively at a high rate, it was early postulated some metabolic alteration may take place in the molecule to make it chemically distinct from the blood glucose and thus increase the diffusion gradient.

Although the experimental data are not convincing, many workers explained glucose absorption on the basis of phosphorylation and then dephosphorylation. In some of the early experiments enzyme inhibitors, such as iodoacetate or phlorizin, did decrease glucose absorption in animals when given with the sugar. Since these compounds were thought to be rather specific phosphorylation reaction inhibitors at that time, the conclusion was easily reached that phosphorylation was a prerequisite to transport of glucose into the blood. Later work has indicated that iodoacetate inhibits many sulfhydryl enzymes and is not specific for phosphorylating enzymes. Furthermore, phlorizin appears to inhibit principally some of the oxidative reactions and not the enzymes involved in the formation of sugar phosphates.

The absorption problem has been approached by newer techniques byLandau and Wilson. These workers studied the absorption of sugars in sacs of hamster intestine *in vitro* by the use of C^{14}-labeled compounds. They concluded that only a small fraction of the absorbed glucose passed through glucose-6-phosphate as an intermediate. In the hamster, then, phosphorylation of glucose may not be necessary for absorption. These workers also confirmed other investigators' work indicating that glucose is not converted into 2 mols of triose phosphate and recondensed to sugar during absorption. Such a theory had some backing at one time and has been reviewed by Wilson. If phosphorylation of glucose is not involved in absorption, as seems to be the case at this time, we are left with no solid explanation to account for the transport mechanism. It is established that "active" transport (transport against a concentration gradient) is involved, since it is suppressed under anaerobic conditions. Passive transport also accounts for some absorption. Quastel has reviewed the absorption of sugars briefly. He previously demonstrated that K^+ ions have a marked effect on absorption rates, and that Na^+ and Mg^{++} are required while Ca^{++} and phosphate are not essential to active transport.

Protein Absorption

The absorption of protein digestion products has been considered the result of a simple diffusion process, since the amino acids are in general sufficiently water-soluble to account for their passage across the intestinal mucosa without the intervention of chemical alteration. At present, however there are indications that the process may be more complex. Researchers for instance, showed that when

DL-alanine was introduced into the lumen of a segment of small intestine, the L, or natural isomer, appeared in the blood in a concentration three times that of the D isomer during the absorption process. These workers feel that the data corroborate previous postulates of a steriochemically specific mechanism in the intestinal wall for the absorption of L-amino acids.

It is now believed that amino acids are absorbed both passively (diffusion) and by active transport. In intestinal sacs of the hamster active transport has been demonstrated for proline, threonine, alanine, glycine, serine, valine, histidine, hydroxyproline, phenylalanine, isoleucine, and leucine. The active absorption of alanine and of phenylalanine is completely inhibited by an equal concentration of L-methionine, while glycine transport is inhibited to the extent of 35 per cent. Quastel states that such findings are consistent with the view that these amino acids may have affinities for a common intestinal absorption mechanism but that the rates of active absorption may not be proportional to the affinities. These may be examples of competitive inhibitions of transport.

It is known that the amino acids readily enter the portal blood with small amounts finding their way into the lymph system. The absorption is a rapid process, since only small quantities of amino acids are present in the intestine at any time. In other words, as they are liberated from proteins by the digestive processes, they are quickly transported into the blood, and the over-all time of amino acid absorption must then be governed largely by the rate of hydrolysis of the protein constituents in the intestine.

It is probable that some small peptides are absorbed; various workers have reported an increased polypeptide-nitrogen content of portal blood during the digestion and absorption of protein.

Researchers established beyond doubt the presence in blood of free amino acids. They employed the "vividiffusion" method. Arterial blood from an animal was led through suitable dialyzing tubes immersed in a saline solution and back into a vein of the animal. Dialysis was allowed to proceed for a number of hours, and the concentration of diffusible molecules was increased in the saline to an extent that a number of amino acids, and other molecules, could be identified in crystalline form after evaporation of the saline. More recently it has been established that the amino acid nitrogen of blood increases in a rather consistent fashion after the ingestion of a given amount of protein. The intravenous administration of moderate amounts of an amino acid mixture leads to a rapid rise in blood amino acid nitrogen without appreciable loss in the urine. The transitory nature of this increased blood level indicates active participation of tissues in removing the amino acids from the blood.

One important problem, which can only be mentioned here, involves the absorption of undigested protein molecules. That both animals and man do absorb native protein has been amply demonstrated experimentally by many workers. Researchers showed that unsplit proteins are able to pass the intestinal wall under normal conditions not only in experimental animals of various ages but also in humans, both young and old. They believe that small amounts of unaltered protein are absorbed regularly.

As we all know, there are many individuals who are allergic to sea foods, strawberries, or even milk, wheat, or other common foods. Supposedly absorption from the intestine of small amounts of proteins of the foods is largely responsible for the many types of allergic reactions exhibited by such individuals.

THE COMPOSITION AND FORMATION OF FECES

During the passage of the intestinal contents through the small intestine the products of digestion, along with many other compounds such as vitamins, water, and mineral salts, are absorbed. When the contents reach the large intestine, the absorption process, with the exception of water, is normally completed. Here more water and NaCl are absorbed, and the remaining material leaves the body as feces. The consistency of the feces depends to a large degree on the water content, or stated differently, on the degree to which the process of water absorption has been carried. Certainly other factors, such as gastrointestinal motility and nature of diet, affect the consistency of the feces also.

Small variations in diet have little or no effect on the nature of the feces. However, an exclusively vegetable diet tends to yield a larger bulk and softer consistency feces, while on a meat diet the feces are harder and the quantity is less. Even during starvation feces are produced, though the quantity is markedly diminished.

The water content of the feces is usually from 60 to 70 per cent by weight. The 20 to 30 per cent dry matter is composed primarily of undigested dietary constituents, such as cellulose material, hair, and seeds, fatty material, mineral matter, and bacteria. The undigested food protein, carbohydrate, and fat amount to very little, since the digestion and absorption of these substances is normally 95 to 98 per cent complete. Practically all the nitrogen present is of bacterial origin.

The normal dark brown colour of the feces is due to bile pigment derivatives. The primary pigment is stercobilin arising from oxidation of the precursor stercobilinogen. Small amounts of bilirubin and biliverdin are sometimes present.

Indole and skatole arising from bacterial enzymatic action on the amino acid tryptophan are the primary compounds responsible for the disagreeable odor of the feces. Hydrogen sulfide and methyl mercaptan (CH_3SH) likewise have displeasing odors, and these substances may be present in considerable amounts at times.

The large intestine is the site of bacterial multiplication. Many billions of bacteria may be excreted daily in the feces. *Escherichia coli* is ordinarily the predominating organism, although many others frequently are found in the feces.

The excrement is generally slightly alkaline; the pH varies from 7 to 7.5 on a normal diet. The inclusion of considerable milk or lactose in the diet may lead to more favorable conditions for certain acidophilus organisms, and the fecal pH may then be lowered.

After the excretion of bile into the intestine, further changes occur with the formation of a number of new compounds. Several of these are established as to identity and chemical structure.

It is generally agreed that bilirubin on entering the intestine is reduced by bacterial enzymes to urobilinogen identical to stercobilinogen. Some of this material is oxidized to urobilin. The addition and removal of hydrogen can occur at various atoms in the rings, so that a number of different compounds are possible. The urobilinogen not oxidized in part absorbed by the portal system, removed by the liver, oxidized there.

THE BIOCHEMISTRY OF HUMAN LIVER

Liver is the most important organ of the human metabolic system sometimes called the chemical power house of human body many enzymes and digestive secretions are produced in the liver. (See Fig. 9.24)

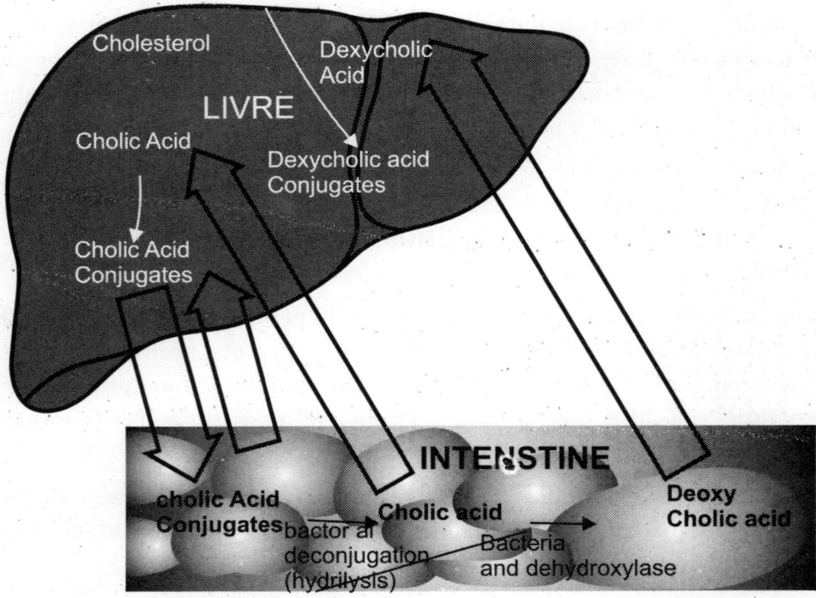

Fig. 9.24 : The Circulation of Bile in Hepatic System

Some of the important secretions of the stated below:

1. Synthesis of Protein, e.g. Albumin
2. Synthesis of Cholesterol
3. Synthesis of Vitamin A, D E and K
4. Synthesis of bilirubin from degraded red blood cells
5. Synthesis of bile and bile salts.

Functions of Liver

The liver has a very important role in the digestive system; Liver is a very intelligent organ plays a very vital role during the fasting state and also in the fed state. In the fed state liver utilizes the glucose and amino acids from food and use them in metabolism. There is therefore an extensive interconversion of complex amino acids to simpler amino acids by transamination other pathways so that the normal pattern of amino acids in plasma is reached as fast as possible just after meals.

During fasting state the liver plays an almost similar role, but only difference is that the nutrients are received from storage tissues for example amino acids from muscles and fats from adipose tissues.

The liver converts the glucogenic amino acids into blood glucose and glycogen, and the fatty acids into very low density lipoproteins (VLDL) that are transported to provide the tissues with fatty acids as fuel. In the conclusion of these general discussion it can be stated that as an important nutrition regulating organ. These metabolic interrelationships are shown in Fig. 9.25.

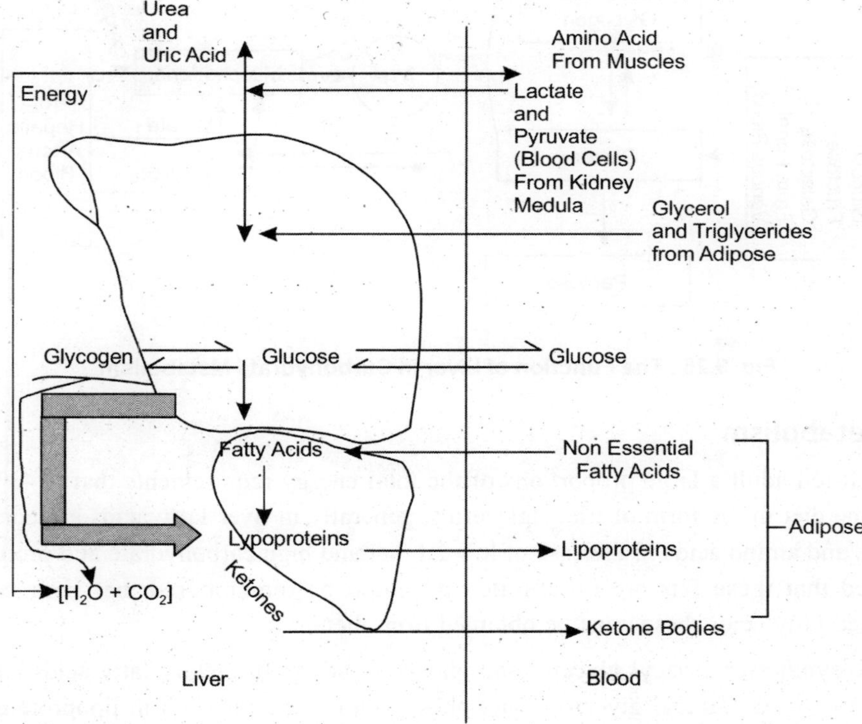

Fig. 9.25 : The Metabolic Agent Exchanging Between Liver and Blood

Liver plays different types of role in the metabolism, to speak specifically the four different types of metabolisms, in which liver plays role respectively:

1. The Carbohydrate Metabolism
2. The Fat Metabolism
3. The Amino Acid Metabolism
4. The protein Metabolism

The Carbohydrate Metabolism

Here liver mainly converts monosaccharide (see carbohydrate metabolism chapter 5). It also converts glucose to glycogen and also glycogen back to glucose, and glucose to pyruvate. Pyruvate is therefore completely oxidized in liver by tricarboxylic acid cycle or Kreb's Acid Cycle. The role of liver is carbohydrate metabolism is shown in Fig. 9.26.

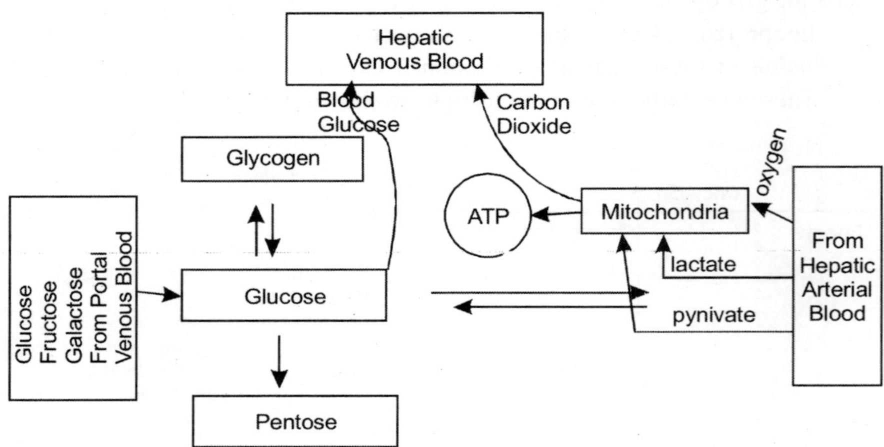

Fig. 9.26 : The Function of Liver in Carbohydrate Metabolism

The Fat Metabolism

In normal well-fed adult a large proportions of the total energy requirements that is 43% to 51% is produced in the diet in the form of triacylglecerols, generally in liver fatty acids are converted from carbohydrates and amino acids in the case of low fat diet and high carbohydrate and protein intake. It must be noted that these fats are unsaturated it should be remembered that liver cannot make polyunsaturated fatty acids these must be obtained from diet.

Liver also synthesizes triacyl glycerol and phospholipids from dietary fatty acids and also from endogenous fatty acids. Triacyl glycerol s and phospholipids are enclosed in lipoproteins (very low density lipoproteins) for transport from liver. Now if there is excess fatty diet then liver may not be able synthesize the required phospholipids and protein at a rate sufficient to maintain lipoprotein synthesis i.e. either the fats will be stored in the adipose tissue in excess for extreme fat diet or in the case of starvation, the loss of the fat from the adipose tissue.

The supply of the precursors for example the essential fatty acids for the phospholipids, and amino acids for proteins, will therefore become critical so the deficiencies of these precursors, in relations to the supply of fatty acids will lead to the decomposition of large amounts of triacyl glycerol in liver, resulting in the condition of fatty liver. The liver also oxidizes fatty acids by β-oxidation to conserve energy at ATP but under conditions of excess production and under utilization the end product of β-oxidation acetyl CoA cannot be further utilized. (See Fig. 9.27)

Acetyl Coenzyme molecules are condensed to form 3-hydroxy methyl glutaryl CoA. Loss of an acetyl CoA as a result of the 3-hydroxy-3-methyl glutaryl CoA lyase causes the formation of acetoacetate. Acetoacetate is dependant 3-hydroxybutyrate dehydrogenase, so that both sides accumulate together.

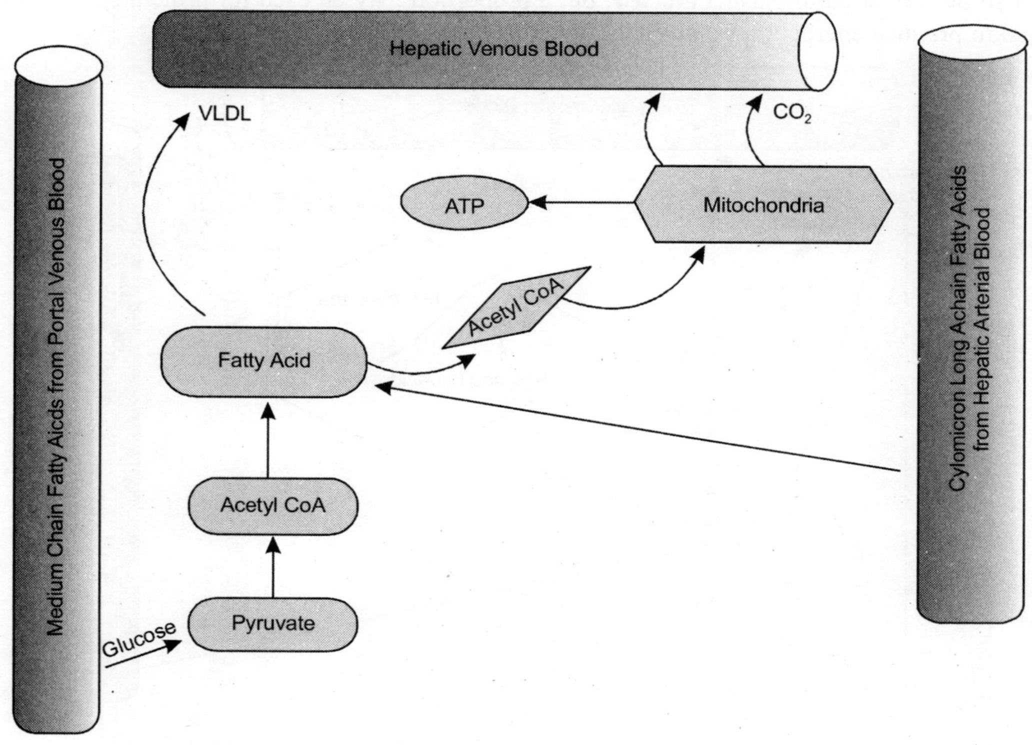

Fig. 9.27 : The Fat Metabolism in Liver

$$CH_3COCH_2COOH + NADH + H^+ \rightarrow CH_3CHOHCOOH + NaD^+$$

The ratio of the acids in the plasma depends on the NADH : NAD$^+$ ratio in mitochondria and is usually three to six in favour of 3-hydroxybutyrate. The acids when produced in moderate quantities, these acids can be utilized by extraheptic tissues such as heart muscle or kidney, but if extensive fatty acid oxidation occurs in the liver, as well happen during starving or diabetes mellitus, large quantities of ketone bodies will be formed. In uncontrolled diabetes, the high concentration of 3-hydroxybutyric acid and acetoacetic acid produced can cause a severe metabolic acidosis.

The Amino Acid Metabolism

The two main sources mainly supply the amino acid, the portal blood, and the muscle.

During the digestion and absorption of protein food via the portal blood and during post absorption and fasting conditions via the extra-hepatic tissues, mainly the muscles. Alanine forms a major proportion of the amino acids from either source enter inside the liver where interconversion can occur as a result of the operation of the tricarboxylic acid cycle and also by transamination. Depending upon the process at that time, the amino acids are received from the portal blood; they may be recirculated following interconversion incorporated into other compounds such as pyrimidines, porphyrins and purines or into liver or plasma proteins. During starvation most of the amino acids received from the hepatic artery will be converted to plasma glucose, but a proportion may be used for protein synthesis or be oxidized to produce energy after conversion to oxoacids.

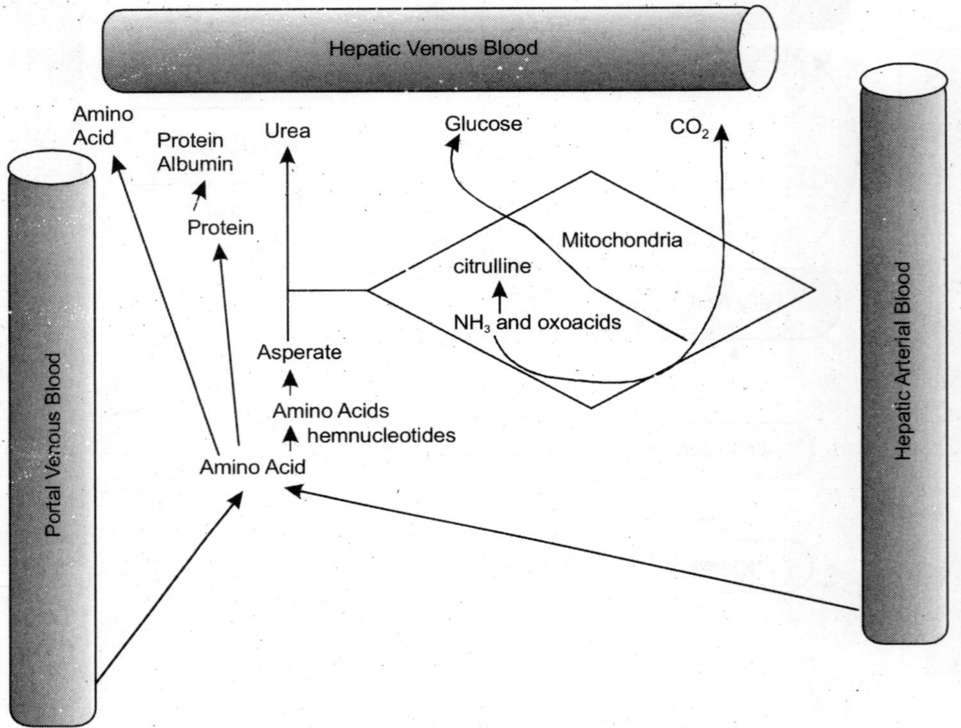

Fig. 9.28 : The Amino Acid and Protein Metabolism in Liver

There is one very important function of live is to synthesize urea from ammonia arising from deamination, quantity of urea produced will therefore depend upon the excess amino acids from the diet not required for protein synthesis and that are being used for energy or during fasting to produce plasma glucose.

The Protein Metabolism

Liver plays a vital role in the protein metabolism, by producing two vital proteins one is the liver tissue protein and plasma protein.

By using conventional pathway liver synthesizes the above mentioned proteins for albumin, α and β globulins the clotting factors prothrombin, factors VII, IX and X in the presence of Vitamin K. Many transport proteins such as tranferrin ceruloplasmin, transcobalmins etc.

LIVER AS A STORAGE ORGAN

The main item of storage in liver is glycogen, the capacity of this storage is almost .25% to .3% of the total weight of the liver during fasting state and during high carbohydrate diet it is about 9.5 to 10%. The stored glycogen is normally converted to blood glucose 12hr after a meal.

Liver also plays an important role in the storage of important nutrients and proteins as discussed previously. Vitamins in large quantities are also stored in the liver.

The liver is also a very good store for the iron, which is largely stored in the form of ferratin in parenchymal cells. In hæmochromatosis the excess iron is stored as hæmosidrin in the Kupffer cells.

THE PANCREAS

The pancreas is also very much involved in the digestive system. The pancreatic juice is the element utilized in the digestion.

The pancreatic juice is conveyed through the common bile duct at the ampulla of Vater so that the bile and the pancreatic juice can well mix in before reaching the intestine.

The Pancreatic juice is mainly composed of three secretions:

1. The electrolyte section (Na^+, K^+)
2. The Organic Secretions and
3. The enzyme protein secretions.

These secretion are controlled by hormones. The cations are mainly Na^+, K^+ and Ca^+. Some anion concentrations is also present, the HNO_3^- whose concentration is normally about 78.5mM and can be increased up to 135mM it has an important regulatory effect on the pH of the intestinal contents, pancreatic juice can neutralize an equal volume of gastric juice.

The Pancreatic juice supplies a quite, large number of precursors of powerful hydrolytic enzymes; the table below shows the list of the pancreatic enzymes with their substrates.

The Composition of Pancreatic Juice

Substrate	Enzymes
Typsinogen Thymotrypsinogen Elastase Carboxy peptidase A and B	Proteins
Amylase	**Starch**
Esterase Lipase Cholesterol esterase Phospholipase	Acyglycerols/ester
Ribonulcease Deoxyriboundclease	Nucleic Acids

THE KIDNEY AND RENAL FUNCTION

Introduction

Water is the most abundant constituent of the body. It is a necessary component of cell structures, though it may be only loosely held by hydrogen bonds. It provides the fluid medium within which the chemical reactions of the body take place and substances are transported. Because of its high specific heat, it absorbs or gives off much heat with relatively small changes in temperature, thereby serving as a sort of temperature buffer system for the body. Since the rate of turnover of water in the body is high, the supply must be continually renewed, and among supplies to the body water ranks next to oxygen as a vital substance.

Electrolytes are distributed in solution throughout all the body fluids, blood, lymph, intracellular fluids, digestive juices, and urine. The cations of these electrolytes are Na^+, K^+, Ca^{++}, and Mg^{++}, with Na^+ and K^+ predomi-nating. Quantitatively, the chief cations of plasma and other extracellular fluids is Na^+, and of intracellular fluids K^+. Among the anions of the body are Cl^-, HCO^{3-}, H_2PO^{4-}, HPO^{4-}, SO^{4-}, protein-, and the anions of many organic acids, usually in small amounts. Quantitatively, the chief anions of plasma and other extracellular fluids are $Cl^-$and HCO^{3-}, while in intracellu-lar fluids they are, in decreasing amounts, inorganic and organic phosphates, protein-, SO^{4-}, and HCO^{3-}. The intracellular fluids of most tissues contain very little Cl^-

Since generally the proteins of tissue fluids and structural tissue proteins exist in the body at pH values above their isoelectric points, they ionize as acids.

The electrolytes of body fluids have various functions, the chief of which is to contribute most of the osmotically active particles, to provide buffer systems and mechanisms for the regulation of pH (acid-base balance), and to give the proper ionic balance for normal neuromuscular irritability and tissue function. A rough general expression of the relation between irrita-bility and the ions of body fluids is as follows:

$$\text{Irritability } \alpha \frac{[Na^+]+[K^+]}{[Ca^{++}]+[Mg^{++}]+[H^+]}$$

which indicates that irritability is directly proportional to the concentrations of Na^+ and K^+ ions and inversely proportional to the concentrations of Ca^{++}, Mg^{++}, and H^+ ions. The relations, however, are not entirely as simple as depicted by this relation. K^+ ions in low concentrations are excitatory, and in higher concentrations are inhibitory, particularly in relation to nerve synapses and myoneural junctions. In addition, the ratio Na^+/K^+ is important in this connection. The relation of K^+ to heart action is of particular interest. At high concentrations K^+ ions cause widespread intracardiac block, first in the auricle, then at the A-V node, and finally in the ventricle. The heart stops in diastole. Low concentrations of K^+ ions impair the contractility of heart muscle.

Under normal conditions fluids constitute about 70 per cent of the body mass. These fluids contain dissolved substances, chiefly electrolytes, which produce an osmotic pressure of about 7.5 atm.

(equivalent to 0.9 per cent NaCl) and provide buffer systems which help maintain the pH within physiologic limits. In maintaining normal conditions in the body it is necessary to have the proper distribution of total fluids between blood, tissue spaces, and tissue cells, and to have the proper composition of these fluids. If too much fluid moves from plasma to the tissue spaces, edema results, while the movement of excessive fluid from the tissue spaces to the plasma causes dehydration. The two chief tasks in maintaining the proper composition of plasma and extracellular fluids and thus, indirectly, the composition of intracellular fluid are:

1. maintenance of the total electrolyte concentration, osmotic pressure, and volume, and
2. maintenance of the proper acid-base balance and pH. These tasks are the functions of the kidneys primarily and the lungs second-arily, which will be considered in detail later.

The membranes of the tissue cells generally are permeable to water, either OH^- or H^+ or both, and HCO_3^- ions. Chloride ions pass readily into and out of certain cells, such as erythrocytes and cells of the gastric mucosa, yet muscle cells contain exceedingly small quantities of Cl^- ions. Cell membranes exhibit restricted permeability to phosphate ions.

In some peculiar manner the quantity of sodium salts (Na^+ ions) in the plasma and lymph determines the volumes of these fluids. Thus, if the intake of salt is greatly increased or if excessive amounts are given intra-venously or subcutaneously, the volumes of these fluids may be increased to the production of tissue edema. Conversely the loss of the large quantities of sodium, as in profuse sweating and severe diarrhea and vomiting decreases the volumes of plasma and other extra cellular fluid which can be maintained. The administration of pure water does not correct the dehydration the dehydration is increased, because water carries more electrolytes into the urine and further increases the deficiency. Here we have the anomalous situation of dehydrating the body by drinking water. This situation is often summed up by the statement that tissues do not hold water without salt. The administration of NaCl both the electrolyte and the water balance. Conversely the administration of potassium salts has a diuretic effect and tends to decrease fluid volume. As previously indicated, Na^+ and K^+ ions in intracellular and extracellular fluids against concentration gradients is accomplished by the expenditure of cellular energy in the active transport across the cell membranes. Conditions which stop or reduce cellular metabolism, such as hypoxia, the presence of enzyme poisons, and low temperatures, tend to equalize the distribution of ions between extra cellular and intracellular fluids. Maintenance of relatively constant volume cellular fluids (plasma plus interstitial fluid plus specialized fluid) is essential to maintenance of normal cellar function, and many of the pathologic conditions with which the physician has to deal have their origins in disturbances of extracellular fluid volume and composition.

THE RENAL FUNCTION

Because the kidneys playa dominant role in regulating the volume and composition of body fluids, some knowledge as to how kidney function is mandatory for the proper understanding of acid-base balance and electrolyte and water balance.

The tissue cells of the body take fluid materials from the surrounding interstitial fluid and pass their waste products into this fluid. In turn, food substances at a higher concentration in interstitial fluid pass into blood plasma. Thus there are concentration gradients which move food materials from plasma to interstitial fluid to cells and waste products from cells to interstitial fluids to the plasma. The waste products are removed by the excretory organs, the lungs the intestine, the skin and primarily the kidneys

The chief waste products of the body are carbon dioxide derived from the oxidation of all organic foods; various nitrogenous substances, such as urea, creatinine, and uric acid, from protein metabolism sulphur and phosphorus compounds, also largely from protein metabolism; and excess quantities of water, sodium, potassium, calcium, and chloride. Many other substances in small amounts are eliminated from the body as waste products.

Carbon dioxide is largely eliminated through the lungs. The intestine excretes most of the calcium, magnesium, and iron, and some other substances. The skin normally eliminates considerable amounts of water by insensible evaporation. Puring active perspiration the skin may eliminate massive quantities of water containing large amounts of sodium chloride, considerable urea, and smaller amounts of other substances. However, it is the kidneys upon which falls the burden of excreting most of the solid waste products (55-70 g per 24 hours) and excess water of the body.

Aside from the gross excretion of solid waste products in urine, the kidneys help regulate the electrolyte and acid-base balance and osmotic pressure of blood by the controlled excretion of Na^+, K^+, Cl^-, HCO_3^-, and HPO_4^-. In addition, the kidneys form ammonia to be used for the neutralization of urinary acids, thereby saving fixed cations to the blood and tissues.

The kidneys apparently play an important role, through formation of the enzyme renin, in the maintenance of blood pressure.

The chief function of the kidneys is the regulated excretion of substances from the blood so as to keep the composition of blood and tissue fluids within the limits compatible with normal tissue metabolism. This is accomplished through operation of the unit mechanisms called **"nephrons,"** of which it is estimated that each kidney possesses something over one million.

The nephron proper consists of a glomerulus which serves as a filter and a subjoined tubule which regulates the composition of urine by selective absorption of substances from the glomerular filtrate and the excretion of substances into it. The tubule is composed of three segments. Directly attached to the glomerulus is the proximal segment, or proximal convoluted tubule, which has the greatest diameter and is composed of irregular epithelial cells having brush like striations at their internal borders. The proximal convoluted tubule passes to the thin intermediate segment, or the loop of Henle, which is made up of very flat or squamous cells. The loop of Henle leads into the distal segment, or distal convoluted tubule, containing rather regular epithelial cells without the brush like striations characteristic of the proximal tubule. The distal tubule passes to the arborized system of collecting tubules leading to the kidney pelvis from which urine flows through the ureter to the bladder.

The blood supply to the nephron is through an afferent arteriole which breaks up within the Glomerulus to form a complex tuft of parallel capillaries which converge into the efferent arteriole. The efferent arteriole in turn divides into a plexus of capillaries closely applied to the basement membrane of the tubule cells.

Glomerular Function

Cushny proposed the theory that urine formation begins with filtration of plasma in the glomerulus to produce protein-free glomerular fluid. Researchers have analyzed the glomerular fluid of frogs and other cold-blooded animals for various constituents and found the concentrations to be the same as in plasma with the exception of proteins and lipids or substances combined with these large aggregates. It was definitely proved, as Cushny had postulated, that glomerular fluid is an ultraflltrate of plasma in these animals. Conclusive indirect evidence has shown such to be the case for the human and other warm-blooded animals.

The glomerular membrane is freely permeable to all the small molecules of the plasma. Also, it permits ready passage of the polysaccharide inulin with a molecular weight of about 5100. Purified egg albumin (molecular weight 40,000) also gets through, but serum albumin (molecular weight 68,000) normally does not. Most of the mesh surface of the glomerular membrane is sufficiently coarse to permit the passage of particles around 20 μ in diameter, and about half of this surface allows particles of about 50 μ to pass through. The nature of the particle also has some influence upon its ability to permeate the membrane.

The filtration rate in the glomerulus is determined primarily by the blood flow and pressure. The blood hydrostatic pressure within the glomerulus, which averages about 60 mm Hg, forces fluid through the membrane. This is opposed by the plasma osmotic pressure, which is about 25 mm higher than the osmotic pressure of the capsular fluid, and by the tension within Bowman's capsule, or the capsular pressure, which averages about 15 mm. Thus the average effective filtration pressure is about 60 - (25 + 15) = 20 mm Hg. Ordinarily obstruction of a ureter causes cessation of urine production when the pressure in the ureter is about 20 mm.

The blood hydrostatic pressure within the glomerulus varies from about 50 to 70 per cent of the mean aortic arterial pressure of 90 mm. Other factors remaining constant, the glomerular filtration rate (GFR) varies with changes in glomerular hydrostatic pressure. This hydrostatic pressure varies directly with pressure changes in the renal artery, when there are no changes in tonus of the renal arterioles. Constriction of the afferent arterioles decreases glomerular hydrostatic pressure, while constriction of the efferent arterioles increases this pressure, with corresponding changes in glomerular filtration rate. Generally, moderate decreases in pressure in the renal artery are compensated for by increased tonus and constriction of the efferent arterioles, thereby decreasing renal blood flow but increasing glomerular hydrostatic pressure and the fraction of plasma filtered. The afferent arterioles may show increased tonus, but to an extent insufficient to prevent an increased glomerular plasma filtration fraction. When the mean arterial pressure decreases to 60-70 mm Hg, urine formation ceases, since the glomerular hydrostatic pressure becomes too low to maintain filtration.

Various nonphysiologic factors may alter the filtration rate. It may be decreased by occlusion of afferent arterioles by emboli (clots), by increased intracapsular pressure due to obstruction in the tubules and larger urinary channels, and by decreased permeability of the glomerular membrane because of infection. Increased filtration is caused by increased blood pressure, decreased plasma protein osmotic pressure, and increased glomerular permeability. Changes in glomerular permeability cause alterations in the sizes of molecules passed through as well as the amount of filtration.

It is estimated that the filter surface area of the glomeruli of the kidneys is normally about 1.56 sq m, and that about 700 ml of plasma or 1200 ml of blood per minute flow over this area. Under normal conditions this produces about 125 ml of glomerular filtrate per minute, for a 70-kg adult.

Glomerular filtration rate (GFR) is the volume of fluid filtered from the renal (kidney) glomerular capillaries into the Bowman's capsule per unit time. The FGR Glomerular filtration rate (GFR) can be calculated by measuring any chemical that has a steady level in the blood, and is freely filtered but neither reabsorbed nor secreted by the kidneys. The rate therefore measured is the quantity of the substance in the urine that originated from a calculable volume of blood. The GFR is typically recorded in units of *volume per time*, e.g. milliliters per minute ml/min. Compare to filtration fraction.

$$\text{GFR} = \frac{\text{Urine Concentration} \times \text{Urine Flow}}{\text{Plasma Concentration}}$$

There are several different techniques used to calculate or estimate the glomerular filtration rate (GFR or eGFR).

Measurement using Inulin

The GFR can be determined by injecting inulin into the plasma. Since inulin is neither reabsorbed nor secreted by the kidney after glomerular filtration, its rate of excretion is directly proportional to the rate of filtration of water and solutes across the glomerular filter. Compared to the MDRD formula, the inulin clearance slightly overestimates the glomerular function. In early stage renal disease, the inulin clearance may remain normal due to hyperfiltration in the remaining nephrons. Incomplete urine collection is an important source of error in inulin clearance measurement

Creatinine-based Approximations of GFR

In clinical practice, however, **creatinine clearance** or estimates of creatinine clearance based on the serum creatinine level are used to measure GFR. Creatinine is produced naturally by the body (creatinine is a break-down product of creatine phosphate, which is found in muscle). It is freely filtered by the glomerulus, but also actively secreted by the peritubular capillaries in very small amounts such that creatinine clearance overestimates actual GFR by 10-20%. This margin of error is acceptable considering the ease with which creatinine clearance is measured. Unlike precise GFR measurements involving constant infusions of inulin, creatinine is already at a steady-state concentration in the blood and so measuring creatinine clearance is much less cumbersome. However, creatinine estimates of GFR have their limitations. All of the estimating equations depends on a prediction of the 24-hour creatinine excretion rate, which is a function of muscle mass. One of the equations, the Cockcroft and Gault equation (see below) does not correct for race, and it is known that African Americans, for example, both men and women, have a higher amount of muscle mass than Caucasians; hence, African Americans will have a higher serum creatinine level at any level of creatinine clearance. A common mistake made when just looking at serum creatinine is the failure to account for muscle mass. Hence, an older woman with a serum creatinine of 1.4 may actually have a moderately severe degree of renal insufficiency, whereas a young muscular male, particularly if African American, can have a normal level of renal function at this serum creatinine level. Creatinine-based equations should be used with caution in cachectic patients and patients with cirrhosis. They often have very low muscle mass and a much lower creatinine excretion rate than predicted by the equations below, such that a cirrhotic patient with a serum creatinine of 0.9 may have a moderately severe degree of renal insufficiency.

Creatinine Clearance C_{Cr}

One method of determining GFR from creatinine is to collect urine (usually for 24-hours) to determine the amount of creatinine that was removed from the blood over a given time interval. If one removes, say, 1440 mg in 24 hours, this is equivalent to removing 1 mg/min. If the blood concentration is 0.01 mg/mL (1 mg/dL), then one can say that 100 mL/min of blood is being "cleared" of creatinine, since to get 1 mg of creatinine, 100 mL of blood containing 0.01 mg/mL would need to have been cleared.

Creatinine clearance (C_{Cr}) is calculated from the creatinine concentration in the collected urine sample (U_{Cr}), urine flow rate (V), and the plasma concentration (P_{Cr}). Since the product of urine concentration and urine flow rate yields creatinine excretion rate, which is the rate of removal from the blood, creatinine clearance is calculated as removal rate per min ($U_{Cr} \times V$) divided by the plasma creatinine concentration. This is commonly represented mathematically as

$$C_{Cr} = \frac{U_{Cr} \times V}{P_{Cr}}$$

Example: A person has a plasma creatinine concentration of 0.01 mg/ml and in 1 hour produces 60ml of urine with a creatinine concentration of 1.25 mg/mL.

$$C_{Cr} = \frac{1.25 \text{ mg/mL} \times \dfrac{60 \text{ mL}}{60 \text{ min}}}{0.01 \text{ mg/mL}} = \frac{1.25 \text{ mg/mL} \times 1 \text{mL/min}}{0.01 \text{ mg/mL}} = \frac{1.25 \text{ mg/min}}{0.01 \text{ mg/mL}} = 125 \text{ mL/min}$$

Commonly a 24 hour urine collection is undertaken, from empty-bladder one morning to the contents of the bladder the following morning, with a comparative blood test then taken. The urinary flow rate is still calculated per minute, hence:

$$C_{Cr} = \frac{U_{Cr} \times 24 - \text{hour volume}}{P_{Cr} \times 24 \times 60 \text{ mins}}$$

To allow comparison of results between people of different sizes, the C_{Cr} is often corrected for the body surface area (BSA) and expressed compared to the average sized man as mL/min/1.73 m^2. While most adults have a BSA that approaches 1.7 (1.6-1.9), extremely obese or slim patients should have their C_{Cr} corrected for their *actual* BSA.

$$C_{Cr-corrected} = \frac{C_{Cr} \times 1.73}{BSA}$$

BSA can be calculated on the basis of weight and height.

The creatinine clearance is not widely done any more, due to the difficulty in assuring a complete urine collection. When doing such a determination, to assess the adequacy of a complete collection, one always calculates the amount of creatinine excreted over a 24-hour period. This amount varies with muscle mass, and is higher in young people vs. old, in African Americans vs. Caucasians, and in men vs. women. An unexpectedly low or high 24-h creatinine excretion rate voids the test. Nevertheless, in cases where estimates of creatinine clearance from serum creatinine are unreliable, creatinine clearance remains a useful test. These cases include "estimation of GFR in individuals with variation in dietary intake (vegetarian diet, creatine supplements) or muscle mass (amputation, malnutrition, muscle wasting), since these factors are not specifically taken into account in prediction equations."

Estimated Values

A number of formulae have been devised to estimate GFR or C_{cr} values on the basis of serum creatinine levels.

Estimated Creatinine Clearance rate (eC_{Cr}) using Cockcroft-Gault formula

A commonly used surrogate marker for estimate of creatinine clearance is the Cockcroft-Gault formula, which in turn estimates GFR: It is named after the scientists who first published the formula, and it employs serum creatinine measurements and a patient's weight to predict the creatinine clearance. The formula, as originally published, is:

$$eC_{Cr} = \frac{(140 - \text{Age}) \times \text{Mass (in kilograms)} \times [0.85 \text{ if Female}]}{72 \times \text{Serum Creatinine (in mg/dL)}}$$

This formula expects weight to be measured in kilograms and creatinine to be measured in mg/dL, as is standard in the USA. The resulting value is multiplied by a constant of 0.85 if the patient is female. This formula is useful because the calculations are simple and can often be performed without the aid of a calculator.

When serum creatinine is measured in μmol/L:

$$eC_{Cr} = \frac{(140 - \text{Age}) \times \text{Mass (in kilograms)} \times \text{Constant}}{\text{Serum Creatinine (in } \mu\text{mol/L)}}$$

Where *Constant* is *1.23* for men and *1.04* for women.

One interesting feature of the Cockcroft and Gault equation is that is shows how dependent the estimation of CCr is based on age. The age term is (140-age). This means that a 20-year-old person (140-20 = 120) will have twice the creatinine clearance as an 80-year old (140-80 = 60) for the same level of serum creatinine (120 is twice as great as 60). The C-G equation also shows that a woman will have a 15% lower creatinine clearance than a man at the same level of serum creatinine.

Estimated GFR (eGFR) using Modification of Diet in Renal Disease (MDRD) formula

The most recently advocated formula for calculating the GFR is the one that was developed by the *Modification of Diet in Renal Disease Study Group*. Most laboratories in Australia and The United

Kingdom now calculate and report the MDRD estimated GFR along with creatinine measurements and this forms the basis of Chronic kidney disease. The adoption of the automatic reporting of MDRD-eGFR has been widely criticised.

The most commonly used formula is the "4-variable MDRD" which estimates GFR using four variables: serum creatinine, age, race, and gender. The original MDRD used six variables with the additional variables being the blood urea nitrogen and albumin levels. The equations have been validated in patients with chronic kidney disease; however both versions underestimate the GFR in healthy patients with GFRs over 60 mL/min. The equations have not been validated in acute renal failure.

For creatinine in mg/dL:

$$eGFR = 186 \times \text{Serum Creatinine}^{-1.154} \times \text{Age}^{-0.203} \times [1.210 \text{ if Black}] \times [9,742 \text{ if Female}]$$

For creatinine in μmol/L:

$$eGFR = 32788 \times \text{Serum Creatinine}^{-1.154} \times \text{Age}^{-0.203} \times [1.210 \text{ if Black}] \times [0.742 \text{ if Female}]$$

Creatinine levels in μmol/L can be converted to mg/dL by dividing them by 88.4. The 32788 number above is equal to $186 \times 88.4^{1.154}$.

A more elaborate version of the MDRD equation also includes serum albumin and blood urea nitrogen (BUN) levels:

$$eGFR = 170 \times \text{Serum Creatinine}^{-0.999} \times \text{Age}^{-0.176} \times [0.762 \text{ if Female}] \times$$
$$[1.180 \text{ if Black}] \times [0.742 \text{ if Female}]$$

Where the creatinine and blood urea nitrogen concentrations are both in mg/dL. The albumin concentration is in g/dL.

These MDRD equations are to be used only if the laboratory has NOT calibrated its serum creatinine measurements to isotope dilution mass spectroscopy (IDMS). When IDMS-calibrated serum creatinine is used (which is about 6% lower), the above equations should be multiplied by 175/186 or by 0.94086Since these formulae do not adjust for body mass, they (relative to the Cockcroft-Gault formula) underestimate eGFR for heavy people and overestimate it for underweight people. (see Cockcroft-Gault formula above).

Estimated GFR (eGFR) using the CKD-EPI Formula

The CKD-EPI (Chronic Kidney Disease Epidemiology Collaboration) formula was published in May 2009. It was developed in an effort to create a formula more precise than the MDRD formula, especially when actual GFR is greater than 60 mL/min per 1.73 m^2.

Researchers pooled data from multiple studies to develop and validate this new equation. They randomly divided 10 studies which included 8254 participants, into separate data sets for development and internal validation. 16 additional studies, which included 3896 participants, were used for external validation.

The CKD-EPI equation performed better than the MDRD (Modification of Diet in Renal Disease Study) equation, especially at higher GFR, with less bias and greater accuracy. When looking at

NHANES (National Health and Nutrition Examination Survey) data, the median estimated GFR was 94.5 mL/min per 1.73 m2 vs. 85.0 mL/min per 1.73 m2, and the prevalence of chronic kidney disease was 11.5% versus 13.1%.

The CKD-EPI equation, expressed as a single equation, is:

$$eGFR = 141 \times \min(SCr/k, 1)^{a} \times \max(SCr/k, 1)^{-1.209} \times 0.993^{Age} \times [1.108 \text{ if Female}] \times [1.159 \text{ if Black}]$$

where SCr is serum creatinine (mg/dL), k is 0.7 for females and 0.9 for males, a is -0.329 for females and -0.411 for males, min indicates the minimum of SCr/k or 1, and max indicates the maximum of SCr/k or 1.

A clearer version may be as follows: For creatinine (IDMS calibrated) in mg/dL:

African American Female

If serum creatinine (Scr) <= 0.7

$$eGFR = 166 \times (SCr/0.7)^{-0.329} \times 0.993^{Age}$$

If serum creatinine (Scr) > 0.7

$$eGFR = 166 \times (SCr/0.7)^{-1.209} \times 0.993^{Age}$$

African American Male

If serum creatinine (Scr) <= 0.9

$$eGFR = 163 \times (SCr/0.9)^{-0.411} \times 0.993^{Age}$$

If serum creatinine (Scr) > 0.9

$$eGFR = 163 \times (SCr/0.9)^{-1,209} \times 0.993^{Age}$$

White or other race Female

If serum creatinine (Scr) <= 0.7

$$eGFR = 144 \times (SCr/0.7)^{-0.329} \times 0.993^{Age}$$

If serum creatinine (Scr) > 0.7

$$eGFR = 144 \times (SCr/0.7)^{-1.209} \times 0.993^{Age}$$

White or other race Male

If serum creatinine (Scr) <= 0.9

$$eGFR = 141 \times (SCr/0.9)^{-0.411} \times 0.993^{Age}$$

If serum creatinine (Scr) > 0.9

$$eGFR = 141 \times (SCr/0.9)^{-1.209} \times 0.993^{Age}$$

This formula was developed by Levey et al.

Estimated GFR (eGFR) using the Mayo Quadratic Formula

Another estimation tool to calculate GFR is the Mayo Quadratic formula. This formula was developed in an attempt to better estimate GFR in patients with preserved kidney function. It is well recognized that the MDRD formula tends to underestimate GFR in patients with preserved kidney function.

The equation is: GFR = exp(1.911 + 5.249/SCr - 2.14/Scr^2 - 0.00686 * Age -0.205 (if female))
If SCr < 0.8 mg/dL, use 0.8 for SCr

Estimated GFR for Children using Schwartz Formula

In children, the Schwartz formula is used. This employs the serum creatinine (mg/dL), the child's height (cm) and a constant to estimate the glomerular filtration rate:

$$eGFR = \frac{k \times Height}{Serum\ Creatinine}$$

Where *k* is a constant that depends on muscle mass, which itself varies with a child's age:

In first year of life, for pre-term babies K=0.33 and for full-term infants K=0.45

For infants between ages of 1 and 12 years, K=0.55.

The method of selection of the K-constant value has been questioned as being dependent upon the gold-standard of renal function used (i.e. creatinine clearance, inulin clearance etc) and also may be dependent upon the urinary flow rate at the time of measurement.

Importance of Calibration of the Serum Creatinine level and the IDMS Standardization Effort

One problem with any creatinine-based equation for GFR is that the methods used to assay creatinine in the blood differ widely in their susceptibility to non-specific chromogens, which cause the creatinine value to be overestimated. In particular, the MDRD equation was derived using serum creatinine measurements which had this problem. The NKDEP program in the United States has attempted to solve this problem by trying to get all laboratories to calibrate their measures of creatinine to a "gold standard", which in this case is isotope dilution mass spectroscopy (IDMS). At the present time in late 2009 not all labs in the U.S. have changed over to the new system. There are two forms of the MDRD equation that are available, depending on whether or not creatinine was measured by an IDMS-calibrated assay. The CKD-EPI equation is designed to be used with IDMS-calibrated serum creatinine values only.

Cystatin C

Problem with creatinine (varying muscle mass, recent meat ingestion, etc.) have led to evaluation of alternative agents for estimation of GFR. One of these is cystatin C, a ubiquitous protein secreted by most cells in the body (it is an inhibitor of cysteine protease) that is freely filtered at the glomerulus and not reabsorbed. After filtration, cystatin C is normally catabolized in the renal tubules and none appears in the urine, so urine collection methods cannot be used. However, equations have been developed linking estimated GFR to serum cystatin C levels. Some proposed equations (2009) have combined creatinine and cystatin. None are available clinically at this time, but this is likely to change in the near future.

The normal range of GFR, adjusted for body surface area, is similar in men and women, and is in the range of 100-130 ml/min/1.73M^2. In children, GFR measured by inulin clearance remains close to about 110 ml/min/1.73M^2 down to about 2 years of age in both sexes, and then it progressively decreases. After age 40, GFR decreases progressively with age, by about 0.4 - 1.2 mL/min per year.

Risk factors for kidney disease include diabetes, high blood pressure, family history, older age, ethnic group and smoking. For most patients, a GFR over 60 mL/min/1.73M^2 is adequate. But, if the GFR has significantly declined from a previous test result, this can be an early indicator of kidney disease requiring medical intervention. The sooner kidney dysfunction is diagnosed and treated, the greater odds of preserving remaining nephrons, and preventing the need for dialysis.

The severity of chronic kidney disease (CKD) is described by six stages; the most severe three are defined by the MDRD-eGFR value, and first three also depend whether there is other evidence of kidney disease (e.g. proteinuria):

(0) Normal kidney function – GFR above 90mL/min/1.73m^2 and no proteinuria

(1) CKD1 – GFR above 90mL/min/1.73m^2 with evidence of kidney damage

(2) CKD2 (Mild) – GFR of 60 to 89 mL/min/1.73m^2 with evidence of kidney damage

(3) CKD3 (Moderate) – GFR of 30 to 59 mL/min/1.73m^2

(4) CKD4 (Severe) – GFR of 15 to 29 mL/min/1.73m^2

(5) CKD5 Kidney failure - GFR less than 15 mL/min/1.73m^2 Some people add CKD5D for those stage 5 patients requiring dialysis; many patients in CKD5 are not yet on dialysis.

Tubular Function

The quantity of glomerular fluid formed in the kidneys is enormously greater than the amount of urine excreted due to reabsorption of most of the water as the fluid passes through the tubules. This reabsorption of water is indicated by the much greater concentration of dissolved substances in urine than in plasma and glomerular filtrate. The solute concentration in plasma and glomerular fluid is normally around 0.3 osmolar per liter, while in urine the solute concentration may rise to 1.4 osmolar, representing a concentration of nearly fivefold.

The pH of glomerular filtrate is about the same as that of plasma, 7.4, and the filtrate is nearly isosmotic with plasma. As the glomerular filtrate passes through the proximal tubules 80-85 per cent of the electrolytes (Na^+, K^+, Cl^-, HCO_3^-, HPO_4^-, and SO_4^-) and of the water and practically all of the glucose, amino acids, and ascorbic acid are reabsorbed. Thus of 125 ml glomerular filtrate passing through the proximal tubules only some 20-25 ml. reaches the loop of Henle and the distal tubules. This fluid is now essentially devoid of glucose and amino acids (very small amounts of each), is isosmotic, and has a pH of 7.4, since water and electrolytes have been absorbed in the proportions present in glomerular fluid. The reabsorption which takes place in the proximal tubules is referred to as "obligatory absorption", since it is relatively independent of the composition and volume of body fluids and so far as known is not under hormonal control.

As indicated above, normally about 80 per cent of the water of the glomerular fluid is reabsorbed into the plasma by the proximal tubules, and the remaining 20 per cent passes to the distal tubules. When the plasma volume increases because of water intake, the cardiac output and glomerular filtration rate increase so that a smaller proportion than normal of the water is absorbed by the proximal tubules and more passes on to the distal tubules. Conversely, in cases of dehydration and excessive concentration of plasma (and other body fluids) the opposite conditions exist, and less water comes to the distal tubules. Thus, the proximal tubules permit more water to pass when the plasma is too dilute and less to pass when the plasma is too concentrated.

Urine formation is completed as the fluid from the proximal tubules flows through the loops of Henle and distal tubules where the cells, through special mechanisms, reabsorb most of the water, various nonelectrolytes, and especially sodium salts from the fluid, and add such substances as K^+, H^+, NH_4^+, and creatinine (at higher plasma levels only) to it. Of the 20-25 ml of fluid reaching the loops of Henle per minute only 0.5 to 2.0 ml enters the collecting tubules.

The reabsorption of electrolytes, chiefly sodium salts, by the distal tubular cells is under the control of adrenal cortical hormones, aldosterone, and deoxycorticosterone. When the electrolyte concentration and osmotic pressure of plasma fall below a certain level, the adrenal cortex secretes more of these hormones, which increase absorption of sodium salts, thereby restoring a larger proportion of electrolyte to the plasma and relieving the lowered electrolyte concentration and osmotic pressure. When the plasma electrolytes and osmotic pressure rise above normal, the adrenals secrete less hormones, permitting the excretion of more sodium salts and lowering the electrolyte concentration and osmotic pressure of plasma.

In Addison's disease of the adrenals the secretion of cortical hormones may be greatly decreased, causing failure to reabsorb sodium salts in the distal tubules and their loss in the urine with profound disturbance of body fluid volume and composition.

The reabsorption of water by the distal tubules is under the control of the antidiuretic hormone vasopressin, which is released into the blood by the neurohypophysis. When the electrolyte concentration and osmotic pressure of plasma rise above normal, the neurohypophysis is stimulated to re-lease more vasopressin, which causes more reabsorption of water back into the plasma by the distal tubules, with less water excretion in the urine, to aid in relieving the condition. Conversely, when the electrolyte concentration and osmotic pressure fall below normal, less vasopressin is released into the blood, decreasing water reabsorption by the distal tubules and increas-ing excretion in the urine, to relieve the condition.

In the clinical condition diabetes insipidus, frequently associated with lesions of the hypophysis or hypothalamus, there is a deficiency in the secretion of vasopressin and a decrease in the distal tubular reabsorption of water. Many liters of urine with exceedingly low specific gravity (1.002 to 1.006) may be excreted, which is associated with great thirst and water intake. The condition is controlled by administration of vasopressin preparations.

Experimental diabetes insipidus may be caused by section of the hypophyseal stalk or destruction of the supraoptic nuclei, which causes atrophy of the neurohypophysis. The net over-all concentration of urine relative to glomerular filtrate is determined by the degree to which reabsorption of water exceeds reabsorption of total solutes in the glomerular filtrate.

About 180 litres of glomerular filtrate on the average are formed per 24 hours, of which some 178.8 liters (99.4 per cent) are reabsorbed. Table showing the plasma electrolytes below:

The Plasma Electrolytes

	A Plasma Conc (mM/1)	B Filtered A × 180 (mM)	C Found in Urine (mM)	D Reabsorbed B-C (mM)	Per Cent Reabsorbed (D/B×100)
Na	142	25,560	111	25,449	99.6
Cl	103	18,540	119	18,421	99.4
K	5	900	60	840	93.4
HPO_4	1	180	30	150	83.4
SO_4	0.5	90	23	67	74.4

Table above shows the average total quantities of various ions present in glomerular fluid and urine per day and the amounts reabsorbed by tubular action. For example, it will be seen that 25,560 mM of Na^+ are filtered by the glomeruli and 25,449 mM are reabsorbed by the tubules (99.6 per cent).

Kidney clearance

As the blood containing waste products passes through the kidneys, a certain proportion of these substances is removed per unit time, and the amount of anyone substance excreted in the urine during this time is equivalent to the amount of the substance in a definite volume of blood or plasma. Specifically the kidney clearance of a substance represents the minimum volume of blood required to furnish the amount of substance excreted in the urine in one minute. Plasma rather than whole-blood clearances are determined because of the unequal distributions of sub-stances between plasma and cells and the wide variations in the proportions of plasma and cells in blood.

The clearance, C, of a substance may be represented by the equation:

$$c = \frac{uv}{p}$$

where U = the concentration of substance per milliliter of urine, V = the milliliters of urine secreted per minute and P = the concentration of substance per millimeter of plasma.

Suppose a subject puts out 120 ml of urine in an hour containing 900 mg of urea, and that the plasma urea concen-tration is 22 mg per 100 ml. The urinary excretion of urea per minute is 900/60 = 15 mg = 0.125 × 120 = UV, and the concentration of urea in the plasma, P, is 22/100 or 0.22 mg. The urea clearance, C, is:

$$c = \frac{UV}{P} = \frac{0.125 \times 120}{0.22} = \frac{15}{0.22} = 68 \text{ ml}$$

This means that the amount of urea excreted per minute by the kidneys equals the amount of urea present in 68 ml of plasma, and does not mean that all of the urea is extracted from 68 ml of plasma as it flows through the kidney. Now if the glomerular filtrate volume per minute in the subject were the normal 125 ml per minute with a plasma clearance of 68 ml, then 68/125 = 0.54 of the urea was removed from the 125 ml of glomerular filtrate. The fraction of a substance removed from the glomerular filtrate is called the" extraction ratio," and for the example given it is 0.54.

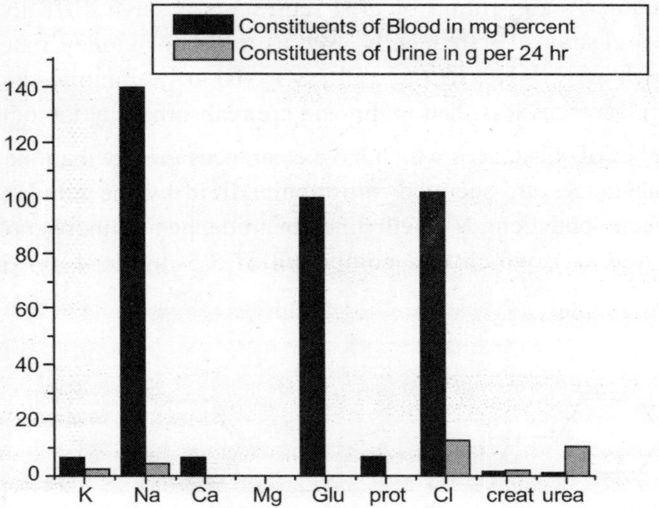

Fig. 9.29 : The Comparative Study of the Constituents of Urine and Plasma

When the rate of urine flow is 2 ml or more per minute in the adult, the clearance for a substance is called" clearance maximum," C_m. At reduced urine flows the clearance does not represent the full capacity of the kidneys.

When the concentrations of a substance in plasma and glomerular filtrate are equal, and the substance is not absorbed or excreted by the tubules, its clearance represents the rate of glomerular filtration. The glomerular filtration rate, as determined by clearance, amounts to around 125 ml per minute per 1.73 sq m surface area for men and a bit less for women.

The polysaccharide inulin (molecular weight about 5000) is most com-monly used for determination of glomerular filtration rates, though other substances, such as creatinine, thiosulphate, and mannitol, may be used. When creatinine is ingested, the clearance is about 175 ml per minute, indicating that 50 ml of plasma have been cleared by tubular excretion (175–125 = 50). Without the ingestion of creatinine the clearance approaches that of inulin. This does not mean the ingested (exogenous) creatinine is handled differently from endogenous creatinine by the tubules, but rather that as plasma creatinine values rise above a certain level due to creatinine ingestion, the tubules begin to excrete creatinine. Consequently, glomerular filtration rates are determined from creatinine clearances at normal plasma levels of creatinine.

Any substance which is not combined with plasma protein and which shows lower clearance than inulin must be reabsorbed by the tubules. For example, urea *Cm* normally ranges from about 64 to 99 ml and averages about 75 ml This is 50 ml less than the glomerular filtration rate, indicating tubu-lar reabsorption. The average proportion of urea absorbed is given by the expression:

$$\frac{\text{Inulin clearance (125)-urea clearance (75)}}{\text{Inulin clearance (125)}} = \frac{50}{125} = 0.4$$

which means that an average of 40 per cent of the urea present in glomerular filtrate is reabsorbed by the tubules. This tubular reabsorption of urea represents passive diffusion with the water being reabsorbed. The urea clearance is frequently used as a test for kidney function. The clearances of many substances, such as Na^{+}, K^{+}, HCO_3^{-}, Cl^{-}, $SO_4^{=}$ HPO_4^{-}, amino acids, glucose, uric acid, and ascorbic acid, show clearances less than inulin and are reabsorbed by the tubules.

There are a number of substances which have clearances greater than the inulin clearance, which means that these substances are secreted into tubular fluid by the tubular cells. Such substances include H^{+}, NH_4^{+}, phenolphthalein, N'-methylnicotinamide, penicillin, phenol red, p-aminohippurate, and iodopyracet (Diodrast), which is a compound of 3,5-diiodo-4-pyridone-N-acetic acid and ethanolamine:

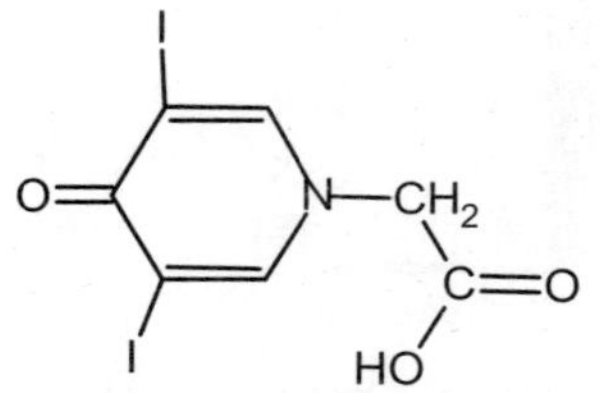

(3,5-diiodo-4-oxopyridin-1(4*H*)-yl) acetic acid

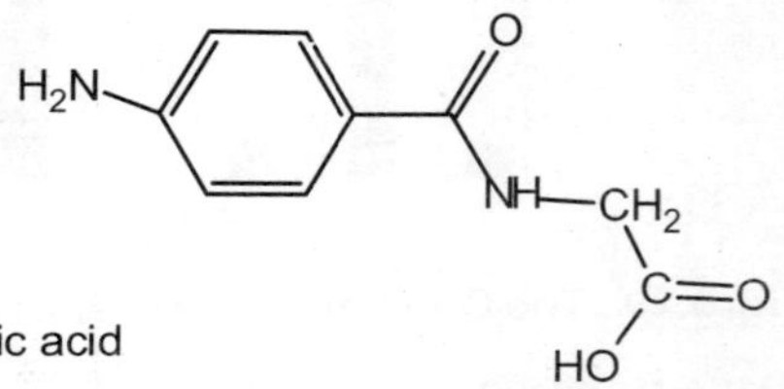

{[(4-aminophenyl) carbonyl] amino}acetic acid

p - Amino hippuric acid

The tubular secretion of iodopyracet and of p-aminohippurate (P AH) is so effective that at low plasma levels of these substances they do not appear in the renal venous blood, indicating complete removal by One passage of blood through the kidney. Their clearances accordingly are a measure of the renal plasma flow per minute and amount to about 700 ml for the average adult, which represe1}ts a renal blood flow of about 1200 ml per minute (about one-fourth of the total cardiac output).

Since p-aminohippurate is less bound to plasma proteins than iodopyracet, does not enter red cells, and is more easily determined than iodopyracet, it is replacing iodopyracet in the determination of renal blood flow. When the Cm values for p-aminohippurate and iodopyracet are 700, this means that 700 – 125 = 575 ml of plasma are cleared of these substances by tubular secretion into tubular fluid. When the glomerular filtration rate is divided by the renal plasma flow, the fraction of plasma filtered through the glomeruli is obtained:

$$\frac{125}{700} = 0.178$$

which means that about one-fifth of the plasma brought to the glomeruli becomes glomerular filtrate.

Transport maximum, T_m, and renal threshold. Transport maximum, T_m, represents the maximum rate at which the tubules can reabsorb or excrete a substance. At normal blood sugar levels (for example, 100 mg per 100 ml of plasma) urine contains essentially no glucose, which means that the tubules are reabsorbing 125 mg of glucose per minute from the glomerular fluid. Suppose the plasma glucose level is raised to 500 mg per 100 ml; then the glomerular filtrate will contain $5 \times 125 = 625$ mg of glucose per minute. Suppose under these conditions the kidney excretion of glucose is 300 mg per minute. This means that the tubules can reabsorb as a maximum 325 mg of glucose per minute, or that the T m for glucose is 325.

In the case of substances excreted by the tubules, such as P AH, the amount appearing in the urine per minute when the plasma level exceeds the excretory capacity represents the excretory T_m.

In the operation of both tubular reabsorption and excretion there are substances which compete with each other for the cellular mechanisms involved indicating that the same entire mechanism or a part of it may function with different substances. For example, substances actively excreted by the tubules such as P AH, iodopyracet, penicillin, etc., may mutually depress their individual urinary excretions when given together in high dosage. Similarly, competition for mechanisms with decreased tubular reabsorption and increased urinary excretion are shown by sugars such as xylose and glucose, by various amino acids, and by sodium chloride and ascorbic acid. It is very interesting that the tubular reabsorption of ascorbic acid is decreased by the simultaneous tubular excretion of P AH.

Renal Threshold

Various substances in plasma, such as glucose, ascorbic acid, Na^+, and Cl^-, do not appear in urine appreciably until their plasma concentrations rise to certain values, and such substances are referred to as "threshold substances." These substances are reabsorbed from glomerular filtrate by the tubules.

Specifically the renal threshold for a substance is the plasma concentration of the substance above which it appears in the urine, a constant glomerular filtration rate being assumed since the efficiency of tubular reabsorption decreases with increased rate of flow in the tubules. The renal threshold generally lies at a much lower plasma level of a substance than the plasma level at which the tubules show T_m.

The glucose renal threshold in the adult shows wide variations, most generally being between 140 and 170 mg per 100 ml of plasma. The plasma concentration above which a substance appears in the urine is referred to as the "threshold of appearance." Another useful concept is the" threshold of retention" at which the urine and plasma concentrations of a substance are equal. When the plasma level of a substance exceeds the threshold of retention, the concentration of the substance in the urine will be higher than in plasma, while if the plasma concentration of the substance is less than the threshold of retention, the urine concentration will be, less than the plasma concentration.

The thresholds of appearance and retention for K^+ in man average about 11 and 11.7 mg per 100 ml of plasma, respectively. Generally the threshold of retention is a bit greater than the threshold of appearance.

Various principles and facts relative to renal operation are brought out in Figures 9.30 and 9.31 Figure 9.30 taken from Homer Smith, diagrammatically shows the relations of the tubules to the excretion of inulin, glucose, urea, and iodopyracet, with clearance values for these substances.

Figure 9.31, taken from Smith, shows the relations of clearances to plasma concentration for a number of substances. The clearance values given represent averages. It has been found that these clearances are portional to body surface area, just as is energy metabolism, the average. surface area being 1.73 sq m. Consequently, the above clearances represent values per 1.73 sq m of body area. The normal clearance for any given individual will be proportionately more or less, depending upon his surface area.

RELATION OF THE KIDNEYS TO HYPERTENSION

Researcher showed that when. the blood flow to a kidney of the dog is considerably reduced (ischemic kidney) by a clamp on the renal artery, hypertension ensues which persists for several weeks and slowly subsides. The condition. is relieved by removal of the clamp or of the ischemic kidney. The

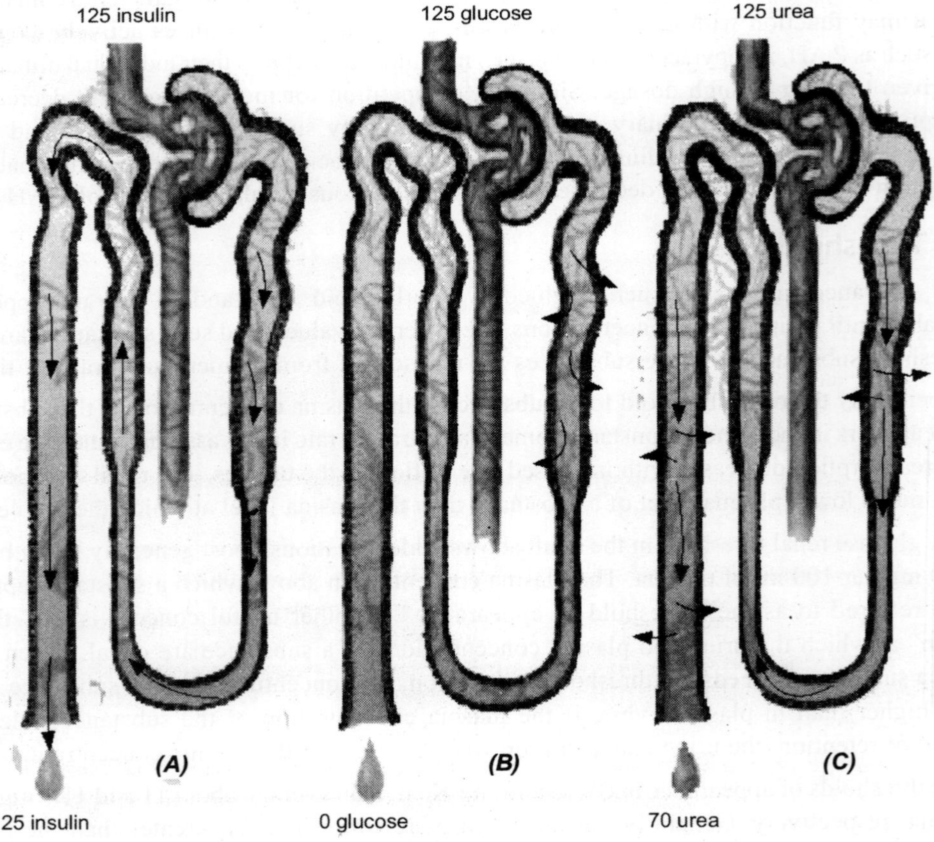

Fig. 9.30 : The Scheme to illustrate the excretion of *(A)* inulin, which is excreted solely by filtration with no tubular reabsorption; *(B)* glucose, which is filtered but at normal plasma level and rate of filtration, is completely reabsorbed by the tubule; *(C)* Urea, which is filtered, but in part escapes from the tubular urine by diffusion

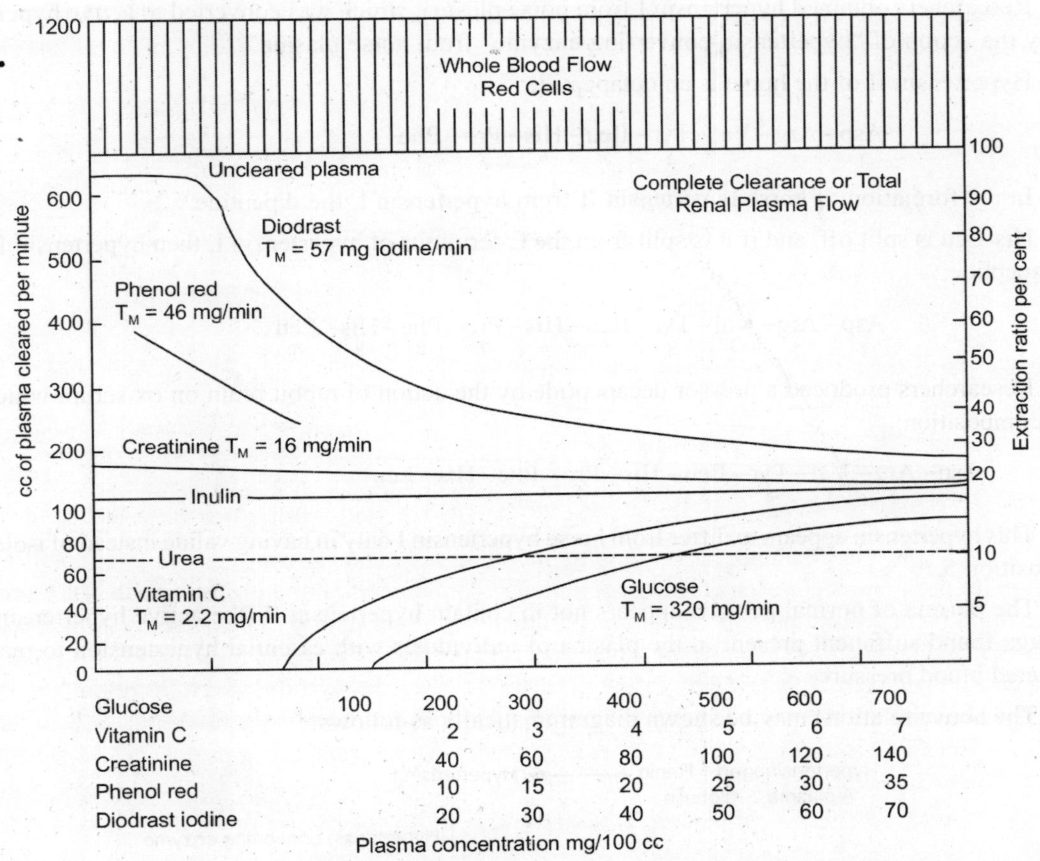

Fig. 9.31 : The Diagrammatic Summary of Excretion of Various Types of Compounds by the Human Kidney. Average normal T_m values are used in the calculation of the clearances at various plasma levels

condition is made more severe when both kidneys are made ischemic, or one is made ischemic and the other removed. That the pressor effect is not of nervous origin has been demonstrated by transplanting the kidney to be made ischemic to the neck or groin, where the same hypertensive effects are observed.

Experimental renal hypertension appears to involve the following mechanisms. The ischemic kidney liberates into the blood an enzyme, renin, which splits hypertensin I, a polypeptide, from hypertensinogen, an α2-globulin formed by the liver. Hypertensin I has no pressor activity; however, an enzyme present in plasma, "hypertensin-converting enzyme," acts upon hypertensin I to form hypertensin II, which is a powerful pressor agent. Tissues in general and kidney and intestine in particular contain a dipeptidase enzyme which destroys hypertensin II.

Different hypertensins have been obtained, depending upon the animal source of renin and of the α2-globulin (hypertensinogen) upon which it acts.

Researchers obtained hypertensin I from horse plasma, which was converted to active hypertensin II by the action of "hypertensin-converting enzyme" from horse plasma.

Hypertensin II of the horse is an octapeptide:

$$\underset{1}{\text{Asp}}-\underset{2}{\text{Arg}}-\underset{3}{\text{Val}}-\underset{4}{\text{Tyr}}-\underset{5}{\text{Ileu}}-\underset{6}{\text{His}}-\underset{7}{\text{Pro}}-\underset{8}{\text{Phe}}$$

In the formation of horse hypertensin II from hypertensin I, the dipeptide.

His-Leu is split off, and if it is split from the C-terminus of hypertensin I, then hypertensin I is the decapeptide:

$$\underset{1}{\text{Asp}}-\underset{2}{\text{Arg}}-\underset{3}{\text{Val}}-\underset{4}{\text{Tyr}}-\underset{5}{\text{Ileu}}-\underset{6}{\text{His}}-\underset{7}{\text{Pro}}-\underset{8}{\text{Phe}}-\underset{9}{\text{His}}-\underset{10}{\text{Leu}}$$

Researchers produced a pressor decapeptide by the action of rabbit renin on ox serum which had the composition:

$$\underset{1}{\text{Asp}}-\underset{2}{\text{Arg}}-\underset{3}{\text{Val}}-\underset{4}{\text{Tyr}}-\underset{5}{\text{Ileu}}-\underset{6}{\text{His}}-\underset{7}{\text{Pro}}-\underset{8}{\text{Phe}}-\underset{9}{\text{His}}-\underset{10}{\text{Leu}}$$

This hypertensin appears to differ from horse hypertensin I only in having valine instead of isoleucine at position 5.

The plasma of normal persons appears not to contain hypertensin II (the active hypertensin), but Skeggs found sufficient present in the plasma of individuals with essential hypertension to maintain elevated blood pressure.

The above relations may be shown diagrammatically as follows:

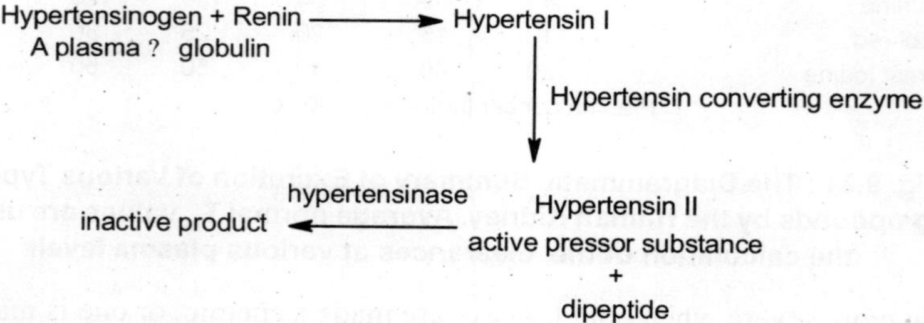

" Angiotonin" is an older term for hypertensin II. Angiotonin more recently is called "angiotensin".

Fluid Compartments of the Body

Water makes up about 70 per cent of the weight of the adult human body. This water is distributed throughout the body as the major component of the intracellular and extracellular fluids. The intracellular fluids amount to about 50 per cent of the body weight. The extracellular fluids represent then about 20 per cent of the body weight. Of the extracellular fluids, interstitial fluid amounts to about 15 per cent and blood plasma to about 5 per cent of the body weight. Relatively small volumes are represented by specialized fluids, such as cerebrospinal fluid, ocular fluids, lymph, synovial fluid, etc., which to a greater or lesser degree are included in determinations of the other fluid compartments.

Measurement of Huid Compartments

The measurement of fluid compartments is based upon the intravenous injection of a known amount of a substance which distributes throughout a given compartment or compartments and, after time for mixing, determination of its concentration in plasma, and correction for the amount excreted or destroyed in the body, so that the quantity of substance in the fluid compartment is known when the sample for analysis is taken (9). The fluid compartment is obtained by dividing the grams of substance present in the fluid compartment by the grams of substance per liter of plasma.

As an example of the calculations involved in determining fluid compartments, consider the following illustration. The sugar alcohol mannitol is excreted entirely by the kidneys. Suppose a subject receives 21 g of mannitol intravenously, and after a time for equilibration, 11 g have been excreted in the urine and the serum contains 0.7 g per litre. The quantity of mannitol in the body at the time of sampling, Q, is 10 g (Q = 21 - 11 = 10 g), and the concentration of mannitol per litre of fluid, C, is 0.7 g. Then the fluid volume or space occupied by the mannitol, *V*, is:

$$V = \frac{Q}{C} = \frac{10}{0.7} = 14.3L$$

When the concentrations, C, are expressed as grams per liter of water, then the volumes obtained represent liters of water. The ideal characteristics of a substance for the measurement of a fluid compartment are:

(a) the substance is evenly distributed throughout the fluid compartment being measured without passage into another fluid phase.

(b) a representative sample of the diluted substance can be obtained for analysis, generally from plasma or serum;

(c) the amount of substance re-tained in the body at the time of sampling can be determined from the amount injected and correction for the quantity destroyed in the body or excreted by any route.

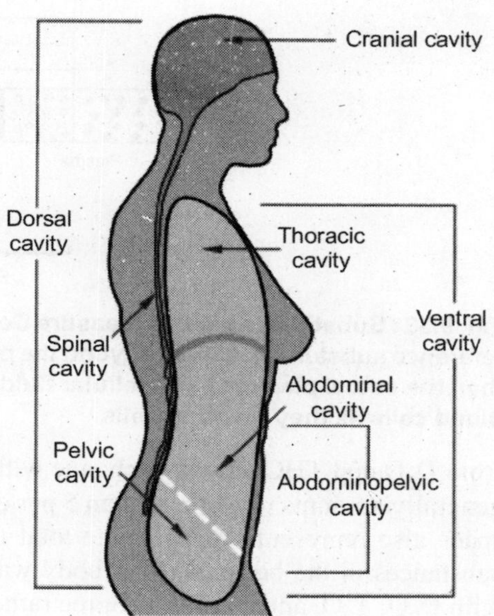

Fig. 9.32 : The Fluid Compartments of the Human body

No substance in practice meets all of these ideal requirements completely, but a number of compounds give very useful approximate values. Fig. 9.33 shows the distributions of various substances in the fluid compartments of the body.

It is customary in the literature to refer to substance space. Deuterium oxide, D_2O, and radioactive tritium oxide, or THO, distribute throughout all body fluids, so that total body fluid volume is obtained by determining D_2O and THO spaces. Some error arises from the fact that deuterium and tritium

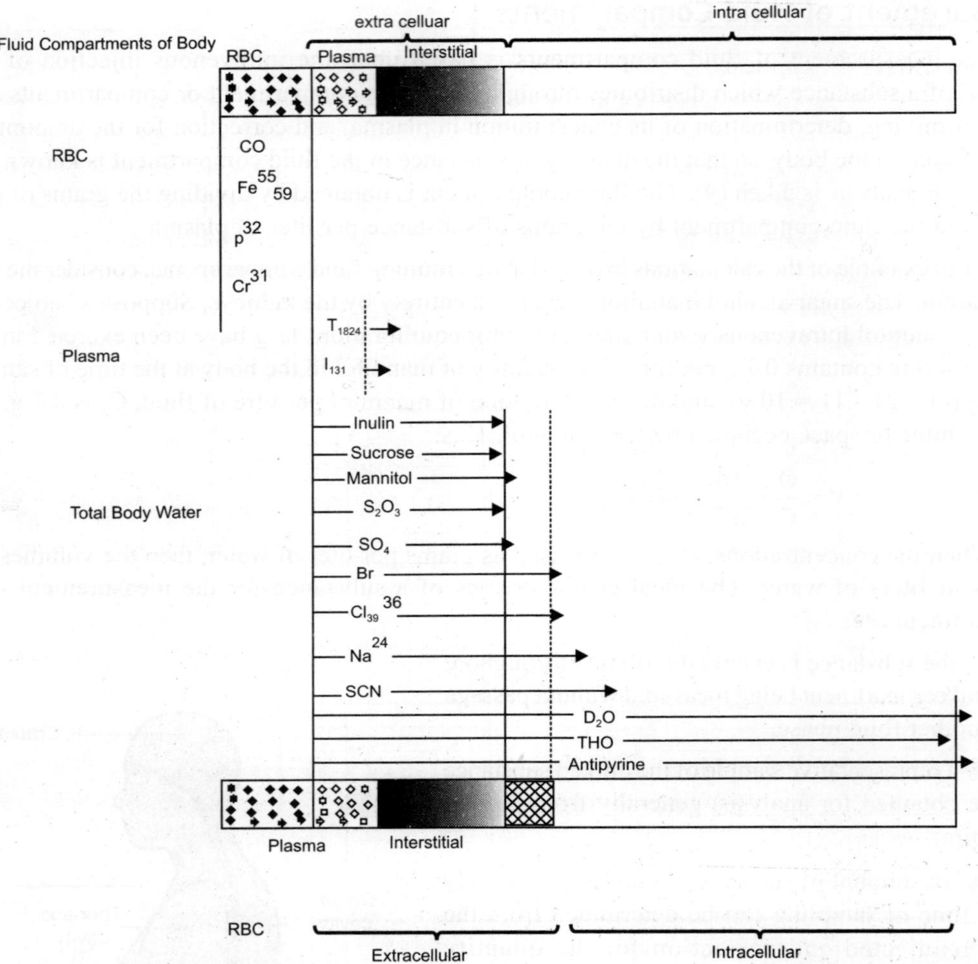

Fig. 9.33 : Substances used to Measure Compartments of the Body Fluids. It is obvious that many of the reference substances diffuse beyond the particular fluid phases, yielding volumes of distribution greater than the actual plasma, extracellular fluid, or total body water. D_2O, THO, and antipyrine penetrate red blood cells as they do other cells

from D_2O and THO slowly exchange with hydrogen atoms of various body substances. However, this generally amounts to no more than 5 per cent and does not seriously affect their usefulness. Antipyrine space also represents approximate total fluid volume. From the dilution of known amounts of these substances in the body the total body water is obtained. The values for total body water determined with D_2O, T_2O, and antipyrine agree rather well with each other and with total body water determined by desiccation.

Extracellular fluid volume appears to be most nearly represented by the inulin space, though the slowness of inulin diffusion into extracellular fluids represents a disadvantage. The thiosulphate space appears to represent the extracellular fluid volume rather well. The plasma volume is generally determined by use of I^{-131}-albumin or the dye Evans blue (T-1824). The intracellular fluid volume is equal to the total fluid volume minus the extracellular volume, or it is represented by the D_2O spacd minus the inulin space. The interstitial fluid volume is equal to the extracellular fluid volume minus the plasma volume (inulin space minus p3l-albumin space). The interstitial fluid represents a fluid buffer between plasma and intracellular fluid. Since plasma is subject to rather sudden variations in composition through absorption from the intestine, the interposition of interstitial fluid between plasma and intracellular fluid aids in maintaining the composition of intracellular fluid more constant than otherwise would be the case. With this arrangement the kidneys have the opportunity to compensate for changes in plasma before they are seriously reflected upon the intracellular fluid.

THE BIOCHEMISTRY OF URINE

At this time the student should review the discussion of renal function in order that he may have a better basis for understanding the composition of urine. The chief waste products of the body are carbon dioxide, derived from the oxidation of all organic foods; various nitrogenous substances, such as urea, uric acid, and creatinine, from protein metabolism; sulphur compounds and phosphates, also largely from protein metabolism; and excess quantities of water, sodium, potassium, calcium, magnesium, and chloride. Many other substances in small amounts are eliminated from the body in urine. Carbon dioxide is largely excreted by the lungs. The intestine excretes most of the calcium, magnesium, and iron, and some other substances. The skin normally eliminates considerable water by insensible evaporation but during active perspiration may excrete massive quantities of water containing large amounts of sodium and chloride, appreciable potassium, considerable quantities of urea, and smaller amounts of other substances. However, it is the kidneys upon which falls the burden of excreting most of the solid waste products of the body (normal average about 75 g per 24 hours for an adult) and excess body water.

The composition of urine is determined by the quantities of substances which must be removed from the body by the kidneys in order to maintain the composition of blood and other body fluids within physiologic limits. Aside from water, the preponderant amount of waste products in urine arises from the metabolism of protein derived from food and tissues. In general, then, the gross amount of substances excreted in urine is proportional to the protein eaten, and the major waste product from protein is urea. During starvation the qualitative composition of uriBe remains essentially unchanged, but the quantitative relations of the constituents are markedly altered. A few substances, such as creatinine, are formed largely as the result of tissue metabolism (endogenous metabolism) and show relatively less variation with diet than do substances, such as urea, which normally are mainly the result of metabolism of food (exogenous metabolism).

The composition of urine in pathologic states generally differs from the normal in the relative amounts of constituents. For example, the urine of a normal person contains an average of only about 0.1 g of glucose per 24-hour sample, while the daily urine of a diabetic may contain 150 g or more. Similarly, only traces of the ketone bodies are present in normal urine, while 50 g or more may be excreted daily by the severe diabetic. Normally less than 100 mg of protein per day are excreted by normal persons, but the urine of the nephrotic patient may contain many grams.

There are two chief objects in studying the composition of urine. One of these is to aid in understanding, through the end products formed and excreted in the urine, the processes of body metabolism, details of which have been considered in the chapters on metabolism. The other object is to obtain evidence, through the presence of abnormal substances or substances in abnormal amounts, that may be of aid in the diagnosis and treatment of pathologic states.

THE DETAIL CONSTITUTION OF URINE

Because the volume of urine relative to contained solutes varies widely, expression of the amount of substance per 100 ml of urine is generally of little value. Rather, it is necessary to calculate the amount of substance contained in the total volume of urine excreted per 24 hours or other definite period of time.

When the quantities of urinary substances per 24 hours are to be determined, the sample is usually collected as follows. Upon arising at a definite time in the morning, the bladder is emptied and the urine is discarded. During the ensuing day and night, and upon arising (at the same time as the day before), all samples of urine are collected in a bottle or flask containing a suitable preservative, such as toluene. This mixed sample represents the urine corresponding to a 24-hour metabolic period. Collections for shorter periods are carried out similarly. Detailed methods for the analysis of urine are to be found in laboratory manuals and works on clinical chemistry.

Since the quantities of most of the constituents of urine depend largely upon the nature and amount of the food eaten, it is impossible to give average values that are of much significance. Table below gives values such as may be found for a 24-hour sample of urine from an individual on a diet containing an average quantity of proteins. In addition, values for the chief nitrogenous substances on high and low protein diets are included.

The Composition of Urine Values are in g per 24 hr

	Ordinary Diet	High protein Diet	Very Low Protein Diet
	1	2	3
Volume, ml	1250	1550	950
Total nitrogen	13.20	23.28	4.20
Urea	24.30	43.80	6.20
Urea nitrogen	11.40	20.40	2.90
Ammonia	0.50	1.00	0.20
Ammonia nitrogen	0.40	0.80	0.17
Creatinine	1.64	1.72	1.60
Creatinine nitrogen	0.61	0.64	0.60
Uric acid	0.60	0.90	0.30

...(Contd.)

	1	2	3
Uric acid nitrogen	0.20	0.30	0.10
Total amino acids	1.5, less than half free		
Amino acid nitrogen	0.2		
Hippuric acid	0.6		
Indican	0.01		
Allantoin	0.015		
Creatine	0.0-0.06		
Undetermined nitrogen	0.60	1.10	0.50
Glucose	0.10		
Nonglucose reducing substances calculated as glucose	1.0		
Phenols	0.20		
Citirc acid	0.30		
Ascorbic cid	0.025		
Oxalic acid (oxalates)	0.015		
Acetone bodies	0.01		
Thiamine	0.0001		
Total sulfur	1.0		
Inorganic sulfates	0.80		
Ethereal sulfates	0.08		
Neutral sulfur	0.12		
Chlorine as chloride	7.0		
Phosphorus as phosphate	1.0		
CO_2 as HCO_3 + H_2CO_2	Varies with urinary pH		
Sodium	5.0		
Potassium	2.5		
Calcium	0.2		
Magnesium	0.15		
Copper	0.00004		
Iron	0.00003		
Iodine	0.00005		
Water	90-95 per cent		

Detailed discussions of the urinary excretion of many of the substances are to be found in the chapters dealing with metabolism, acid-base and electrolyte-water balances, and nutrition. In addition to the components indicated in the table, urine contains many other substances, some of which are poorly defined. The values indicated for the very low protein diet represent the distribution of urinary substances resulting from a diet very high in carbohydrate and containing very little protein. The nitrogen excretion under such conditions is decidedly less than found during starvation, due to the protein-sparing action of a high carbohydrate diet. Table below, shows the composition of the urine of Victor Beaute on different days of a prolonged fast.

The Composition of Urine during Fasting the Values are in g per 24 hr

	Day of Fasting			
	1st	3rd	12th	14th
Total N	10.51	13.72	8.77	7.78
Urea N	8.96	12.26	6.62	5.99
Ammonia N	0.40	0.73	1.05	0.73
Uric acid N	0.12	0.06	0.17	0.17
Purine base N	0.029	0.032	0.023	
Creatinine N	0.42	0.34	0.30	0.24
Creatine N	0.02	0.09	0.09	0.10
Total S	0.614	0.801	0.577	0.536
Total P_2O_5	2.26	2.98	1.55	1.25
Cl	3.2	1.5	0.18	0.24
Ca		0.216		0.096
Mg		0.131		0.037
K		1.33		0.515
Na		0.865		0.096

It is of interest to note the increased rate of tissue protein breakdown on the third day of starvation, causing a sharp rise in nitrogen output. The greatly decreased excretion of cations and chloride in the urine during starvation indicates their conservation to the plasma by the kidneys in order to maintain the ionic balance and osmotic pressure of blood.

Urine specific gravity and volume. The specific gravity of urine normally varies between 1.015 and 1.025 but is subject to wide fluctuations under various conditions. It may fall to 1.003 or lower as the result of copious water drinking and rise to 1.040 or higher because of hemoconcentration due to excessive perspiration. The specific gravity is increased by the excretion of large amounts of sugar in diabetes and of protein in nephrosis. The posterior pituitary secretes an antidiuretic hormone pitressin, which regulates the absorption of water by the tubules. Deficiency of this hormone leads to the condition known as "diabetes insipidus," in which enormous volumes of urine are excreted (up to 30 I per 24 hours) having a specific gravity of 1.001 to 1.003. In such cases the solute concentration of urine may be less than that of blood plasma.

The specific gravity of urine is directly proportional to the solute content and generally varies inversely with the volume. The solute content of urine may be calculated by the use of Long's formula, in which the last two figures of the specific gravity (to the third decimal) are multiplied by the factor 2.6. This gives approximately the solids in grams per liter of urine. For example, suppose we wish to calculate the total solids in 1250 ml of urine with a specific gravity of 1.020. The solids per liter would be:

$$20 \times 2.6 = 52 \text{ g}$$

and per 1250 ml

$$\frac{1250}{1000} \times 52 = 65\text{g}$$

The osmolar concentration of urine may be calculated from the freezing point depression which generally varies from -1.3° to - 2.3°C, but as a maximum may reach - 2.60 C. Since a depression of -1.86° C represents a concentration of 1 osmol, a depression of -2.6° C represents -2.6/ -1.86 = 1.4 osmols per litre.

The Composition of Minimal Urine Excretion for the Excretion of Waste Product

Waste Products Total Milliosmols	Minimal Urine Volume in Millimetres for Excretion	
	At 1.4 Cone Cap	At 0.4 Cone Cap
200	143	500
400	286	1,000
600	429	1,500
800	571	2,000
1,000	714	2,500
1,200	857	3,000
1,400	1,000	3,500
1,600	1,140	4,000

This value of 1.4 osmols represents the maximum concentrating capacity of the normal kidney and may be used in calculating the minimal water requirement for the excretion of a given quantity of waste products. Suppose an individual excretes 1250 ml of urine per 24 hours, showing a freezing point depression of -1.49° C. The osmolar concentration per liter is: -1.49/-1.86 = 0.80, and the milliosmolar concentration is 0.80 × 1000 = 800. The total milliosmols excreted, then, would be 1250/ 1000 × 800 = 1000. Assuming the kidneys of the individual to be capable of concentrating to 1.4 osmols per liter (1400miliosmols), the minimal quantity of urine required for excretion of waste products will be :

$$\frac{\text{total miliosmols of waste products}}{1.4} = \frac{1000}{1.4} = 714 \text{ ml}$$

Similarly, the minimal quantity of urine required for the excretion of any given quantity of waste products for any given kidney concentrating capacity may be calculated. Table above lists these values for different quantities of urine solutes at kidney concentrating capacities of 1.4 and 0.4 osmols per liter. An individual on an ordinary diet generally excretes about 1200 milli-osmols of solutes, which are greatly increased on diets high in protein and salt, and which may fall to 200 on diets high in carbohydrate (very low protein) and low in salt. It can be seen from the table that when the concentrating capacity of the kidneys is severely reduced, as may be the case in diseased states, the volume of water required for excretion of waste products may be very greatly increased. A number of tests for kidney function are based upon kidney concentrating capacity under controlled conditions.

URINE SECRETION AND WORK

In general, the solute concentration of urine is much greater than is plasma. This concentration of urine over plasma is effected through the reabsorption of water and solutes from the glomerular fluid as it passes through the tubules. This concentrating process represents osmotic work done by the kidneys (tubules) upon the glomerular fluid.

The minimum osmotic work, W_{min}, in small calories, involved in trans-ferring solute from a concentration C_1 in plasma to a concentration C_2 in urine is given by the thermodynamic equation:

$$W_{min} = NRT\ln\frac{c_2}{c_1} = 2.3NRT\ \log\frac{c_2}{c_1}$$

in which C_1 and C_2 are expressed as mols per kilogram of water, T = the absolute temperature, N = the molar concentration in urine, R = the gas constant in calories (1.987), In = the natural logarithm, log = the logarithm to base 10, and 2.3 = the factor for the conversion of natural to base 10 logarithms.

This equation may be applied to any constituent of plasma and urine. For example, if the concentration of Cl^-, C_1, in plasma is 0.104 mol per kilogram of water, and in urine C_2 is 0.166 mol per kilogram of water, the mini-mum osmotic work done by the kidneys at 37° C (310° A) is given by the expression:

$$W_{min} = 2.3\times\overset{N}{0.166}\times\overset{R}{1.987}\times\overset{T}{310}\ \log\frac{0.166}{0.104}$$

$$W_{min} = 2.3\times0.166\times1.987\times310\times0.2 = 47\text{ cal}$$

Suppose the concentration of urea in plasma, C_1, is 0.005 mol, and in urine C_2 is 0.333 mol per kilogram of water. Then the minimum work in-volved in its excretion would be:

$$W_{min} = 2.3\times0.333\times1.987\times310\ \log\frac{0.333}{0.005}$$

$$W_{min} = 2.3\times0.333\times1.987\times310\times1.82 = 861\text{ cal}$$

Since the concentration of water is higher in plasma than in urine, the work done is negative. The value of N for pure H_2O is 1000/18 = 55.55. Suppose the concentration of H_2O, C_1, in plasma is about 55.2, and in urine about 54.8. Then the minimum work done would be:

$$W_{min} = 2.3 \times 54.8 \times 1.987 \times 310 \log \frac{54.8}{55.2}$$
$$= 2.3 \times 54.8 \times 1.987 \times 310 \times (-0.0032) = -240 \text{ cal}$$

It is of interest that the concentration of Na^+ in plasma normally is about the same as the concentration in urine so that essentially no net work is required for its excretion.

The concentration of HCO_3- is generally lower in urine than in plasma and contributes about -2 cal to the work of excretion. Normally, then, the excretion of only water and HCO_3- represents negative work calories.

The net over-all work done by the kidneys in forming urine from plasma is the difference between the sums of the positive calories and negative calories-represented by the various substances excreted. This net over-all work is about 800 to 900 cal per litre when the concentrations of solutes are normal. The work of the kidneys varies with the concentrations of solutes in the urine which in turn are affected by the diet (especially NaCl and protein) and water intake. The concentration of Cl^- in urine may be less than in plasma and thus contribute negative calories. With large urine volumes the concentration of water in the urine may become greater than in plasma and contribute positive calories.

It will be noted, as to be expected, that most of the work of the kidneys is performed in the excretion of urea. The efficiency of the kidneys in performing concentration work in the secretion of urine may be expressed by the equation:

$$\text{efficiency} = \frac{\text{concentration work}}{\text{total energy consumption}} \times 100$$

The total energy consumption may be arrived at from the O2 utilization. It is estimated that the total energy consumption of the kidneys in forming II of normal urine is about 65,000 cal, and the minimum concentration work, *W* min, done is around 850 cal. This gives:

$$\text{efficiency} = \frac{850}{65{,}000} \times 100 = 1.3\%$$

By comparison, the efficiency of the salivary glands in forming saliva is 250/5000 × 100 = 5.0 per cent, that of the stomach in forming gastric juice is 900/12,000 × 100 = 7.5 per cent, and that of the pancreas in forming pancreatic juice is 55/75,000 ×100 = 0.073 per cent. The above efficiency values are calculated on the basis of all the energy expended being utilized for one purpose, which generally is not true, since each organ utilizes energy for various purposes. However, these values do indicate the proportion of total energy expenditure which is utilized for the specific purpose of forming a given fluid.

URINE AND PH

The kidneys may secrete urine with a pH value as low as 4.5 and as high as 8.2 under extreme conditions. The mean pH of the normal mixed 24-hour sample is about 6.0.

The titratable acidity of urine (to phenolphthalein) in terms of milliliters of 0.1 *N* alkali required to neutralize the 24-hour sample generally varies from 200 to 500 and averages around 300. It is dependent almost entirely upon the nature of the diet, being high on a high protein diet, which yields much sulphuric and phosphoric acids upon metabolism, and low on a high vegetable and fruit diet, which when metabolized yields a basic residue from oxidation of the potassium and sodium salts of organic acids present.

While the acidity of normal urine is due to a complex mixture of organic acids and acid phosphates (BH_2PO_4), the latter contribute most of the acidity. In severe diabetes mellitus the ketone acids acetoacetic and β-hydroxybutyric may represent major proportions of the titratable acidity. The ingestion of ammonium salts of strong acids, such as NH_4Cl, increases urinary acidity due to the fact that the ammonia is converted to neutral urea in the liver, releasing HCl, which lowers the pH of blood and of the urine formed from it.

The pH of urine may be raised above 7 and the titratable acidity reduced to 0 (alkaline urine) by the ingestion of sodium bicarbonate or salts of organic acids such as sodium citrate which, when metabolized, yield sodium bicarbonate. Urine collected shortly after meals may be alkaline on account of the so-called alkaline tide. Occasionally urine as voided may be ammoniacal and alkaline due to bacterial action upon urea in the bladder or other parts of the urinary tract. The usual titration of urine to phenolphthalein, pH around 9, is considerably above the normal pH of blood, 7.4. Samples of urine with pH values above that of blood may still show appreciable titratable acidity when titrated to phenolphthalein. More significant values are obtained by titrating electrometrically to pH 7.4 with the glass electrode. This procedure is necessary when accurate studies of urinary acidity in relation to blood pH are carried out. In the titration of urine to phenolphthalein neutral potassium oxalate is added to precipitate the calcium as oxalate to prevent its interference with titration of acid phosphates. The chief reaction in the titration is:

$$H_2PO_4 + OH^- \rightarrow H_2O + HPO_4^-$$

If the $H_2PO_4^-$ is present as the Na^+ or K^+ salt, the titration proceeds normally. However, in the presence of Ca^{++} ions the relatively insoluble and unstable salt, $CaHPO_4$, is formed as the pH is raised and the concentration of HP04~ ions increases. The $CaHPO_4$ then slowly undergoes transformation as follows:

$$4CaHPO_4 \rightarrow Ca_3(PO_4)_2 + Ca(H_2PO_4)_2$$

The triple phosphate $Ca_3(PO_4)_2$ precipitates out, leaving the acid salt $Ca(H_2PO_4)_2$ in solution; this makes the solution more acid and requires additional alkali. In effect, the presence of Ca^{++} ions causes the titration of some of the H_2PO_4 to PO_4= instead of stopping at HPO_4^-, the desired stage. Unless the calcium is removed before titration, samples of urine of the same acidity may show widely different values due to this calcium effect.

The acidity of the urine formed is chiefly dependent upon the degree to which the Na^+ ions of the glomerular filtrate are replaced by H^+ ions supplied by the carbonic acid-carbonic anhydrase system of the distal tubules, and also upon the extent to which these cations are substituted by NH_4^+ ions. The reabsorption of HCO_3^- ions into the blood is dependent upon the availability of Na^+ ions for simultaneous reabsorption with the HCOs ions. In an acidosis such as caused by diabetic ketosis, the kidneys are flooded with so much of the Na^+ salts of the ketone acids that sufficient H^+ and NH_4^+ cannot be substituted to conserve the Na^+, and excessive amounts of these ions are lost in the urine, and the body supply is depleted. With depletion of Na^+ ions in the plasma and glomerular filtrate the reabsorption of the basic HCO_3^- ions into the blood is decreased, and the condition of acidosis becomes established. In the acidosis of chronic nephritis the damaged kidney tubules are unable to exchange H^+ and NH_4^+ for Na^+ sufficiently rapidly to conserve the Na^+, the reabsorption of basic HCO_3^- into the blood is decreased, and a condition of acidosis results.

Normally the reabsorption of glomerular fluid HCO_3 back into the blood by the tubules is so efficient that only about 0.1 per cent escapes into the urine. However, after the ingestion of bicarbonate, large quantities can be excreted in the urine with only a slight increase in plasma level. The drug acetazolamide (Diamox) (2-acetylamino-1,3,4-thiadiazole-5-sulphonamide) is a powerful inhibitor of carbonic anhydrase, and when administered decreases the concentration of H_2CO_3 in the distal tubule cells and the capacity to exchange H^+ for Na^+, This results in a copious excretion of Na^+, HCO_3, and water, a rise in urinary pH (urine may become alkaline) and a fall in blood pH: Diamox is used to promote the excretion of Na^+ and water in the reduction of edema due to heart failure. The student will find a discussion of the renal tubular mechanisms involved in regulating urinary acidity.

Turbidity of urine. Normally samples of freshly voided urine are clear. However, less acid urines upon standing may become cloudy owing to the precipitation of insoluble phosphates (calcium and magnesium phosphates). This precipitation is the result of an increase in pH into the zone of formation of the insoluble phosphates resulting from the loss of dissolved CO_2 or to the formation of ammonia by the bacterial decomposition of urea when the sample of urine is not preserved. This precipitate of phosphates readily dissolves upon acidification with acetic acid. Turbidity also may appear in urine upon standing due to the separation of a cloud (nebecula) composed chiefly of mucoid or nucleoprotein and epithelial cells. Grossly pathologic urine containing pus may be densely turbid.

Odor of urine. The odor of normal urine, characteristic and faintly aromatic, is due to a number of substances, among which are certain volatile organic acids in small amounts. Some of the odor has been attributed to a neutral ill-smelling substance called "urinod", C_6H_8O. The urine voided after eating asparagus possesses a typical odor said to be due to the presence of methylmercaptan, CH_3SH, which may arise from an unstable precursor in urine. The ingestion of substances such as turpentine, tolu, and copaiba give excretory products which impart rather characteristic odors to urine. When urine undergoes so-called alkaline fermentation in which urea is hydrolyzed to ammonia, it develops a strongly ammoniacal odor. This may be associated with a putrid odor, if pus and decomposing tissue cells are present.

Colour of urine. The color of normal urine shows a wide variation from the light straw color of highly dilute urine to the reddish yellow color of concentrated urine. Acid urines are generally darker than alkaline urines. The colour of normal urine is due chiefly to pigment called "urochrome", which appears to be a compound of urobilin and a peptide. Fresh urine contains little if any free urobilin, though upon storage at ordinary temperature or heating free urobilin, a brown substance, is liberated, causing the urine to become darker in color. Also, a colorless reduction product of urobilin, urobilinogen, is present in urine and is oxidized to the colored urobilin uponexposure to air. Urobilinogen originates through bacterial action upon bile pigments in the intestine. Urochrome is present in urine partly as the colorless reduction product, urochromogen, which is oxidized to urochrome upon exposure to air.

Urobilin is also excreted in the feces, where it is generally known as "ster-cobilin", to the extent of 100 to 200 mg per day. The urobilin of urine, as urochrome and urobilinogen, generally amounts to 70 to 80 mg per day, all but a few milligrams of which are represented by urochrome. The total urobilin of feces and urine is considered to be a measure of the rate of hemoglobin breakdown in the body.

The amount of urochrome excreted per day appears to vary with the metabolic rate and accordingly is increased in fevers. The vitamin riboflavin is excreted in urine and contributes to its yellow colour.

Hemoglobin, myoglobin, and bile pigments may be present in urine as the result of pathologic conditions or injury and contribute to its color. In the rare condition of alcaptonuria the urine contains. homogentisic acid, which, upon making the urine alkaline, is oxidized to a black pigment. Melanotic tumors may cause the excretion of melanin precursors in the urine and give it a brown or black color.

The ingestion of certain foods and drugs may cause characteristic changes in the color of urine. Rhubarb, cascara, and senna color urine a reddish brown which turns to blood red upon addition of alkali. Methylene blue causes pale green to greenish blue coloration. Phenol and some of its derivatives lead to dark green or brownish black colors, especially upon standing exposed to air.

NORMAL CONSTITUTION OF URINE

The origins of most of the usual constituents of urine have been considered in the chapters on metabolism, and the following discussion is concerned primarily with additional facts relative to the chemistry of these constituents and to their importance as urinary components.

Urea: Urea is the principal end product of protein (amino acid) metabolism in the human, mammals in general, and in certain lower forms. In the human it usually represents 80 to 90 per cent of the total urinary nitrogen. In general, the proportion of nitrogen as urea increases as the total urinary nitrogen increases, and *vice versa.* On very low protein diets urea may represent 60 per cent or less of the total nitrogen.

The quantity of urea excreted, in general, is proportional to the total protein metabolism, whether this protein represents food protein or the protein of tissues undergoing catabolism. The excretion of urea is decreased in certain liver diseases, such as acute yellow atrophy and cirrhosis, in which the capacity to form urea is decreased. In cases of severe acidosis the amount of urea excreted may be considerably reduced because of diversion of amino nitrogen to ammonia formation. Excretion of urea also may be decreased in nephritis when the ability of the kidneys to excrete it is severely impaired. This may cause greatly increased concentrations of urea in the blood (uremia) and other body fluids.

Ammonia: The mechanism of ammonia formation by the kidneys and the regulation of the amount formed have been considered in detail in the chapters dealing with protein metabolism and acid-base balance, to which the student is referred.

Ammonia, normally, is the second most important nitrogenous substance of urine quantitatively. Ordinarily, 2.5 to 4.5 per cent of the total urinary nitrogen is composed of ammonium salts. On the average this represents about 0.7 g per day.

Since both urinary ammonia and urea are derived from the amino groups of the amino acids, for a given quantity of nitrogen excreted an increase in the amount of the one leads to a decrease in the amount of the other.

As previously pointed out, ammonia is formed in the kidney to be substituted for Na^+ and K^+ so that they may be conserved to the plasma. Accordingly, the quantity of urinary ammonia per day varies from practically 0 in alkalosis to 8 to 10 g in cases of severe diabetic acidosis.

Urinary ammonia may be markedly decreased in cases of severe nephritis in which the capacity of the kidneys to form it has been impaired. This reduces the capacity of the kidneys to conserve base and contributes to the development of acidosis.

The quantity of ammonia in the urine may be enormously increased through hydrolysis of urea by bacteria in the bladder (cystitis) or other parts of the urinary tract. This bacterial production of ammonia from urea in normal urine may take place if the samples are stored without preservative (alkaline fermentation).

Uric acid and other purines: Uric acid is the chief end product of purine metabolism in man, the higher apes, and the Dalmatian dog. The quantity of uric acid in human urine is generally from 0.5 to 1.0 g per 24 hours, though the amount is subject to wide variations. On a purine-free diet the uric acid excretion may fall to 0.1 g per day, while on a high purine diet the daily excretion may rise to 2 g.

Uric acid is very sparingly soluble in water (0.06 g in 100 ml of hot water) and is insoluble in alcohol and ether. Through its enol form uric acid acts as a weak dibasic acid and forms both neutral and acid salts. The neutral lithium and potassium salts are the most soluble urates. The acid salts of the alkali metals are rather insoluble. Ammonium urate is difficultly soluble. The alkaline earth urates are insoluble. The acid salts of sodium and potassium form most of the sediment which separates from concentrated acid urine upon standing and cooling.

Crystals of uric acid are readily precipitated from urine by acidifying with HCl. Such preparations are colored reddish brown by associated urine pigments, which may be removed by solution in concentrated H_2SO_4 and precipitation of the uric acid by careful dilution with water or alcohol. Uric acid separates from pure solution as transparent, colorless, rhombic plates, while from urine it gives a variety of colored crystal forms.

The two *pKa* values (- log *Ka)* of uric acid are approximately 5.7 and 9.8.

In very acid urines the proportion of free uric acid is considerable, while in urines of average pH (6 ± 0.5) the sodium and potassium acid salts, which are much more soluble than uric acid, predominate. Alkaline urines contain a large proportion of the soluble sodium and potassium salts. Acid urates tend to separate from urines more acid than normal when permitted to stand, and free uric acid may separate from strongly acid urines.

Uric acid reduces alkaline copper and silver but not bismuth solutions. It also reduces sodium phosphotungstate in alkaline solution and arseno-phosphotungstate in the presence of cyanide to give blue colors. The latter reaction is commonly used in the quantitative estimation of uric acid.

When uric acid is oxidized with concentrated nitric acid and the mixture made alkaline with NH_4OH, a reddish purple colour appears because of the formation of ammonium purpurate, also called "murexide". The murexide test is given by various purines:

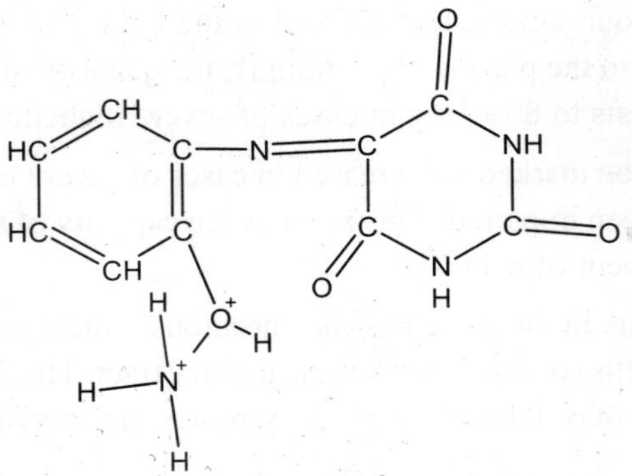

ammonium pėrpurate (murexide)

Uricase

$+ H_2O + ^1/_2 O_2$

$+ CO_2$

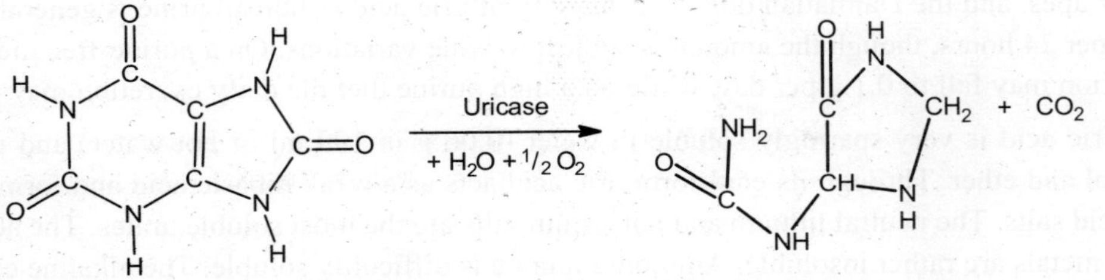

7,9-dihydro-1H-purine-2,6,8 (3H)-trione
uric acid

1-(5-oxoimidazolidin-4-yl) urea
Allantoin

This reaction with uricase has been made the basis of a specific quantitative method for uric acid, the principle of which is to run colorimetric estimations with arsenophosphotungstic acid on samples before and after treatment with uricase, which specifically destroys the uric acid and leaves the other substances which give a color with the reagent. Uric acid is represented by the difference between the uric acid equivalents of the color values before and after treatment with uricase.

The excretion of uric acid in pathologic conditions varies widely. Conditions involving the breakdown of large quantities of nuclear material such as is found in the destruction of leucocytes in leukemia may cause the excretion of large amounts of uric acid (10 g or more). In gout the uric acid content of blood may rise to 15 mg per cent, apparently partly due to decreased kidney elimination. The uric acid of the urine tends to decrease prior to an attack of gout and to increase during the crisis and recovery, sometimes to twice the normal value.

Small amounts of various purines other than uric acid are excreted in urine. These substances include I-methylguanine, N2-methylguanine, 1-methylhypoxanthine, 8-hydroxy-7-methylguanine, guanine, hypoxanthine, adenine, 7-methylguanine, 1-methylxanthine, 7-methylxanthine, and 1,7-dimethylxanthine (paraxanthine). The latter three compounds are present only after the ingestion of coffee, tea, etc. The total excretion of these purines averages around 30 mg per day and varies widely in a group and in an individual case. The excretion is little affected by diet, indicating an endogenous source.

Creatinine and Creatine

The formation and metabolism of these substances have been considered in the chapter on protein metabolism:

$-H_2O$

Strong acid and heat

(N-methylcarbamimidamido) acetic acid
Methyl guanidino acetic acid (creatine)

2-imino-1-methylimidazolidin-4-one
Anhydrous creatine (creatinine)

The quantity of creatinine excreted in urine by the normal adult generally is from 1.2 to 1.7 g per 24 hours and depends partly upon the amount of creatinine (from meats and soups) in the diet, since this is excreted un-changed in the urine. The quantity excreted on a creatinine-free diet is practically constant for a given individual over long periods of time and is independent of the total nitrogen output. Researcher defines the creatinine coefficient as the milligrams of creatinine, or creatine plus creatinine (when creatine is present), excreted per kilogram of body weight per day. The creatinine coefficient averages 20 to 26 in normal men and 14 to 22 in women. It varies in proportion to muscular development, is not significantly altered by variations in diet (when these do not involve the ingestion of much preformed creatinine), exercise, urine volume, and many pathologic conditions.

Prolonged illness and old age, both of which are associated with loss of muscular tone and activity, may cause marked decreases in the creatinine coefficient.

The statement is commonly made that creatine is absent from the urine of normal adult males. However, using improved methods of analysis, have shown that normal adult males regularly excrete a small amount of creatine. Researchers found that 60 to 150 mg of creatine per day are excreted, which on the average amounts to about 6 per cent of the total creatinine output. Adult females show much greater variation in creatine excretion, although in the great majority of cases it amounts to two or two and a half times as much as for adult males. In about one-fifth of the females the creatine excretion is similar to that for the males.

The excretion of creatine is greatly increased in infants and children, during starvation, and particularly in patients with certain muscular dystrophies. Creatinine crystallizes as colorless, glistening, monoclinic prisms. It is soluble in water and alcohol. Its nitrogen atoms are weakly basic, and it forms salts only with strong acids. It forms an important compound with zinc chloride which has the formula $(C_4H_7N_3O)_2ZnCl_2$. Creatinine reduces alkaline copper solutions and forms an insoluble cuprous-creatinine salt. It is responsible for some of the reducing action of urine, and consequently may cause errors in urinary sugar determinations by certain methods.

Creatinine reacts with picric acid in alkaline solution to form a red substance, which is the basis for the method commonly used for its quantitative determination. The nature of this coloured product and considers it a complex of creatinine, picric acid, sodium hydroxide, and water in the molecular proportions of 2-1-3-3, respectively. When creatinine is treated with alkali and sodium nitroprusside, a red color changing to yellow is produced. The addition of acetic acid to the yellow solution, followed by heating, produces a green color, which finally changes to blue.

Researchers have described a highly specific method for the quantitative determination of creatinine based upon the carmine colour formed when creatinine reacts with 1,4-naphthoquinone-2-potassium sulphonate.

Creatine is converted to creatinine by heating its solutions with strong acid. The difference in creatinine before and after heating urine and other biologic fluids with acid gives the amount of creatine present, calculated as creatinine. The method is not very accurate.

Amino Acids

The total amino acid excretion in urine per 24 hours normally amounts to about 1.5 g, of which more than half is in peptide or other combinations from which the free amino acids are liberated by acid hydrolysis (8).

The quantity of urinary amino acids may be increased in diseases asso-ciated with excessive tissue destruction such as severe diabetes, typhoid fever, and acute yellow atrophy of the liver. In the latter condition the capacity of the liver to deaminize amino acids is impaired sufficiently to cause their excessive accumulation in blood and excretion in the urine.

Hippuric Acid

Hippuric acid, so called because it was first found in the- urine of horses, is a peptide of benzoic acid and glycine, benzoylglycine:

Hippuric acid represents a detoxication product of benzoic acid. Benzoic acid is present in many fruits and vegetables, especially in cranberries and prunes. Some food products, such as ketchups, are preserved with 0.1 per cent of sodium benzoate and contribute to the benzoic acid intake. Benzoic acid also may be formed by the oxidative breakdown of phenylalanine through bacterial action in the intestine.

The amount of hippuric acid excreted per day in the urine averages about 0.7 g and ranges from 0.1 to 1.0 g. In the dog hippuric acid formation occurs in the kidneys, but in man and the rabbit synthesis appears to take place largely in the liver. The capacity of an individual to synthesize hippuric acid after a test dose of sodium benzoate is used as a clinical test of liver function.

Hippuric acid is readily soluble in alcohol and hot water and is sufficiently soluble in ether to permit extraction from aqueous solution with this solvent.

Allantoin

Allantoin occurs, in varying amounts, in the urine of prac-tically all mammals. The urines of man, anthropoid apes, and the Dalmatian dog contain very small amounts, while those of other mammals contain allantoin in quantity, where it represents the chief end product of purine metabolism. Allantoin is formed in the liver by the action of the enzyme uricase upon uric acid (see uric acid above), and the quantity of the substance in urine varies with the amount or activity of this enzyme. Extirpation of the liver of a dog causes allantoin excretion to be changed to uric acid excretion.

Human urine generally contains 10 to 25 mg of allantoin per 24-hour specimen.

Proteins

Normally, human urine contains exceedingly little protein (20-80 mg per 24 hours), which is composed of a small quantity of insoluble nebecula (mucoid or nucleoprotein), albumin, and enzymes. However, in various kidney diseases large quantities of protein may be excreted in the urine, as much as 20 g per day in cases of nephrosis. The kinds of protein that may be found in urine under various pathologic conditions include serum albumin and globulin, nucleoprotein, fibrin, myoglobin, hemoglobin, and related substances, proteoses and peptones, and Bence Jones protein. Of these proteins serum albumin and globulin 'are most frequently present in urine, with albumin predominating, apparently. because of its smaller molecular size and more ready passage through the glomerular membrane. Protein in the urine is usually referred to as " albuminuria". Nucleoproteins in appreciable amounts may appear in the urine in conditions involving inflammation of the urinary epithelia such as cystitis (in the bladder) and pyelitis (in the kidney). Bence Jones protein, considered a low-molecular-weight globulin, may occur in the urines of patients with multiple myeloma, which is a tumorlike growth in bone marrow, and at times in the urines of leukemia patients. Bence Jones protein is characterized by precipitating when the urine is warmed to 40 to 60° C and dissolving almost completely when the urine is further heated to boiling. Upon cooling the precipitate reappears. The excretion of proteoses and peptones in the urine may occur in conditions involving destruction of much tissue protein, such as pneumonia, carcinoma, and diphtheria. Blood cells and other constituents may be present in urine because of hemorrhage into the urinary tract (hematuria). Hemoglobinuria, the presence of hemoglobin in urine, results when appreciable quantities of hemoglobin are released into the blood (hemoglobinemia) through hemolysis of the red cells. Hemoglobin dissolved in plasma is treated as a foreign protein and excreted in the urine. However, it is a threshold substance and is retained by the kidneys until the plasma concentration rises above about 150 mg per cent. Myoglobin, or muscle hemoglobin, may occur in the urine after severe muscle trauma.

Functional proteinurias, not related to diseased states, may be observed after violent exercise, cold baths, and as a result of standing upright (postural proteinuria). Postural proteinuria is probably due to interference with the blood supply to the kidneys, causing venous stasis, since the condition disappears when the individual is lying down. The amounts of protein excreted in functional proteinurias are generally small.

The subject of proteinuria is large and complex, and for detailed dis-cussions the student should consult books dealing with clinical pathology.

Carbohydrates in Urine

Normal urine contains a variety of carbohydrate like substances which reduce sugar reagents. When calculated as glucose, these ordinarily amount to 1.0 to 1.5 g per 24 hours. Much of this material consists of non-glucose substances present in foods or formed from them during cooking. The quantity of glucose present generally averages about 140 mg per 24-hour specimen, though it is subject to considerable variation. The presence of appreciable quantities of glucose in urine is known as "glucosuria" (or "glycosuria"). The most severe glucosurias are seen in severe diabetes mellitus, when the urine may contain 10 per cent or more of glucose. In such conditions both the urine volume and specific gravity are high. Such glucosurias are due to elevated blood sugar levels (hyper-glycemia) exceeding the renal threshold. In renal diabetes the blood sugar is normal, but the renal threshold is below normal, so that some glucose is excreted in the urine, particularly after meals. Emotional glucosuria is the result of the rapid breakdown of liver glycogen through adrenaline action, causing a hyperglycemia which exceeds the kidney threshold.

Pentoses such as arabinose may occur in urine (pentosuria) as the result of ingestion of large amounts of foods rich in pentoses such as cherries, grapes, prunes, and plums. This type of pentosuria is alimentary. In cases of idiopathic or essential pentosuria, which represents an "inborn error of metabolism", the major pentose excreted is L-xylulose. Researchers have identified both D-ribulose and L-xylulose, in very small amounts, as constituents of normal human urine. D-Ribulose is formed in the pentose cycle of carbohydrate metabolism, and L-xylulose in the metabolism of D-glucuronic acid.

Lactose is frequently present in the urine of lactating women as the result of the passage of the sugar from the mammary gland into the blood. Since lactose as such cannot be utilized by the body, it is excreted as a foreign substance in the urine. It is important to differentiate lactose from glucose in such cases, since a lactosuria is of no significance whereas the presence of glucose may indicate diabetes mellitus. The sugars are best differentiated by reduction tests before and after fermentation with bakers' yeast which removes glucose but not lactose. If both sugars are present, as may be the case, quantitative determination of fermentable sugar indicates the proportions of each. Lactose may be qualitatively detected in urine through the osazone and mucic acid tests.

Galactosuria has been observed following the ingestion of large amounts of galactose in nursing infants with gastrointestinal disturbances, and in both infants and adults with liver disease.

Fructosuria or levulosuria may occur in severe cases of diabetes mellitus, in which fructose accompanies glucose. Alimentary fructosuria may result after the ingestion of large quantities of fructose by patients with liver disease. Essential levulosuria or fructosuria, in which fructose is excreted in the urine from birth, is a rare condition which may be considered as an "inborn error of metabolism." In some cases there is complete inability to metabolize ingested fructose; the sugar is quantitatively excreted in the urine. Insulin has no effect upon the condition, and carbohydrates other than fructose are metabolized normally.

Fat in Urine

Fat may be present in the urine of normal persons after a high fat meal (lipuria), or after the ingestion of large amounts of fatty oils such as cod liver oil. Lipuria may occur in cases of severe diabetes mellitus, in which the plasma lipid concentration is markedly elevated, in nephrosis, in cases of traumatic injury to the bone marrow, and when there has been extensive crushing of superficial fat. Lipuria has been reported in cases of alcohol and phosphorus poisoning. Obstruction of lymph flow in the thoracic duct may cause distension and rupture of lymph vessels of the kidney and bladder with resulting lymphuria or chyluria, and the presence of lipids in the urine. Fat may appear in the urine as the result of fatty degeneration of epithelial cells and leucocytes in conditions such as pyelonephritis and nephrosis.

Glucuronic acid. Glucuronic acid occurs in urine conjugated with a wide variety of substances under both normal and pathologic conditions and after the administration of certain drugs. These conjugates represent detoxication products of the body.

The glucoside conjugates are formed from aromatic alcohols and phenols such as naphthol, phenol, and borneol, while ester conjugates are formed particularly from aromatic acids such as benzoic and phenylacetic.

Both free glucuronic acid and ester conjugates reduce alkaline sugar reagents, the latter being hydrolyzed to glucuronic acid by the alkali. Glucoside conjugates are not hydrolyzed by alkali and show reduction only after acid hydrolysis. Glucuronic acid and its conjugates are not fermentable. Peculiarly, free glucuronic acid is dextrorotatory while its conjugates are levorotatory.

The total glucuronic acid content of urine normally amounts to 0.5 to 1.0 g per day and is present chiefly as conjugates of phenol, indoxyl, skatoxyl, and the estrogenic hormones.

The urinary glucuronates may be greatly increased by the administration of substances such as antipyrine, camphor, chloral hydrate, borneol, menthol, aspirin, morphine, phenolphthalein, most of the sulfonamide drugs, and turpentine. Decreases in glucuronate excretion have been observed in the rat associated with vitamin deficiencies. Riboflavin deficiency, however, appears to cause increased excretion.

Glucuronic acid and glucuronates give a color reaction with naphtha-resorcinol when heated with strong HCl, which is used for their qualitative detection and quantitative estimation.

Ketone bodies: The ketone bodies found in urine are derived chiefly from the metabolism of fatty acids and consist of acetone, acetoacetic acid, and β-hydroxybutyric acid. Acetoacetic acid is the primary ketone body from which β-hydroxybutyric acid is formed by reduction and acetone by decarboxylation. The liver is the site of formation of the ketone acids.

A normal person on a mixed diet generally excretes less than 100 mg of ketone bodies per 24 hours. Typical ketonuria results whenever the rate of production of ketone bodies in the liver exceeds the capacity of the body tissues to oxidize them. This occurs on very high fat diets, during starvation when a large amount of body fat is metabolized, in diabetes mellitus, in both normal and toxemic pregnancies, during ether anesthesia, and often in conditions of alkalosis. In severe diabetic acidosis 75 g or more of ketone bodies may be excreted in the 24-hour urine. Sodium Nitroprusside reacts with acetoacetic acid in the presence of NH_4OH and $(NH_4)_2SO_4$ to give a permanganate color (Rothera's test), which is commonly used as a test for the substance in urine. Acetone also reacts, but less readily. Acetoacetic acid also gives a Bordeaux red colour with ferric chloride.

Acetone gives iodoform when treated with alkaline iodine solution. In testing urines for acetone the latter should first be separated by distillation. β-Hydroxybutyric acid, when oxidized with hydrogen peroxide and treated with a mixture of ferric and ferrous chlorides, gives a rose color (Black's reaction). The acid must be separated from urine before applying the test.

Citric acid: Citric acid is formed continually in carbohydrate metabolism through operation of the tricarboxylic or citric acid cycle, and small amounts, 0.2 to 1.2 g per 24 hours, are excreted in the urine. The excretion is increased in conditions of alkalosis. Citric acid excretion is increased after the administration of estrogens. A marked increase follows hepatectomy.

Lactic acid: Lactic acid accumulates in the tissues and blood under all I conditions in which the supply of oxygen to the tissues is deficient, and small amounts escape in the urine. The urinary lactic acid increases in cases of pneumonia, eclampsia, ether anesthesia, carbon monoxide poisoning, acute yellow atrophy of the liver, and epileptic attacks. As expected, severe muscular exercise may markedly increase urinary lactic acid.

Oxalic acid: Normal urine contains 10 to 30 mg of oxalic acid as oxalates per 24-hour sample. Although it appears that some oxalic acid may arise from metabolic processes, the larger proportion generally is derived from the oxalates present in foods such as asparagus, spinach, rhubarb, lettuce, apples, and other vegetables and fruits. Oxalic acid separates from urine as insoluble calcium oxalate. Many urinary calculi are largely calcium oxalate. The oxalic acid content of urine is increased in diabetes mellitus, in certain liver diseases, and in various conditions involving deficient tissue oxidations.

Normal urine generally contains 50 to 200 mg of lactic acid per 24-hour specimen

Oxaluric acid: H_2N-CO-NH-CO-COOH, a combination of oxalic acid and urea, occasionally is present in traces in normal urine. It breaks up into urea and oxalic acid upon hydrolysis.

Sulphur compounds: Sulphur is excreted in the urine in three combinations: inorganic sulphate, organic ester or ethereal sulphate, and so-called neutral sulphur of the urinary sulphur arises from the metabolism of protein, specifi-cally from the S-containing amino acids, cystine, cysteine, and methionine, and varies with the protein intake and the rate of tissue protein breakdown. The total urinary sulphur varies ordinarily from about 0.8 to 1.4 g per 24 hours, and averages about 1.0 g. Generally the urinary N: S ratio ranges from 13 to 16 on a high meat diet and during fasting. Inorganic sulphate sulphur generally makes up the greater proportion of urinary sulphur and ordinarily amounts to 0.7 to 1.0 g per day. It normally repre-sents from 85 to 95 per cent of the total sulphur excreted. The sulphate sulphur excretion may be diminished in conditions of renal functional impairment and is increased on high protein diets and when there is excessive tissue breakdown such as occurs in high fevers.

Ethereal sulphate sulfur excretion in the urine generally ranges from 0.06 to 0.12 g per day. It consists of the sodium and potassium salts of sulfuric acid esters of phenols such as indoxyl, skatoxyl, phenol, and cresol. These ethereal sulphates represent detoxication compounds of phenols and are formed in the liver. Some of the phenolic substances originate through bacterial action in the intestine, and some appear to be formed in tissue metabolism. Indoxyl and skatoxyl are formed practically entirely by the putrefactive decomposition of tryptophan in the intestine. Both substances are esterified with sulfuric acid in the liver and excreted in the urine as sodium and potassium salts. The amount of indican, potassium indoxyl sulphate, excreted in the urine is taken as a rough index of intestinal putrefaction.

Indican is oxidized to the dye indigo blue by hypochlorite (Jaffe test) and by ferric chloride in strong HCl solutions (Obermayer test): these tests are used for the detection of indican in urine. The neutral sulfur fraction of urine, ordinarily ranging from 0.08 to 0.16 g per day, is composed of a heterogeneous mixture of sulfur compounds. These include cystine, methionine, oxyproteic acids, urochrome, thiosulphates, thiocyanates, bile acids, and taurine anft-its derivatives. Urinary neutral sulfur is increased in cystinuria, melanuria, and in obstructive and hepatocellular jaundice due to the excretion of bile acids.

Chloride: Next to urea, chlorides make up the chief solid constituent of urine, generally amounting to 6 to 9 g per day of Cl^-, and equivalent to 10 to 15 g of NaCl, which is the chief chloride present. However, the daily excretion varies widely with the dietary intake. The excretion of chlorides in the urine may be markedly decreased by excessive perspiration owing to loss in the sweat. During fasting, chloride toward increasing the acidity and decreasing the phosphate content of the urine. Urinary phosphate excretion is increased in diseases of the bones such as rickets, osteomalacia, and periostosis. It is also increased by the excessive amounts of parathyroid hormone in cases of parathyroid hyperfunction. Conversely, it is decreased in parathyroid hypofunction. Decreases in urinary phosphates may be observed in pregnancy as the result of utilization of maternal phosphate in the formation of fetal bones, in acute infections, and in nephritis, when the capacity of the kidneys to eliminate phosphates is decreased.

Sodium and Potassium: The quantity of sodium excreted in the urine normally ranges from 4 to 5 g per day, although it varies widely and generally parallels the chloride excretion. Urinary potassium ordinarily amounts to 2.5 to 3.0 g per day, and the ratio K:Na is generally about 3:5. During fasting, however, because of lack of NaCl intake and the release of much more K^+ than Na^+ in the catabolism of tissues, more potassium than sodium is excreted in the urine. The urinary excretion of potassium relative to sodium is increased in acute febrile conditions involving tissue destruction and, like sodium excretion, is increased in acidosis. In Addison's disease with adrenal cortical deficiency, the excretion of sodium is increased and of potassium decreased.

Calcium and magnesium: Ordinarily the greater proportions of calcium and magnesium are excreted in the feces. Generally, the urinary excretion of calcium lies between 0.1 and 0.3 and of magnesium between 0.1 and 0.2 g per day. As indicated above in the discussion of phosphates, increased alkalinity of intestinal contents causes precipitation of increased amounts of insoluble calcium and magnesium phosphates which are lost in the feces, with less of these elements being absorbed into the blood and excreted in the urine. Conversely, increased acidity of the intestinal contents causes more calcium and magnesium to be absorbed and excreted in the urine. The urinary excretion of both calcium and magnesium normally is primarily dependent upon the nature of the diet. It is of some interest that whileihe excretion of calcium in osteomalacia is increased as the result of bone decalcification, the excretion of magnesium may be decreased. The injection of $MgSO_4$ into rats has been found to increase by several times the urinary calcium excretion, presumably due to mobilization from bones.

THE SEDIMENTS OF URINE

Urinary sediments are obtained by centrifuging the urine or permitting it to stand. These precipitates ordinarily are composed of so-called organized and unorganized sediments. Organized sediments may contain erythrocytes, epithelial cells, pus cells, bacteria, animal parasites, spermatozoa, urethral filaments, tissue debris, and contaminating foreign substances.

The organized sediments of urine are examined microscopically in the clinical laboratory as an aid to the diagnosis of various urinary tract diseases.

The most common unorganized sediments contain calcium oxalate and phosphate, magnesium ammonium phosphate (triple phosphate), uric acid, and ammonium and sodium urates. Less common constituents are calcium Sulphate and carbonate, magnesium phosphate, amino acids (cystine, leucine, and tyrosine), xanthine, melanin, bilirubin, indigo, and hippuric acid. These substances separate from urine under conditions in which their solubilities in urine are exceeded.

Calcium phosphate separates from urines more alkaline than pH 6, and calcium oxalate from urines of any pH. Ammonium magnesium phosphate, ammonium urate, and calcium carbonate separate from urines made both alkaline and ammoniacal by "ammoniacal fermentation" due to bacterialaction. Uric acid and sodium acid urate separate from the more acid urines. Urinary calculi, concretions found pathologically, generally in the kidney and bladder, are of much medical importance. They are of simple and compound forms, the former being composed of a single compound and the latter of two or more substances. Calculi generally consist of a central nucleus surrounded by a succession of concentric layers, which often but not always represent different components. Urinary calculi often appear secondary to urinary tract infections. Clumps of pus and epithelial cells and bacteria may serve as foreign body nuclei upon which calcium phosphate and magnesium ammonium phosphate precipitate due to the alkaline ammoniacal medium caused by bacterial decomposition of urea. The majority of stones arise over long periods of time in the absence of infection and are composed of a number of substances, generally with one predominating. Calculi are generally classified according to their principal constituent. The most common types in man are uric acid and urate calculi, phosphatic calculi, and calcium oxalate calculi. Calcium carbonate calculi are frequently found in herbivorous animals, but very rarely in man. Cystine calculi occur rarely, generally making up less than 1 per cent of calculi. Xanthine calculi are more rare than those of cystine. Indigo and cholesterol calculi are extremely rare, as are urostealith calculi, which are composed principally of fat and fatty acids. Fibrin calculi may be formed by blood coagulation in the urinary tract.

To identify urinary sediment cells is a difficult task. Some cells have characteristic features, therefore easily identifiable. On the other hand, some cells, even with sophisticated stains, remain a challenge. The diversity of cells that one can meet is notable. Cells found in urine can belong to the reticuloendothelial system (leucocyte, macrophage) or to the epithelial system. Urinary cells can originate from the kidney or from the lower urinary tract, from the superficial lining or from a deeper source. In an adult male, cells can also originate from the prostate and the urethra.

Urine is not a favorable media for maintaining cell structures. Most cells undergo rapid changes (degeneration, rupture, vacuolization, granulation of the cytoplasm) that profoundly affect the visual aspect of the cell. These changes are more pronounced if the cell comes from the higher urinary tract because agression due to osmotic variation adds to the energy lack and other causes of morphological change. The best preserved cells are usually bladder cells, while a well preserved proximal renal tubular cell is exceptional. Some pathologies, like inflammation, metaplasia, neoplasia, are known to affect the cellular aspect.

All experienced urinary sediment microscopists have seen some cases of cells, so special, that identification was a best guess based on the sediments context. This situation, while being frustrating, should not be considered as a failure. Identification of all the figures seen in the urinary sediment is quite impossible; so,one must focus his effort on element of clinical value.

MICRO-ORGANISMS

Bacteria

The urinary infection is the most frequently observed anomaly in urinary microscopy. The presence of many white blood cells and bacteria is characteristic of this situation. On the other hand, urine specimens for routine examination are not usually obtained through a sterile technique with the result that old specimens can have a lot of bacteria with only a few leukocytes. The presence of numerous squamous cells can indicate an external genital source for the bacteria. In these two situations, a positive nitrite result may indicate a urinary tract infection but a safe diagnosis is made only by a positive culture obtained with a midstream specimen.

Bacteria associated with urinary tract infection are mostly bacillus *(E. coli)* but other shapes cannot be ruled out. Bacteria-coated urothelial cells are frequent in cystitis. This situation is different from Clue Cells, which are vaginal squamous cells coated with a coccobacillus (Gardnerella Vaginalis), forming a crust over the cell. Under the microscope, these latter cells have a granular aspect with a blunted border.

Yeast

Like bacteria, the presence of yeast in the urine sediment may indicate an infection. A frequently seen yeast in urine is Candida. Identification of this organism is relatively easy because of its usual club shape. In the majority of cases, only the isolated cells are seen but, in some cases, budding pseudohyphea may be observed. Other forms of yeast can be seen in the urine specimens. With a wet uncolored sediment, some of these forms are difficult to differentiate from red blood cells or some other elements. However, yeast contains DNA that can be demonstrated with an usual stain like the Sedistain™. This dye preparation stains yeast cells in blue. In a number of specimens, the presence of yeast is the result of a contamination with vaginal secretion. Yeasts are often observed in specimens that contain sugar. It is important to be careful with these specimens because a yeast infection is a frequent finding with diabetic patients. Yeast containing casts have a very high clinical value; these are pathognomonic of pyelonephritis.

Parasite

The parasite that is the more frequently seen in urine is Trichomonas. Usually, this cell comes from genital secretions contaminating the specimen. But Trichomonas should be mentioned because cases of vesical and prostate colonisation by this organism have been reported in the literature. Identification of the living cell is quite easy owing to its spectacular motility. Identification of immobile cells is less obvious. Special stain can be use for Trichomonas. These stains can be found in the parasitology literature. Other parasites can be seen in the urine sediment. But these situations are rare and are mostly seen in exposed population. Identification of these parasites should be transferred to the parasitology department. This laboratory section has the tools and the expertise to make a proper identification.

Viral Infection

Some cellular manifestations of viral infection can be seen in the urine sediment.To see these manifestations, the specimen must be stained. In some cases, phase contrast microscopy can be used but, a good staining procedure on a cytospin specimen gives a preparation that is easier to read. Identification of infected cells is within the scope of the cytology department. Suspected specimens should be transfered to the cytology lab.

- Herpes simplex is the most observed viral manifestation. The infected transitional cells will show a large eosinophilic inclusion surrounded with a clear halo.
- Cytomegalovirus is characterized by a bird's eye inclusion.
- Infection with the polyoma virus gives a cellular transformation formely named "Decoy cells". The enlarged nucleus is invaded by a huge basophilic inclusion.

Spermatozoa

Urinary spermatozoa is a contamination arising from sexual activity. With a male subject, these represent a residual drainage while with a female, these have a vaginal contamination source. Some labs do not report spermatozoa. The problem with such a policy is that the labs rarely know all the elements to make a truly wise decision. Most cases are due to a normal human "activity" that should stay personal. But some cases, not necessarily known to the lab, are a result of reprehensible activity. Reporting the sperm cells found in a young girl's specimen is an easy decision but, cases of abuse are not always so obvious. We think that the lab should report all cases of spermatozoa and let the clinician decide what to do with the result.

Mucus

Mucus is a frequent finding of the urinary sediment. The exact function of mucus is unknown. Some think that this substance is a protection against bacterial infection. This action is done by coating the bacterial's pilis, essential to colonization of the lower urinary tract wall. The mucus coated bacteria are eliminated through miction. Mucus can also protect the lower urinary tract against irritating chemical agents.

Mucus forming cells are found scattered all over the urinary tract from the ascending section of the loop of Henle to the bladder. Consequently, mucus can originate from the kidney or from the lower urinary tract. Mucus originating from the kidney is made of Tamm-Horsfall protein. This explains the frequent association of mucus threads and casts. In elderly patients, mucus is a frequent finding and seems to originate from the lower urinary tract. In the majority of cases, presence of mucus threads is a benign situation. An irritating factor could stimulate mucus secretion.

Contaminants and Artifacts

The number of contaminating elements found in the urine sediment is surprising. Some of these artifacts are unavoidable, such as glass fragments, bubbles. Other artifacts are accidental, such as fibers and hair.

Starch or Talc Powder

Since the advent of systematic use of latex gloves by the hospital's staff, the presence of starch crystals or sometimes talc has become very frequent. The crystals are birefringent with a maltese cross interference pattern when seen under crossed polarized filter. The aspect of these crystals under bright field allows one to easily distinguish them from the birefringent fatty droplets.

RARE CRYSTALS

Several crystals are found only rarely in the urine. Often these crystals are difficult to differentiate from some particular shapes of a "usual" crystals. Before considering the possibility of an exotic crystal, it is necessary to exhaust the possibilities of a rare shape of a usual one. Some rare crystals are found mainly in acidic specimens, while others are found in alkaline media. Again. this represents only a tendency.

Cystine

The cystine is seen as colorless hexagonal plates. The solubility of the cystine is much larger in alkaline urine, with the result that the former is rarely found in alkaline urine. These colorless crystals can be difficult to distinguish from the hexagonal plate form of uric acid crystals. But, in this case, the examination of the crystals under polarized light will probably show birefringent crystals with a polarization color interference.

Cystine Crystals is a Clinically Significant Finding

Leucine et Tyrosine

Crystals of the amino acids leucine and tyrosine are very rarely seen in a urinary sediment. These crystals can be observed in some hereditary diseases like tyrosinosis and the "Maple syrup disease", but these conditions are very rare. The majority of cases where one finds these crystals are in patients with a serious hepatic problem, often in a terminal stage. In these cases, a concurrent presence of leucine and tyrosine is observed.

Leucine

The leucine crystals are seen as yellow spheres with concentric and radial strias. These crystals can sometimes be mistaken with cells, the central part simulating a nucleus. Under polarized light, the former presents a maltese cross interference pattern.

Tyrosine

Tyrosine crystallizes as brown neddles, isolated or forming a dense rosette.

Bilirubin

Bilirubine crystallizes in the urine as fine needles that regroup in a clump or as red brown spheres. The clinical meaning is the same as the bilirubin detected with the dipstick.

Cholesterol

Cholesterol crystallizes as thin rectangular plates with one of the corners (sometimes two or more) having a square notch. These crystals are very slightly birefringent. The cause of the presence of crystallized cholesterol is obscure. These crystals are seen in degenerative kidney diseases and are thought to have an identical clinical meaning as oval fat bodies. The presence of these crystals is normally accompanied by a heavy proteinuria. These crystals are very rare.

Hemosiderin

In a case of intravascular hemolysis, a part of the free hemoglobin passes through the glomerule. The former is then reabsorbed by the tubular cells. The hemoglobin is then concentrated and transformed slowly to a deep red brown pigmented granules, the hemosiderin. These hemosiderin granules can be seen free, inside tubular cells, and embedded in a cast. The free granules agglomerate forming amorphous red brown deposit. A staining procedure, based on the Roux reaction (Prussian blue), is possible for the urinary sediment.

Ammonium Biurates

The ammonium biurates, also called acid ammonium urates, crystallize as a sphere with strias that reminds a dried apple. Several crystals will show characteristic ox-horn projections. The crystals are strongly birefringent. Ammonium biurates are rarely seen in a fresh specimen. The former are found in old specimens that turned alkaline.

Calcium Phosphates

The calcium phosphate crystals are also named dicalcium phosphate or hydroxyl-apatite. Its mineral name is brushite. This substance crystallizes as a long prism with one sharped end. These crystals are slightly birefringent.

Calcium phosphate crystals are found with triple phosphates and their clinical meaning is identical.

Calcium Carbonates

The calcium carbonate crystallizes as very small spheres. These spheres can be found alone, in pair as dumbbell shape or in four units taking a cross shape. These are strongly birefringent. Calcium carbonate crystals are rare, probably because they are difficult to distinguish from amorphous phosphates. Some authors report as calcium carbonate what is recognized by others as amorphous phosphates. The reason is that this crystal is found mixed with amorphous phosphates thus forming a combined crystalluria of homogeneous appearance. The clinical meaning of the calcium carbonate is the same as amorphous phosphates.

Calcium Sulphates

Calcium Sulphate crystallizes as thin plates with sharp ends. The plate can be isolated or forming a rosette. These crystals are of little clinical meaning.

Food and Water Management

Introduction

Although various aspects of foods are discussed in several other chapters, it is the purpose here to bring together, in brief form, the composition and certain pertinent facts regarding a few nutritionally important, naturally occurring foods and food products derived from them. There is perhaps little virtue in learning the specific composition of a variety of foods, but it has long been a source of amazement to the writers how many beginning medical students are able to state a more satisfactory approximate composition of solder than of cow's milk.

DAIRY PRODUCTS

The principal dairy products used in this country are whole milk, evaporated milk, dried milk (whole or skim), cream, butter, cheese, and ice cream. Under each of these general terms we find various types and speciality products. There are innumerable varieties of cheese available for instance, and several grades of milk and cream. It is not the purpose to discuss the various types of cheeses, etc., but to treat the dairy products from a general viewpoint.

Table 10.1 gives the approximate composition of a number of foods obtained from cow's milk. The values in the table are subject to considerable variation. The fat content of milk varies with the breed of cow from 3 to 6 per cent. Cream is marketed with a fat content of as low as 18 per cent and as high as 40 per cent or even over. Butter without added salt has a mineral content of only 0.1 or 0.2 per cent. Salt is added in quantities of from 1 to 5 per cent in manufacture. The fat content varies usually between 80 and 85 per cent because of the amount of water and salt worked into the butter.

Table 10.1: The Approximate Composition of Some Typical Dairy Products

	Water (per cent)	Protein (per cent)	Lactose (per cent)	Fat (per cent)	Mineral Ash (per cent)	Cal/100 g
Whole milk	87	3.4	4.8	3.5	0.7	65
Evaporated milk	74	6.8	9.7	7.9	1.4	140
Dried milk, whole	3.5	26.0	38.0	27.0	6.0	495
Dried milk, skim	3.5	36.0	52.0	1.0	8.0	360
Cream	52-74	2-3	2.5-4.8	20-40	0.5	200-388
Butter	16	1.0	0.4	82	2.5	730
Cheese, cheddar	35	26	2.0	34	3	400
Cheese, cottage	72	21	4.0	1.0	1.8	108
Ice cream	61	4	21	13	1.0	217

The greatest variations are probably found among the hundreds of types of cheeses. Here the fat content may run from less than one per cent (cottage cheese made from skim milk) to 40 per cent in a cream cheese. The protein also varies from 20 to 40 per cent, depending largely on the amount of water allowed to remain in the final product. The composition of ice cream produced commercially at the present time is subject to some variations. Local and other government requirements must be met (especially in regard to fat content), and economically it is not generally feasible to produce ice cream with more fat than is required. Most states require a minimum of 10 or 12 per cent fat. Of importance to the consumer is the fact that ice cream today is sold on a weight and not a volume basis; since it has a large quantity of air incorporated into it.

Milk

No other single food has as many nutritional virtues as milk. It is not, however, a perfect food, since it lacks iron and copper among other essential nutrients. The proteins of milk are casein, lactalbumin, and lacto-globulin. About 80 per cent of the total is due to casein, a phosphoprotein of high biologic value.

Table 10.2 gives the approximate amino acid composition of casein. The composition of milk fat is governed to some extent by the breed and the diet of the cow, period of gestation, etc. In general, the fat contains fatty acids from C4 through C2o. It is unusual to find appreciable quantities of the low molecular weight acids in natural fats. The approximate fatty acid content of butter fat is as follows: butyric 3.0, caproic 1.5, caprylic 1.0, capric 1.8, lauric 6.5, myristic 21, palmitic 19, stearic 15, oleic 30, and linoleic plus linolenic 0.4-all in per cent of total fatty acids. Other lipids in the "fat" include cholesterol and phos-pholipids in small quantities.

Lactose constitutes the principal carbohydrate of milk. It is a disaccharide composed of glucose and galactose, and though it is an excellent carbohydrate from a biologic standpoint, it appears to have little special virtue compared to sucrose, for instance.

The mineral constituents of milk are present in a rather unique proportion. Early workers simulated this composition for a mineral mixture to be used in synthetic experimental animal diets, and, indeed, such mixtures are widely used for this purpose today. The calcium-phosphorus ratio is such that excellent utilization of these elements is possible in man and animals, and the calcium content is higher than in practically any other important food. Table 10.3 lists the approximate mineral composition of whole milk and of milk ash.

Table 10.2: The Approximate Composition of Casein calculated to 16 % nitrogen

Amino Acids	Per Cent
Arginine	4.1 ± 0.2
Histidine	2.5 ± 0.3
Lysine	6.9 ± 0.7
Tyrosine	6.4 ± 0.4
Trytophan	1.8 ± 0.2
Phenylalanine	5.2 ± 0.5
Cystine	0.36 ± 0.04
Methionine	3.5 ± 0.3
Serine	6-7
Threonine	3.9 ± 0.1
Leucine	12.1
Isoleucine	6.5
Valine	7.0
Glutamic acid	22.8
Aspartic cid	6.3
Glycine	0.5
Alanine	5.6
Proline	8.2
Hydroxyproline	2

Table 10.3: The Approximate Mineral Composition of the Whole Milk and of Milk Ash

Element	Whole Milk (per cent)	Milk Ash (per cent)
Ca	0.12	16
P	0.09	12
Mg	0.02	1.8
Na	0.048	5.4
K	0.13	21
Cl	0.11	13

Small amounts of sulphur, iron, copper, zinc, manganese, etc., are also found in milk. The copper and iron content of milk is low; both humans and experimental animals develop a dietary anemia on an exclusive milk diet if care is observed to obviate metallic contamination. A milk diet is regularly used to produce anemia in rats for experimental study.

Milk is a good source of vitamin A. It contains very little vitamin D, unless it is fortified to produce one of the types of "vitamin D milk". The ascorbic acid concentration is low, and pasteurization destroys around one-half the original content. Milk has rather low concentrations of the B vitamins, but because of the relatively large amounts consumed by the majority of children; this source is quite significant in the overall intake.

Since all dairy products contain milk in one form or another, it is not necessary to discuss further the nutritive value or the composition of these. Human milk differs markedly from cow's milk in a number of ways. The protein content of human milk is far lower, while the lactose is much higher. Table 10.4 compares the general informant feeding, cow's milk generally is diluted with water to bring the protein content down and then fortified with carbohydrate so that the composition of mother's milk with respect to these constituents is approached.

Table 10.4: Comparison of Human Milk and Cow's Milk

	Approximate composition					Cal/100 g
	Water (per cent)	Protein (per cent)	Lactose (per cent)	Fat (per cent)	Mineral Ash (per cent)	
Human milk	87	1.5	7.2	3.6	0.2	67
Cow's milk	87	3.5	4.8	3.5	0.7	65

NON-VEG FOODS

Meat and fish, because of their fine flavor, have been sought after by man since the beginning of history. There is no doubt that they play a most important part in our nutrition. The principal nutrient supplied is protein, although from the quantity standpoint many meat cuts contain a higher percentage of fat. In general, the proteins of meat and of fish are of good biologic value, and especially important as supplements to the proteins of grains, many of which, by themselves, are rather poor biologically.

Table 10.5 lists the approximate, partial composition of some typical meat and fish products. Here again the values in the table are subject to variations; this is especially true of the figures for fish. In the case of salmon" the fat content may be a small fraction of 10 per cent or a good deal more, depending on variety, time of year, and where the fish are caught.

Muscle meat, in itself, is not a complete food; it is seriously deficient in calcium, and the high phosphorus content makes the calcium-phosphorus ratio out of balance. Ascorbic acid, the fat-soluble vitamins, and certain B vitamins are likewise deficient. In general there IS practically no carbohydrate present.

Table 10.5: Approximate Composition of Some Meat and Fish Products

Product	Fresh basis			Cal/l00 g
	Water (per cent)	Protein (per cent)	Fat (per cent)	
Beef roast	62	16	21	253
Veal roast	65	18	15	207
Lamb roast	60	16	22	263
Beef liver	70	20	3.5	111
Salmon steak	65	20	10	170
Halibut filet	74	21	5	130

EGGS

For most people, eggs are a palatable and enjoyable food. From a nutritive standpoint, the egg stands with dairy products and meat. The proteins of egg are high on the scale of biologic values. The calcium, phosphorus, and available iron content are especially noteworthy. Certain of the vitamins are supplied in highly significant amounts. Although the egg contains very little carbohydrate, the lipids (fat, phospholipids, and cholesterol) constitute a worthy supplement to our diet. The approximate composition of the whole egg (edible part), white and yolk, is given in Table 10.6. The shell amounts to about one-tenth of the egg weight.

Table 10.6 : Approximate Composition of Hens' Eggs

	Whole Egg	Yolk	White
Water	73	49	85
Lipid	12	33	Trace
Protein	12.5	16	11.5
Carbohydrate	0.5	0.5	0.6
Ash	1.0	1.5	0.5
Calories per 100 g	160	360	47

In the white of the egg are found the proteins ovalbumin and ovoglobulin, and the yolk contains the phosphoprotein ovovitellin. Together these proteins offer an excellent supply of all the essential amino acids for growth, repair, and reproduction in animals, and, as far as we know, for human beings as well.

The fat of the egg yolk is readily and thoroughly digested and absorbed. It contains appreciable amounts of linoleic acid, one of the essential fatty acids. Lecithin and cephalin are the principal phospholipids of the yolk.

Few foods supply as much available iron as do eggs. It appears to be present practically all in inorganic form.

Eggs are a good source of vitamins A and D only if the hen eats feed high in these nutrients. In the case of vitamin A, the various carotene precursor pigments, if present in the diet, are transferred to the yolk. Since grain is one of the important constituents of the laying hen's ration, various B vitamins, especially riboflavin, from it are found in significant amounts in eggs.

Acid and Alkaline Residue Foods

All naturally occurring foods contain some acid and some base-forming elements. In certain foods one or the other may predominate to a slight or to a marked extent. In general, the organic matter of foods does not leave an acid or an alkaline residue on oxidation in the body.

The Grains

Since wheat is by far our most important grain crop, the following discussion will be limited to this food commodity. About 30 per cent of the protein intake of the average American diet comes from wheat flour. Some years ago our government made the enrichment of white flour with iron, thiamine, riboflavin, and niacin a compulsory measure. The amounts added are such that the final concentration in white flour approaches that of whole wheat flour with respect to these nutrients. Table 10.7 shows this very well.

The minimum government standards at the present time for enriched white flour are in milligrams per pound: thiamine 2, riboflavin 1.2, niacin 16, and iron 13. The addition of calcium is optional.

White flour contains approximately the following in percentage: water 10 to 12, protein 10 to 12, fat 1 to 2, carbohydrate 72 to 76, and ash 0.5. Whole wheat flour has slightly more protein, fat, and ash, while the available carbohydrate is somewhat lower. Thus, aside from the valuable protein source,

Table 10.7 : Comparison of Composition of White, Enriched, and Whole Wheat Flours

	White Flour	Enriched Flour (mgjlb) (mg/lb) (mg/lb)	Whole Wheat Flour
Calcium	82.0	82.0[1]	140.0
Phosphorus	400.0	400.0	1600.0
Iron	3.2	13.0-16.5	17.5
Thiamine	0.35	2.0-2.5	2.25
Riboflavin	0.15	1. 2-1. 5	0.55
Niacin	3.5	16.0-20.0	22-29
Pantothenic acid	2.2-2.7	2.2-2.7	4.5-5.4
Pyridoxine	0.9-1.1	0.9-1.1	1.8-2.2

this food offers an abundance of carbohydrate. Because of the enrichment program and the relatively large amounts consumed, wheat flour products constitute an important dietary adjunct to our intake of various members of the vitamin B complex.

The primary proteins of wheat are glutenin (a glutelin) and gliadin (a prolamin). Together they form the so-called wheat gluten. The biologic value of wheat proteins is lower than for milk, egg, or meat proteins. The essential amino acid lysine is especially deficient in wheat proteins, so that on supplementation with this amino acid or with another protein liberally

Most proteins contain sulphur and many also contain phosphorus. On metabolism of the protein, these elements remain as sulfuric and phosphoric acids and must be neutralized by ammonia, calcium, sodium, and potassium for excretion by the kidney. From this it follows that high protein foods in general are acid residue foods. This is also true of most grains.

In fruits and most vegetables, on the other hand, the organic acids are present partly as salts (K, Na, Ca, Mg), depending on the pH and in accordance with the Hasselbalch equation. Since the organic acids are oxidized to CO_2 and water, the alkaline metallic elements remain and must be neutralized by body acids. Exceptions to this statement can be found in the case of benzoic and oxalic acids, which occur in a few foods and are not oxidized to any great extent in the body, thus necessitating cations for their excretion.

Quantitatively the acidity or alkalinity of a given food is expressed as ml of 0.1 *N* acid or alkali equivalents to the residue of 100 g of the food. On this basis lean beef, wheat flour, and eggs have an acid residue of around 10. Many common fruits and vegetables have an alkaline residue varying from around 2 or 3 to 10 or 15.

WATER MANAGEMENT

In a lean 70-kg human adult the total free water is about 49 L (70 per cent of body weight). The extracellular water amounts to about 14 1, of which some 31 are in plasma and III in interstitial fluid and lymph. The intracellular water amounts to about 35 L (49 - 14).

The quantity of water required by the body is determined by the amount needed to give the proper volume and osmotic concentration to the body fluids, and to compensate for water lost by excretion through the kidneys, skin, lungs, and intestines.

Body water is derived from drinking water and beverages, the water content of foods eaten, and the water formed in the oxidation of foods which amounts to 0.6 g, 0.4 g and 1.07 g per gram of carbohydrate, protein, and fat oxidized, respectively. This oxidative or metabolic water is roughly proportional to the calorific value of foods, amounts to 100-140 g per 1000 calories, and varies with the proportions of the different foods utilized.

The amount of water required depends upon the water lost from the body, which fluctuates widely under different conditions.

Urinary Excretion

The urine volume of a normal adult varies greatly, but generally it ranges from 1000 to 2000 ml and averages about 1200 ml. The urine volume decreases with increased loss of water through the skin, lungs, and intestines, and *vice versa*. The volume also varies with diet. A very high protein diet produces large quantities of urea and other substances requiring much water for kidney elimination, while this is not the case with diets high in carbohydrate and fat. The amount of water required for the elimination of waste products in the urine varies with the concentrating capacity of the kidneys. The kidneys normally can concentrate urine to a specific gravity of 1.035 as a maximum, and have an obligatory requirement of about 600 ml of water per day to excrete the average load of waste products. In cases of severe kidney disease the concentrating capacity of the kidneys may fall so low that urine of a specific gravity of only 1.005 is formed, and in such conditions the obligatory water requirement is enormously increased.

Urinary excretion of water beyond the water obligatorily required for the elimination of waste products is referred to as "facultative water excretion" or "loss".

Pulmonary Excretion

The quantity of water exhaled by the lungs per day normally averages about 400 ml, but it may be greatly increased in fevers with hyperventilation. This water loss from the lungs is all obligatory.

Fecal Excretion

Ordinarily the quantity of water lost in the feces averages about 200 ml per day, but it may be greatly increased in cases of diarrhea and lead to severe dehydration. At the same time, large quantities of electrolytes containing an excess of $BHCO_3$ over BCl may be lost, leading to total electrolyte deficiency and acidosis. This water loss is obligatory.

Loss from the Skin

The loss of water from the skin by insensible cutaneous evaporation, when there is no visible sweating, averages about 400 ml per day. This is an obligatory loss. Loss by sweating varies from 0 to as much as 14 liters per day during severe exercise in a hot, humid environment. The concentrations of electrolytes in sweat vary greatly; $[Na^+]$ from 12 to 120, $[Cl^-]$ from 8 to 80, and $[K^+]$ from 5 to 30 meq per liter. Thus it is seen that prolonged copious sweating may lead to both severe dehydration and electrolyte deficiencies, especially Na+ and Cl^- deficiencies. The loss of Na^+, in particular, reduces the volume of plasma and other extracellular fluids which can be maintained, and thus tends to accentuate the dehydration and stabilize it until the deficiency is corrected by taking both salt and water. In severe dehydration the tissue cells lose K^+ and gain Na^+, which upsets the normal ionic balance and function.

Men doing Hard Labor

At elevated temperatures and sweating profusely long have been known to suffer severe muscle cramps and nausea. This condition is aggravated by drinking pure water, because the diuresis increases the electrolyte deficit, but it is relieved by saline solution.

The loss of water through the skin (and lungs) in patients with fever may be large, though there is no obvious sweating. The loss of water by the body increases about 13 per cent for each degree C increase in temperature.

Water Balance

The water balance of an average adult in a temperate environment is outlined in Table 10.8, as modified from Wolf (10).

From the table it will be seen that for water balance an average obligatory intake of 1600 ml is required to meet the obligatory loss of 1600 ml. The facultative loss, all through the kidneys, balances the facultative intake.

Table 10.8 : Showing the Water Balance for Normal Human Adult Values represent ml/24 hrs

Water Source	Intake		Organ of Excretion	Output	
	Obligatory	Facultative		Obligatory	Facultative
Drinking					
water and beverages	600	600	Kidneys	600	600
Food					
water	700	0	Skin	400	0
Oxidative			Intestines	200 (fecal)	0
water	300	0	Lungs	400	0
Subtotal	1600	600		1600	600
Total	2200			2200	

The water requirement is roughly equal to 1 ml per metabolic calorie. The values in the table are for an adult at light work. A greatly decreased water intake is required during starvation because of reduced kidney excretion. During starvation appreciable water is provided by the oxidation of tissue components.

Water balance is achieved in the body through control of intake by the stimulus of thirst, and control of renal excretion by the anti-diuretic hormone (vasopressin). A "thirst center," which is medially located in the vicinity of the third ventricle, causes the sensation of thirst when the extracellular fluids become hypertonic or are decreased in volume.

The hormone vasopressin of the posterior pituitary regulates absorption of water in the distal tubules of the kidney. Vasopressin secretion is stimulated by increased osmolarity of the extracellular fluids, causing more reab-sorption of water in the tubules and less excretion in the urine.

In the case of water deprivation and body dehydration, water deficiency first appears in the plasma, which is directly connected with the channels of excretion. As the plasma becomes more concentrated and its osmotic pressure increases, water passes to the plasma from the interstitial fluid. Then, as the interstitial fluid becomes more concentrated, water passes into it from the intracellular fluid. Thus, since water passes all membranes freely, in water deprivation all compartments lose water, with maintenance of equal osmotic pressures in them. However, since the intracellular compartment contains the most water, it loses the most. As the tissue cells become dehydrated, they also lose potassium to the extracellular fluid, and this potassium is then excreted in the urine. At least some of this cellular potassium is replaced by sodium, which upsets the normal cellular ionic balance and functions of the cells. The loss of cellular water induces thirst, which is the main symptom in the first two days of water deprivation. Continued dehydration of muscle and nerve cells causes functional abnormalities leading to weakness and confusion. Death usually occurs when about 20 per cent of the total body water has been lost. It seems probable that much of the effect of water deficiency upon the body is related to dehydration of cellular proteins with changes in their physiologic functioning.

Although a certain concentration of sodium salts in body fluids is necessary for the maintenance of normal fluid volumes, excessive amounts lead to dehydration. The upper limit of salt excretion by the kidneys is about 300 meq per liter (1.75 per cent NaCl), and when salt solutions stronger than this are taken, the excess salt is excreted along with the water necessary' for its excretion, leading to dehydration. The harmful effects of drinking sea water are related to this dehydrating effect.

The intravenous injection of large volumes of glucose solution without salt is equivalent to the injection of pure water so far as effects upon body fluids are concerned. The glucose is metabolized, leaving the water, which dilutes the body fluids and causes diuresis with salt loss. The salt deficiency leads to excess water excretion and dehydration.

When the water content of body fluids is appreciably increased without proportionate. increase in electrolytes, a condition of water intoxication with nausea and muscle cramps occurs. This is often seen in persons drinking large volumes of water after severe sweating during which the body fluid and electrolytes have been depleted. Normally the kidneys are very efficient in the excretion of excess water and maintenance of body fluid concentrations.

The maintenance of water equilibrium in the infant is less efficient than in the adult. The body of the infant contains about 80 per cent water, as compared with 70 per cent in the adult. An infant weighing 7 kg will contain about 5.6 1 of body water, of which some 4.2 1 are intracellular and 1.4 1 extracellular fluid. His daily output and input of water at equilibrium is about 0.7 1, which is 50 per cent of his extracellular fluid. On the other hand, the normal adult at water equilibrium has a turnover equivalent to only about 18 per cent of his extracellular fluid. Thus, it is obvious that failure to take in or retain water, and excess loss of body fluids through diarrhea, vomiting, and sweating more quickly cause dehydration in the infant than in the adult. The loss of body fluids by the infant also causes a proportionately greater loss of body electrolytes than in the adult, with more acute deficiencies and derangements of ionic balances.

ELECTROLYTES OF BODY FLUIDS

While the electrolyte contents of extracellular fluids can generally be obtained by direct analysis, the electrolytes of intracellular fluids must be determined indirectly and are subject to many errors, with the result that our knowledge of the composition of intracellular fluid is incomplete. As an example of the approach to the problem, suppose the electrolyte content of the heart muscle of an animal is to be determined. A known amount of a substance which distributes throughout extracellular fluid, such as thiosulphate, is injected intravenously. After a short time to permit equilibration, samples of plasma and heart muscle are taken and analyzed for thiosulphate and other electrolytes. The thiosulphate and electrolyte concentrations in plasma and interstitial fluid are assumed to be essentially the same. From the dilution factor of thiosulphate in plasma and the plasma electrolyte concentrations, the volume of extracellular fluid in the tissue and the electrolytes present in this amount of extracellular fluid can be calculated. When these electrolyte values are subtracted from the electrolyte values obtained on the whole wet muscle tissue, the values for intracellular electrolytes are obtained. The total water of the tissue sample is obtained by desiccation, and this minus the extracellular water of the tissue represents the tissue intracellular water. Then the concentrations of electrolytes in the intracellular water may be calculated.

The total electrolyte equivalent of a body fluid may be readily obtained by determining the total cation equivalent present and doubling this value, since electrical neutrality must be maintained. For example, if the total cation equivalent of plasma is found to be 155 meq per liter, then the total electrolyte equivalent is 310.

The electrolyte composition of plasma, exclusive of protein salts, is very similar to that of interstitial fluid, while the composition of intracellular fluid is quite different. Table 10.9 shows the detailed electrolyte composition of normal plasma, and Table 10.9 the approximate composition of muscle intracellular fluid.

Table 10.9 : The Electrolyte Composition of the Normal Human Plasma

	Mg/100 ml	Mg/1	mM/1	meq/1	Milliosmols* per Litre
Cations					
Na^+	327	3,270, ÷ 23 =	142	142	142
K^+	19	190, ÷ 39 =	5	5	5
Ca^{++}	10	100, ÷ 40 =	2.5, × 2 =	5	2.5
Mg^{++}	3.6	36, ÷ 24 =	1.5, × 2 =	3	1.5
Total cations			151	155	151
Anions					
Cl^-	365	3,650, ÷ 35.5 =	103	103	103
HCO_3^-	165	1,650, ÷ 61 =	27	27	27
HPO_4^-	9.6	96, ÷ 96 =	1, × 2 =	2	1
SO_4^-	4.8	48, ÷ 96 =	0.5, × 2 =	1	0.5
Organic acids⁻			6	6	6
Protein⁻			2	16	2
Total anions			139.5	155	139.5

*Milliosmols = millimols. In case of electrolytes such as NaCl and $CaCl_2$, 1 millimol = 2 and 3 milliosmols, respectively, becuse of dissociation.

While the values for K^+, Mg^{++}, and Na^+ of intracellular muscle fluid can be determined with considerable accuracy, the value for Ca^{++} is uncertain because of difficulty in determining the very small amount present. Potassium ions make up by far the greater proportion of cations in intracellular fluid, followed by MgH and Na^+. The MgH concentration of intracellular fluid is many times greater than in interstitial and other extracellular fluids. It is very probable that the distribution of ions, particularly cations, is not uniform throughout living cell protoplasm, due to combination with proteins, nucleoproteins, and. other anionic substances concentrated in specific cell regions and structures are related to the differences in protein content of the fluids and the resultant Donnan distributions.

Table 10.10: The Approximate Composition of Musccle Intracellular Fluid Values are mili Equivalents per liter of cell water

Cations	smeq/l	Anions	meq/l
Na^+	10	HCO_3^-	8
K^+	148	Phosphates plus other	
Ca^{++}	2	non-protein ions	136
Mg^{++}	40	$Protein^-$	56
Total	**200**		**200**

While the Donnan principles certainly must apply in the distribution of diffusible ions between interstitial and intracellular fluids, the operation of active ionic transport across the membranes separating these fluids obscures the Donnan effects. The fact that various ions, such as Ca^{++} and MgH, form nondiffusible complexes with the proteins of extracellular and intra-cellular fluids also may make uncertain calculations of the ionic distribution ratios, since the proportion of the substance in the freely diffusible ionic form is unknown.

The various fluids of the body-plasma, interstitial fluid, and intracellular fluid to maintain osmotic equilibrium despite variations in electrolyte distribution. It will be noted that the concentration of cations in intra-cellular fluid is appreciably higher than in interstitial fluid, yet these fluids are in osmotic balance. Much of the anion equivalence of intracellular fluid is supplied by protein, each protein molecule bearing about eight negative charges or equivalents. Then $8K^+$ associated with nondiffusible protein, if completely ionized, would yield only nine osmotically active particles:

$$K_8 \text{ protein} \leftrightarrows \text{protein}^{8-} + 8K^+$$

However, in the interstitial fluid, where Na^+ is the chief cations, the concen-tration of protein is relatively low and most of the cation is associated with univalent anion (Cl^- and HCO^{3-}). In this case $8Na^+$ associated with these ions would yield 16 osmotically active particles:

$$8NaCl - 8Na^+ + 8Cl^-$$

Thus it is evident that intracellular fluid must contain more cations (and total electrolyte equivalence) than interstitial fluid for the fluids to be in osmotic balance. Figure 10.1 shows the electrolyte compositions of gastro-intestinal secretions as compared to plasma. It will be noted that while the total electrolyte concentrations are very similar, there are wide variations in the distribution of ions in the fluids, which give to them their individual characteristics. These fluids are isosmotic.

CONTROL OF THE OSMOTIC PRESSURE, VOLUME AND COMPOSITION OF EXTRACELLULAR FLUID

The control of the quantity and composition of extracellular fluid is essential to control of the intracellular fluids upon which normal cellular and organ function depend. Most of the clinically observed disturbances involving electrolyte and fluid balance originate in the extracellular fluid, and therapeutic measures are formulated upon the basis of correcting these disturbances in the extracellular fluid. This concern with the extracellular fluid is a matter of practical necessity because of the difficulties in obtaining and analyzing intracellular fluids.

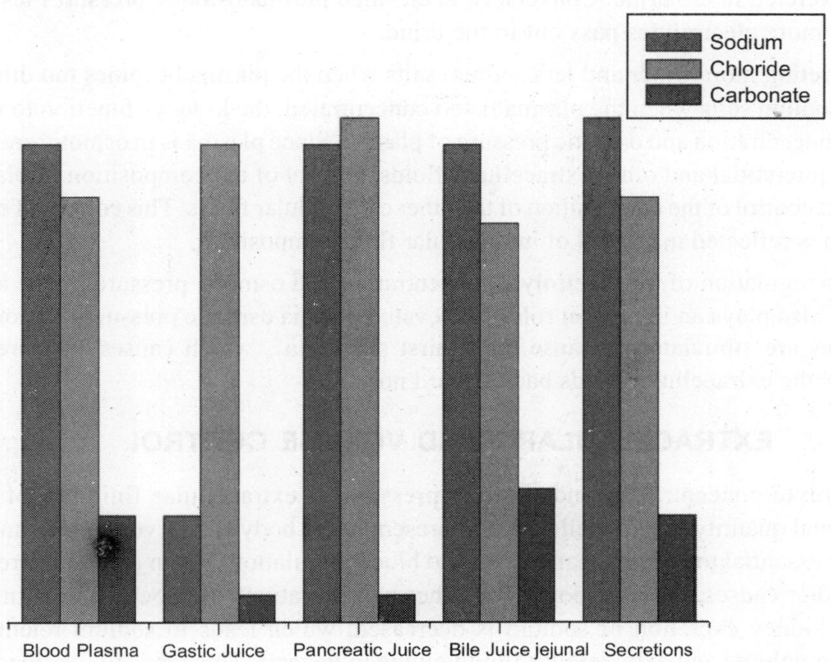

Fig. 10.1 : The Electrolyte Composition of Gastrointestinal Secretions

CONTROL OF TOTAL ELECTROLYTES AND OSMOTIC PRESSURE

The central nervous system is especially sensitive to appreciable changes in the osmotic pressure of its intracellular fluid, which is in balance with and regulated by the osmotic pressure of the surrounding extracellular fluid. Death may result from either water intoxication or deprivation due to hypotonicity and hypertonicity of brain intracellular fluids, respectively.

The electrolyte concentration and osmotic pressure of extracellular fluid are regulated by the kidneys, which excrete more electrolyte and less water or more water and less electrolyte, as required, to maintain the composition of plasma and other extracellular fluids. The major proportion of extracellular fluid electrolytes is composed of Na^+ and Cl^- ions, and regulation of the composition of extracellular fluid by the kidneys is chiefly effected through variation in the excretion of these ions. The amount of water excreted by the kidneys is determined by the amount of antidiuretic hormone (ADH) or vasopressin coming to them from the neurohypophysis. When the plasma osmotic pressure increases above normal, the neurohypophysis is stimulated to release more hormone, which increases water reabsorption in the distal kidney tubules, with less excretion in the urine. When the plasma osmotic pressure falls below normal, less hormone is released, and more water is excreted, to increase plasma osmotic pressure.

The amount of sodium salts excreted by the kidneys is controlled by adrenal cortical hormones aldosterone and deoxycorticosterone. At lowered plasma osmotic pressures the adrenals are stimulated to release more hormones, which increase the reabsorption of sodium salts in the distal kidney tubules, with less being excreted in the urine. Conversely, at elevated plasma osmotic pressures less hormones are secreted, and more electrolytes pass out in the urine.

Thus, by excreting more water and less sodium salts when the plasma becomes too dilute and less water and more sodium salts when the plasma is too concentrated, the kidneys function to regulate the total electrolyte concentration and osmotic pressure of plasma. Since plasma is in osmotic and electrolyte equilibrium with interstitial and other extracellular fluids, control of the composition of plasma by the kidneys is in effect control of the composition of the other extracellular fluids. This control of extracellular fluid composition is reflected in control of intracellular fluid composition.

In addition to regulation of the electrolyte concentration and osmotic pressure by the kidneys, the thirst mechanism also plays an important role. At elevated plasma osmotic pressures osmoreceptors in the- hypothalamus are stimulated to cause the "thirst sensation", which causes an increased water intake that dilutes the extracellular fluids back toward normal.

EXTRACELLULAR FLUID VOLUME CONTROL

While the electrolyte concentration and osmotic pressure of extracellular fluid are of paramount importance, the total quantity of extracellular fluid present in the body is also very important. A normal plasma volume is essential to normal heart action and blood circulation. When greatly decreased due to hemorrhage or other causes, tissue hypoxia and other manifestations of shock appear. In congestive heart failure the kidney excretion of sodium is decreased, which leads to sodium retention and an increase in plasma volume and extracellular fluid volume in general. This tends to overload the failing heart and cause tissue edema. The expansion of extracellular fluid volume is associated also with an increase in intracellular fluid volume, so that the volumes of all body fluid compartments are increased.

Plasma volume is primarily dependent upon the concentration of plasma proteins, particularly the albumin fraction, which determines the effective osmotic pressure of plasma and the capacity to maintain the proper fluid volume distribution between the vascular system and interstitial fluid spaces. Marked plasma protein depletion, especially of the albumin fraction, results in decreased plasma volume. When the plasma volume is reduced by hemorrhage, interstitial fluid moves into the vascular system to help maintain plasma volume and circulatory efficiency. However, this represents an emergency compensation only, since the protein content of the plasma is low. Such situations are best corrected by intravenous administration of whole blood, or plasma or concentrated albumin solution, to restore the normal osmotic balance and volume.

The total extracellular fluid volume of the body in some unknown manner is a function of the total sodium content of the body and fluctuates directly as the sodium content. While the kidney mechanisms rather closely regulate the electrolyte concentration and osmotic pressure of extracellular fluids, these mechanisms are relatively insensitive to changes in extracellular fluid volume.

When hypertonic salt solutions, such as sea water, are ingested, water passes from the intracellular fluid to expand the extracellular fluid volume until osmotic balance is attained. Unless compensated for by the intake of water, severe intracellular dehydration and concentration may occur, with damage to the central nervous system in particular, and with fatal results. If isotonic NaCl solution is administered, there is a proportionate increase in extracellular fluid volume without the diuresis following the administration of water alone. Excess salt is excreted in the urine for several days, and the extracellular fluid volume gradually returns to normal. When a person is on a salt-free diet, the excretion of Na^+ and Cl^- by the kidneys gradually decreases until these ions are practically absent from the urine. Under such conditions the kidneys retain the proper amounts of sodium and water to maintain the electrolyte concentration and osmotic pressure of extracellular fluids within the normal range, but since the total body sodium is depleted, the volume of the extracellular fluids is proportionately decreased. This decrease in extracellular fluid volume is the reason for the clinical use of low-sodium diets in patients with congestive heart failure and hypertension.

The direct relation of the sodium content and the tonicity of extracellular fluid to the volume and tonicity of intracellular fluid is the result of the osmotic balance between these fluids, which is maintained primarily by the transfer of water from the intracellular to the extracellular fluid when the osmotic concentration of extracellular fluid is increased and the reverse transfer of water from extracellular to intracellular fluid when the osmotic pressure of extracellular fluid is decreased.

As pointed out, in clinical conditions with low extracellular sodium (hyponatremia) the amount of sodium necessary to be given to correct the deficit must be calculated on the basis of total body water and not upon only the extracellular fluid volume, despite the fact that the sodium does not enter the cells appreciably. When salt is added to the extracellular compartment to increase its tonicity, a definite volume of water is transferred from intracellular to extracellular fluids to establish osmotic equilibrium, and sufficient salt must be given to bring the concentration in this water to normal as well as to correct the deficit in the extracellular fluid.

Suppose analysis of a patient's serum shows a deficit of 10 mM of sodium per liter (also 10 mM of associated anion, considered as Cl^-, per liter), and assume that from the weight of the patient, 80 kg, a volume of 481 total body water is estimated (0.60×80). Then the serum sodium and the extracellular fluid sodium can be increased 10 mM by giving the patient 480 mM of NaCl

The above situation may be better understood if the processes taking place are analyzed. A sodium and chloride deficit each of 10 mM per liter represent a deficit 20 milliosmols (mOsm) per liter. Assuming a normal mOsm concentration of 310 per liter, the initial concentration in the extracellular fluid of the patient was 290 mOsm per liter. When NaCl is added to the extracellular fluid, its osmotic pressure increases and fluid passes from the intracellular compartment.

The initial extracellular fluid volume of the patient was 16 1 (0.20×80), and the milliosmol deficit 20 per litre, or a total of $20 \times 16 = 320$. The intracellular fluid volume was 32 1 (48 ~ 16) and represented $32 \times 290 = 9280$ total milliosmols. In order to increase the concentration of the intracellular fluid to normal 310 mOsm per liter, the volume must be reduced to 29.935 I ($9280/310 = 29.9351$). This is equivalent to the transfer of $32 - 29.935 = 2.065$ I of water from the intracellular to the extracellular compartment. The NaCl required to provide the normal concentration to this volume of water is $310 \times 2.065 = 640$ mOsm.

Thus, it is evident that to raise the concentration in the 16 1 of extra-cellular water from 290 to 310 mOsm per liter (16 × 20 = 320), and that of the 2.0651 of water transferred from intracellular to extracellular fluid from 0 to 310 mOsm per liter (2.065 × 310 = 640), 960 mOsm of NaCl (320 + 640) must be administered to the patient. Since each minimol of NaCl equals 2 mOsm, this represents 480 mM of NaCl, the quantity calculated as necessary to correct the deficit on the basis of total body water.

Another approach here is as follows: The volume of extracellular water theoretically would become 18.065 1, and the concentration 4640/18.065 = 256.85 mOsm per liter, leaving a deficit of 310 – 253.85 = 53.15 mOsm per liter, or a total deficit of 18.065 × 53.15 = 960 mOsm = 480 mM of NaCl, which should be given to the patient.

Figure 10.2 illustrates diagrammatically the fluid and electrolyte relations in a case such as discussed above.

In cases of dehydration where there is a deficit of water with increased sodium, the water deficit may be calculated from the ratio of the elevated serum sodium concentration to the normal serum sodium concentration. The increase in serum sodium concentration is inversely proportional to the total body water volume.

$$\frac{\text{Normal Serum}\left[Na^{+}\right]}{\text{Elevated Serum}\left[Na^{+}\right]}=\frac{\text{Body water volume}}{\text{Normal body water volume}}$$

Or

$$\text{Body water volume}=\frac{\text{Normal Serum}\left[Na^{+}\right]}{\text{Elevated Serum}\left[Na^{+}\right]}\times\text{Normal body water volume}$$

Then the water deficit is :

Normal body water volume – Body water volume

For example suppose the 80Kg subject above with a normal total body water of 49 L has a serum sodium value of 160 meq per liter as compared with normal value of 138 meq per litre. Then his body water is given by:

$$\text{Body water volume}=\frac{138}{160}\times 48=41.37\text{L}$$

Body water deficit = 48 – 41.37 = 6.63 L

VARIATIONS IN EXTRACELLULAR FLUID VOLUME AND ELECTROLYTES

If one considers only the volume and electrolyte concentrations of extracellular fluids, there are six possible alterations of volume and/or concentration.

	Intracellular	Extracellular
I.	Water = 32 litres Electrolytes = 290 mOsm/1 Electrolyte deficit = 20 mOsm = 10 mM NaCl/L Total electrolytes = 32 × 290 = 9280 mOsm Total electrolyte (NaCl) deficit = 32 20 = 640 mOsm	Water=16 litres Electrolytes = 290 mOsm/1 Electrolyte deficit = 20 mOsm/1 Total electrolytes = 16 × 290 = 4640 mOsm Electrolyte deficit = 16 × 20 = 320 mOsm
II	Water = 29.935 litres $\frac{9280}{29.935}$ = 310 mOsm per liter, which is normal	Water = 18.065 litres Electrolytes = 4640 mOsm = $\frac{4640}{18.065}$ = 256.85 mOsm/L Total electrolyte deficit = (310–256.8) × 18.065 = 53.15 × 18.065 = 960 mOsm = 480 mM NaCl required

Fig. 10.2 Diagram I shows the initial distribution of water and electrolytes in intracellular and extracellular compartments in a hypothetical patient weighing 80 kg with an extracellular deficit of 10 mM sodium chloride (20 mOsm) per litre of water. Diagram II shows the shift of water from intracellular to extracellular compartments when NaCl is given to the patient and the amount required bringing the mOsm concentration up to the normal value of 310.

Changes in extracellular fluid volume are indicated by changes in the hematocrit value of blood and of the concentrations of hemoglobin and plasma proteins. An expansion of volume decreases hematocrit and the concentrations of hemoglobin and plasma proteins, while a contraction of volume has the opposite effect. Isotonic expansion of extracellular fluid, in which both water and salt are in normal proportions but the volume is above normal, is observed after parenteral administrations of saline solution in excess, which may lead to edema of the lungs and extremities. The hematocrit, hemoglobin, and plasma protein values are low. Intracellular fluid volume is unchanged, since there is osmotic balance between extracellular and intracellular fluids. Kidney excretion of both sodium salts and water is increased in compensation if function is normal. Serum sodium concentration is normal in isotonic expansion.

Hypertonic expansion results from the accumulation or retention of sodium without a proportionate increase in water, but with an over-all expansion of extracellular. fluid volume. Since the extracellular fluid is hypertonic, water passes from the intracellular fluid until osmotic balance is attained, resulting in the expanded extracellular volume and cellular dehydration. This condition may be observed as the result of drinking hypertonic salt solution, such as sea water, which contains nearly four times as much sodium as extracellular fluid. The hematocrit, hemoglobin, and plasma protein values are low. Kidney excretion of both sodium salts and water is increased in compensation if function is normal. Serum sodium concentration is high.

Hypotonic expansion of extracellular fluid, in which there is an accumula-tion of water without a proportionate increase in salt, is observed occasionally when large volumes of salt-free fluids, such as five per cent glucose solution, are given to patients with decreased renal function. Since the extracellular fluid is hypotonic, water passes into the intracellular fluid and increases its volume until osmotic balance is reached. The hematocrit, hemoglobin, and plasma protein values are low. Insofar as kidney function permits, compensation is affected by increased excretion of water and decreased excretion of salt. Plasma sodium concentration is low.

Isotonic contraction, in which there are proportional decreases in both extracellular salt and water, occurs as the result of loss of isotonic fluids from the body (gastrointestinal secretions) in conditions such as diarrheal disease and pernicious vomiting. Since the volume of these secretions per day is large (saliva 1500 ml, gastric juice 2500 ml, pancreatic juice 700 ml, bile 500 ml, intestinal mucosal secretions 3000 ml, total 8200 ml) as compared to the total extracellular volume (14,000 ml), the continued loss of a considerable proportion of these secretions results in severe extracellular dehydration and total electrolyte depletion (chiefly Na^+ and Cl^-).

In isotonic contraction cardiovascular disturbances arise because of decreased plasma volume, and the kidneys fail to excrete normal quantities of nitrogenous waste products, as evidenced by increased blood nonprotein nitrogen. The volume of urine decreases, and unless plasma volume, blood pressure, and kidney function are restored by administration of salt and water, kidney function may cease and the patient become comatose and die of circulatory collapse.

In isotonic contraction intracellular fluid volume remains normal, because extracellular fluid is isotonic. Hematocrit, hemoglobin, and plasma protein values are high. The kidneys tend to compensate by excreting both less salt and less water. Plasma sodium concentration is normal.

Hypertonic contraction occurs when there is excessive loss of water relative to the loss of sodium salts, such as is found in persons deprived of water or who because of illness or inattention fail to take water, in persons subjected to copious sweating without adequate intake of water (and salt), and in persons with diabetes mellitus or insipidus when the large loss of water in the urine is not balanced by water intake.

In hypertonic contraction water passes from intracellular to extracellular fluid to establish osmotic equilibrium at a higher tonicity than normal for both fluids; that is, both intracellular and extracellular dehydration exist. Since the volume of intracellular is more than twice that of extracellular fluid, in hypertonic contraction the major proportion of the water is lost from the intracellular compartment. The chief clinical manifestations are due to the effects of cellular dehydration in the central nervous system.

In hypertonic contraction the hematocrit, hemoglobin, and plasma pro-tein values are high. The kidney excretion of salt is increased and that of water decreased if function is normal. Serum sodium concentration is high.

Hypotonic contraction results when excessive salt relative to water is lost from the body. This happens in the adrenal cortical hormone deficiency of Addison's disease (deficient sodium reabsorption in distal kidney tubules) and may be observed after traumatic brain injury (cerebral salt-wasting syndrome). In some patients with chronic renal disease the capacity of the kidneys to reabsorb sodium salts is markedly decreased, which leads to excessive salt loss relative to water, and hypotonic contraction ensues.

In hypotonic contraction water passes from the extracellular to the intra-cellular fluid to establish osmotic equilibrium, which further contracts the extracellular volume. The chief clinical manifestations are due to the decreased plasma volume and its effects upon cardiac function and circulation.

In hypotonic contraction the hematocrit, hemoglobin, and plasma protein values are high, the kidney excretion of sodium salts is decreased, and that of water increased. The plasma sodium concentration is low. From the above discussion it will be seen that the extracellular fluid volume cannot be judged from the concentration of serum sodium but is indicated by values for the hematocrit, hemoglobin, and plasma proteins.

ACID-BASE BALANCE

General Considerations

The acid-base balance of the body is concerned with the metabolism of H^+ ions or protons. the mechanisms of their origin in the body, and the ways in which they are handled and excreted so as to maintain the pH of the body fluids relatively constant. While the pH of extracellular fluids generally is maintained around 7.4, values for intracellular fluids obtained largely by staining with intravital dyes-which also serve as pH indicators, and which are subject to error-show more variation and generally average lower than the pH of extracellular fluids. For example the following intracellular pH ranges have been found for rat tissues: stomach parietal cells 6.4-6.6, duodenal cells 6.8-7.0, peripheral liver cells 7.1-7.5, central liver cells 6.7-7.0, Kupffer cells 6.4-6.5, bone 6.5-7.3.

The chief acids of the body are H_2CO_3, HHb, $HHbO_2$, and other proteins; various organic acids, such as lactic, pyruvic, and citric acids; organic phosphates; and phosphoric and sulfuric acids. These are the substances which are the H + ion or proton donors.

The chief bases of the body, which are the H+ ion or proton acceptors, are the anions of the weak acids, such as HCO_3^-, Hb-, and HbO_2-; protein in general; $HPO_4^=$; and the anions of organic acids and organic phosphates.

The buffer systems of the body accordingly are made up of the weak acids and their salts (anions):

$$\begin{array}{l} \\ \text{Proton Acceptors} \rightarrow \\ \text{Proton Donors} \rightarrow \end{array} \frac{\overset{B^+}{HCO_3^-}}{H_2CO_3}, \frac{\overset{B^+}{Hb^-}}{HHb}, \frac{\overset{B^+}{HbO_2^-}}{HHbO_2}, \frac{\overset{B^+}{\text{Protein}}}{\text{H.Protein}} \text{etc}$$

where B^+ = a cation such as Na^+ or K^+.

When the pH of body fluids is below normal, the ratio of proton acceptors (bases) to proton donors (acids) is below normal; and when the pH of body fluids is above normal, the ratio of proton acceptors to proton donors is above normal.

When protons are added to body fluids, they are taken up by the buffer bases (anions) to form the buffer acids; and when there is a deficit of protons, they are given to the body fluids by the buffer acids, the process tending always to regulate and minimize the change in $[H^+]$ or pH:

$$H^+ + \text{buffer bases} \underset{H^+ \text{ decrease}}{\overset{H^+ \text{ increase}}{\rightleftharpoons}} \text{buffer acids}$$

The ratios of the buffer salts to buffer acids change in proportion to the change in $[H^+]$ or pH, and the pH change within a fluid is represented by the Henderson-Hasselbalch equation for anyone of the buffer systems present, such as the $BHCO_3/H_2CO_3$ buffer system:

$$pH = pKa + \log \frac{B^+HCO_3^-}{H_2CO_3}$$

While the buffer systems of body fluids represent the first mechanism involved in the regulation of pH, the major role in this regulation is played by the kidneys, which excrete more or less H^+ ions (acid) in the urine as the $[H^+]$ of the plasma increases and decreases. This action of the kidneys is importantly supplemented by the respiratory mechanism, which operates to decrease respiration and retain more $CO_2(H_2CO_3)$ in the body fluids when the $[H^+]$ is below normal (pH elevated, alkalosis) and to increase respiration and decrease the $CO_2(H_2CO_3)$ in body fluids when the $[H^+]$ is above normal (pH lowered, acidosis).

It should be pointed out that according to older concepts, the cations Na^+, K^+, Ca^{++}, and Mg^{++} are considered to be the bases of the body; the older and much of the recent literature use this concept. This older view considers that the Na+ ion of a buffer salt, such as NaHCO., neutralizes a strong acid HA:

$$NaHCO_a + HA \rightarrow NaA + H_2CO_3$$

which, of course, is not true. The reaction is as follows:

$$Na^+ HCO_a^- + H^+A^- \rightarrow H_2CO_3 + Na^+ + Cl^-$$

in which HCO_3^- combines with H+ from the acid HA to form the weakly dissociated acid H_2CO_3, which breaks up into H_2O and CO_2, and the CO_2 is eliminated; that .is, the anion of the weak acid HCO_3^- is the proton acceptor or base which neutralizes the H^+ ions. A cation such as Na^+, of course, must be associated with the HCO_3^- to maintain electrical neutrality, but it does not neutralize acids.

The total equivalence of all the cations present in a tissue or body fluid must equal the total equivalence of all the anions present. Then twice the cation equivalence equals the total electrolyte equivalence.

The total cation equivalence may be readily determined, and in the older literature it is considered to be the total base equivalence, while the anion equivalence is considered to represent the acids. According to the modern (Bronsted) concept, the total base of a body fluid or tissue is represented by all of the anions of the weak acids present which combine with H^+ (HCO_3^-, protein-, Hb^-, $HPO_4^=$, etc.). Since considerable of the cation equivalence is balanced by CI- and other anions which do not neutralize H^+, the base equivalence is less than the cation equivalence to this extent. The total cation equivalence of an extracellular fluid minus the Cl^- equivalence closely approximates the total base equivalence, because this difference essentially represents the equivalence due to HCO_3^-, protein- Hb^-, $HPO_4^=$, etc., associated with the cations not balanced by Cl^-.

The metabolic processes of the body continually form acidic substances which without operation of the buffer systems, kidneys, and lungs would rapidly lower the pH to a fatal level.

The oxidation of the carbon of foods in the human adult with average exercise (2500 calories) forms about 400 L of CO_2 per day, which represents about 90 equivalents of acid (H_2CO_3). This CO_2 is largely eliminated through the lungs, but it is buffered and transported in extracellular fluids with exceedingly little change in pH.

The nucleoproteins, phosphoproteins, and phosphatides of foods and tissues, when broken down, liberate phosphoric acid. The oxidation of sulphur-containing amino acids, such as cystine and methionine, in cells yields sulfuric acid. Lactic and pyruvic acids in considerable quantities escape from muscles during strenuous exercise.

Certain foods, such as the juices of fruits and vegetables, contain citric and other organic acids and the potassium and sodium salts of these acids. When metabolized in the body, the organic acid components are oxidized to CO_2 and water; and the cations 'K^+ and Na^+' are released as $BHCO_3$, which increases body base. Thus the metabolism of a mol of monosodiurn citrate forms a, molecule of $NaHCO_3$.

In general, the acid load formed by metabolic processes) in the body greatly exceeds the base load, and, it is the handling of this large amount of acid which is of primary concern in the normal regulation of the acid-base balance.

Certain substances used as drugs have decided effects. upon the acid-base balance of the body. For example, NH_4Cl is taken to the liver, where the ammonia component is converted to neutral urea and HCI is released, which must be neutralized. Of course, the ingestion of $NaHCO_3$ leads directly to an increase in body base (HCO_3^-) and may cause alkalosis.

Various pathologic states tend to upset the acid-base balance by increasing the acid components. In severe diabetes mellitus the liver forms the relatively strong organic acids β-hydroxybutyric and acetoacetic in large amounts, which escape into the extracellular: fluids, As much as 50-100 g of these acids and their salts may be excreted in the urine' per day.

In nephritis the kidneys may be unable' to excrete much of the acid load formed in the body~ which accumulates and leads' to severe acidosis, In pernicious vomiting, as observed in duodenal and pyloric; obstruction, and in gastric lavage, HCl in large amounts may be lost from the stomach. Since this is formed from Cl^- of plasma and H_2CO_3 in the gastric cells by a reaction such as the following:

$$B^+Cl^- + H_2CO_3 \rightarrow \underset{\text{To plasma}}{B^+ HCO_3} + \underset{\text{into stomach}}{HCl}$$

The loss of a mol of HCl from the stomach leaves a mol of $BHCO_3$ to increase: the base of plasma and extracellular fluid and leads to base excess and alkalosis. Normally the HCl formed in the stomach passes into the intestine, where it is neutralized by $BHCO_3$ secreted from plasma; in pancreatic: and intestinal juices to reform BCl, which is absorbed back into the blood stream to re-establish the original acid-base balance.

In severe cases of diarrhea the loss of pancreatic and intestinal secretions containing $BHCO_3$ is large and since this. $BHCO_3$ comes from, plasma and other extracellular fluid, the acid-base balance is upset, to cause acid excess or acidosis.

Both Na^+ and K^+ (mainly Na^+) and small amounts of Ca^{++} and Mg^{++} are present in gastric juice, pancreatic juice, and intestinal secretions. Prolonged severe vomiting lead not only to alkalosis', but to; total electrolyte deficiency (mainly Na^+) and decreased extracellular fluid volume (isotonic contraction). Also, severe diarrheal disease causes not only acidosis but, in addition, total electrolyte deficiency' (mainly Na^+) and, deereased, extracellular fluid volume (isotonic contraction).

Thus it is evident that the body must be provided with mechanisms regulating the acids and, basis of the blood and other body fluid with, respect to both the quantities present and; also the ratios of bases to acids. which determine the pH. These control processes, as previously indicated, are supplied by the buffer systems of the body and renal and respiratory mechanisms.

RESPIRATORY REGULATION OF ACID-BASE BALANCE

The ventilation of the lungs or the minute volume of air passing into and out of the lungs determines the CO_2 tension in the alveolar air, and this tends to regulate the amount of dissolved CO_2 (H_2CO_3) in the blood and other body fluids. An increase in ventilation due to either an increased rate or depth of breathing or both (hyperpnea) lowers the CO_2 tension of alveolar air and causes equilibrium with dissolved CO_2 in the blood and other body fluids to be established at a lower concentration than normal, thus tending to raise the pH. Conversely, decreased lung ventilation due to slow and shallow breathing raises the CO_2 tension of alveolar air and causes equilibrium with dissolved CO_2 in the blood and other body fluids to be established at a higher concentration than normal and thus tends to lower the pH of these fluids. Since the respiratory apparatus is stimulated by increased H^+ ion concentration (lowered pH) and increased pCO_2 and is depressed by decreased H^+ ion concentration (increased pH) and decreased pCO_2, the rates of pulmonary ventilation and alveolar pCO_2 are automatically controlled in the direction of maintaining normal pH relations in the blood and other body fluids. Thus, in an alkalosis, in which the pH is above normal and the ratio HCO_3^- /H_2CO_3 is high, the respiratory apparatus is depressed, lung ventilation is decreased, the pCO_2 of alveolar air is increased, and the H_2COa in blood and other fluids is increased, tending to lower the ratio HCO_3^- /H_2CO_a and the pH toward normal. In an acidosis, with lowered pH and a lowered HCO_3^-/H_2CO_a ratio, the respiratory apparatus is stimulated, lung ventilation is increased, alveolar pCO_2 is decreased, and body fluid H_2CO_a is decreased, tending to increase the ratio HCO_3^-/H_2CO_a and pH of blood and other fluids toward normal.

RENAL REGULATION OF THE ACID-BASE BALANCE

The pH of the urine is determined by the substances which the kidneys must remove from the blood in order to help keep blood pH within physiological limits. Urine normally is a strongly buffered solution, particularly because of the large amounts of phosphates excreted, though it contains smaller quantities of many buffer systems. The pH of urine normally varies between 4.8 and 7.4. It appears that the kidneys are unable to excrete urine more acid than pH about 4.5 and more alkaline than pH 8.2. On ordinary diets the pH of urine is generally a bit on the acid side, and the quantity of acid titratable to a plasma pH of 7.4 is appreciable. This titratable acidity between the pH of urine and of plasma represents the amount of acid removed from the blood by the kidneys in helping maintain a normal plasma pH. In severe alkalosis the pH of urine may exceed that of blood, and the concentration of urinary HCO_3 may be three or four times that of normal plasma. In such conditions the urine shows a titratable alkalinity rather than acidity. Ordinarily urine acidity is titrated to phenolphthalein after the addition of sodium oxalate to precipitate calcium. Normally a 24-hour sample of urine shows the equivalent of 20 to 40 ml of 1 *N* acid. In severe acidosis such as found in diabetes, the value may rise to 150 ml of 1 *N* acid.

The pCO_2 of urine is generally that of arterial blood, about 40 mm. This represents about $0.03 \times 40 = 1.2$ mM of dis-solved CO2 per liter.

At a constant pCO_2 of 40 mm it is possible to calculate the quantities of HCO_3 contained in urine at different pH values. At the lower range of urinary pH (4.5) practically no HCO_3 is present (0.03 mM/L), while at the upper range of pH (8.2) much is present (151 mM/L). Thus we see that practically no base as HCO_3 is excreted in the HCO_3/H_2CO_3 buffer system at the lowest urinary pH, while a great deal is excreted at the higher pH values. At urinary pH values of 7.5, 7.0, and 6.5 there are 30, 9.6, and 3 mM of HCO_3 per liter of urine, respectively. The quantities represent 2.5, 0.8, and 0.25 g of $BHCO_3$ figured as Na HCO_3 per liter, respectively.

The phosphates constitute the most important buffer of ,normal urine quantitatively. This buffer system is composed of $HPO_4{=}/H_2PO_4^-$. While the blood contains relatively little phosphate buffer because of kidney efficiency in excreting phosphate, much phosphate buffer passes through the blood into the urine as a result of the metabolism of phosphorus compounds.

At a plasma pH of 7.4 the ratio of B_2+ HPO_4= to B+ $H_2PO_4^-$ is 5, which represents about 83 per cent $B2^+$ $HPO_4^=$ and 17 per cent B^+ $H_2PO_4^-$. The excretion of B^+ $H_2PO_4^-$ in the urine represents the removal of acid from the plasma and tends to raise plasma pH, while the .excretion of B_2+ $HPO_4^=$ tends to lower plasma pH. When urine of pH 4.8 is formed, around 99 per cent of the phosphate excreted is B+H2POc, while at a urine pH of 8.2 about 97 per cent B2+ $HPO_4^=$ is excreted.

The proportions of acid and salt (base) of an acid excreted in the urine at a given pH may be calculated from the *pK* values using the Henderson-Hasselbalch equation. While the values obtained at extreme salt-acid ratios are not accurate, they show the approximate relations. Table 10.11 gives such calculations for important urinary acids. It is obvious that when the kidneys excrete a very acid urine, much HCO_3^- (base) and Na^+ of the plasma is conserved to help prevent a drop in plasma pH and maintain the electrolyte concentration. The amount of $B^+HCO_3^-$ conserved is equal to the equivalents of urinary acid obtained by titrating the urine with alkali to the pH of plasma. When the kidneys excrete urine at a high pH $B^+HCO_3^-$ is removed from the plasma to compensate a rise in plasma pH.

This conservation of plasma HCO_3^- by the excretion of acid urine is exceedingly important in cases of severe diabetes when large quantities of β-hydroxybutyric and acetoacetic acids are formed and accumulate in the tissues and blood. These are strong organic acids which are neutralized by reaction with HCO_3^- and other buffer bases, thereby lowering the pH. At plasma pH these acids exist 100 per cent as anions (salts). However, if they are excreted by the kidneys in urine at a pH of 4.5 to 4.8, more than half is present as free acid, This means a proportionate removal of H^+ ions from plasma and an increase in HCO_3^- and other plasma bases.

From the above discussion we may conclude that the excretion of acid and base by the kidneys is adjusted to help maintain normal plasma pH. As the plasma pH rises above normal owing to excess HCO_3^- and other bases, the kidneys excrete more of these bases in the urine to lower plasma pH.

Conversely, when the plasma pH is below normal, the kidneys increase the excretion of plasma acids to raise the plasma pH.

Table 10.11: The Relation of Urinary pH to percent of Salt and Acid Buffer Components Excreted in urine

Urine pH	Buffer Systems in Urine							
	$B^+HCO_3^-$	H_2CO_3	$B_2^+HPO_4^-$	$B^+H_2PO_4^-$	B^+HB^-	HHB*	B^+Ac^-	HAc*
4.5	2.4	97.6	0.63	99.37	9.1	90.9	83.4	16.6
6.0	44.2	55.8	16.6	83.4	97.6	2.4	99.4	0.6
7.0	88.8	11.2	66.6	33.4	99.8	0.2	100	0
7.4	95.2	4.8	83.3	16.7	100	0	100	0
8.0	98.8	1.2	95.2	4.8	100	0	100	0
8.2	99.2	0.8	97	3	100	0	100	0

*HB- and HHB = β-hydroxybutyrate ion and β-hydroxybutyric acid, respectively; Ac- and HAc = acetoacetate ion and acetoacetic acid, respectively. In diabetes mellitus about 75 per cent of these acids in urine is β=hydroxybutyric acid.

Another very important way in which the kidneys conserve Na^+ and HCO_3^- is formation of ammonia and substitution of NH_4^+ for Na^+ in the urine. In severe diabetic acidosis the equivalent of 500 ml of 1 *N* NH_3 may be excreted per day, so that as much or more Na^+ and HCO_3^- are conserved to the body through ammonia formation in the kidney as by acid excretion by the kidney.

The mechanisms of sodium reabsorption and bicarbonate and acid formation in the kidney. The H^+Na^+ and $K^+H^+Na^+$ exchanges. The student will find excellent discussions of these very important phases of electrolyte and acid-base balance

As indicated above, the kidneys play the dominant role in regulating both the electrolyte concentration and acid-base balance of extracellular fluids. The processes involved have been studied by many investigators over a long period of time, but Pitts and associates worked out the modern concept explaining the detailed mechanisms by which the kidneys operate in performing these functions.

The glomerular filtrate contains the electrolytes and acids and bases present in plasma. The chief cation present is Na^+, and the chief anions are Cl^-, HCO_3^-, and HPO_4^-; that is, the chief electrolytes in glomerular fluid may be represented as Na^+ Cl^-, Na^+ HCO_3^-, and Na^{2+} HPO_4. The main buffers are the Na^+ HCO_3^-/H_2CO_3 and Na^{2+} $HPO_4^=/Na^+$ $H_2PO_4^-$ systems. At a pH of 7.4 about 95 per cent of the CO_2 is present as Na^+ HCO_3^-, and about 83 per cent of the phosphate is present as Na_2^+ $HPO_4^=$.

Normally, as the glomerular filtrate passes into the tubules most of the water is reabsorbed, and the greater proportion of the Na^+ is taken up by the tubular cells in exchange for H^+ formed in the tubular cells (from H_2CO_3) this Na^+ is returned to the plasma in association with HCO_3^- formed from H_2CO_3 in the tubular cells. The formation of H+ and HCO_3^- from H_2COg in the tubular cells (both proximal and distal) is catalyzed by the enzyme carbonic anhydrase. The principles involved are shown in the diagram of Figure 10.3.

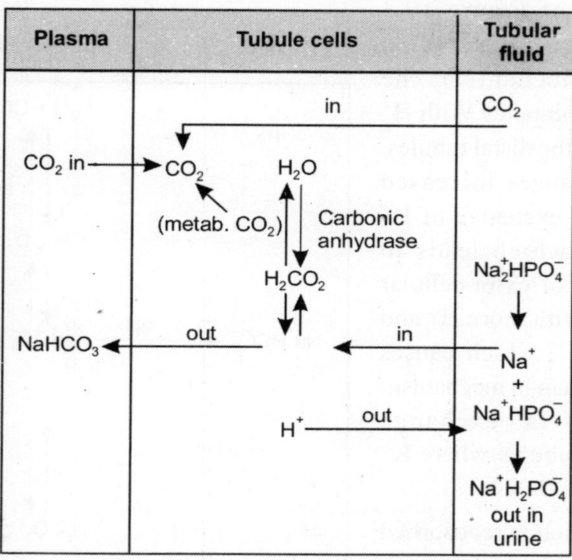

Fig. 10.3 : The Principals involved in the Glomerular Mechanism in the Tubule Cells

While the anions of buffer salts other than phosphate ($HPO_4^=$) may accept H^+ in the H^+-Na^+ exchange and secretion of acid into the tubular fluid, normally it is $HPO_4^=$ which is the chief H^+ acceptor. This exchange amounts to the over-all reaction:

$$\underset{\text{Tubule Cell}}{H_2CO_3} + \underset{\text{Tubular fluid}}{Na_2^+ HPO_4^-} \rightarrow \underset{\text{Into Plasma}}{Na^+ HCO_3\text{-}} + \underset{\text{Out in Urine}}{Na^+ H_2PO_4\text{-}}$$

by which half the sodium of the basic salt $Na^{2+} HPO_4^=$ is converted to plasma $Na^+ HCO_3^-$, and the other half is excreted in the urine as the acid salt Na+ $H_2PO_4^-$. Thus, Na^+ and HCO_3^- are returned to the plasma and H^+ is excreted in the urine to help maintain normal acid-base balance and electrolyte concentration.

Researchers showed that a subject made acidotic with NH_4Cl excreted much more acid in the urine when also given phosphate. This was due to increased amounts of $HPO_4^=$ in the tubular fluid to accept H+ from the tubular cells and promote the H^+-Na^+ exchange.

In severe diabetic acidosis, in which large amounts of the relatively strong ketone acids, β-hydroxybutyric and acetoacetic, are formed, their sodium salts in high concentration are present: in the glomerular filtrate and pass to the tubules. In such a condition the H^+ – Na^+ exchange mechanism operates with sodium β-hydroxybutyrate and sodium acetoacetate so that much of the Na^+ is returned to the plasma as $Na^+HCO_3^-$ and proportionate amounts of ireeJ3-hydroxybutyric and acetoacetic acids are excreted in the urine.

As shown in Figure 10.3, the CO_2 for the formation of H^+ and HCO_3^- in the tubular cells is supplied by the general CO_2 pool (plasma), by CO_2 entering ,from the tubular fluid, and by CO_2 derived from the metabolism of tubular cells.

While the diagram of Figure 10.3 indicates only H^+ exchanging with Na^+ in the H^+Na^+ exchange mechanism, the evidence shows that K^+ competes With H^+ in the exchange for Na^+ in the distal tubules. Administration of K^+ causes increased .exchange of K^+ and less exchange of H^+ or Na^+ in the tubules, which leads to acidosis. In K^+ depletion of extra-cellular fluid the reverse occurs, with more H^+ and less K^+ exchange for Na^+, which causes alkalosis. The H^+-Na^+ exchange mechanism in reality is the H^+K^+-Na^+ exchange mechanism in the distal tubules where K^+ excretion takes place.

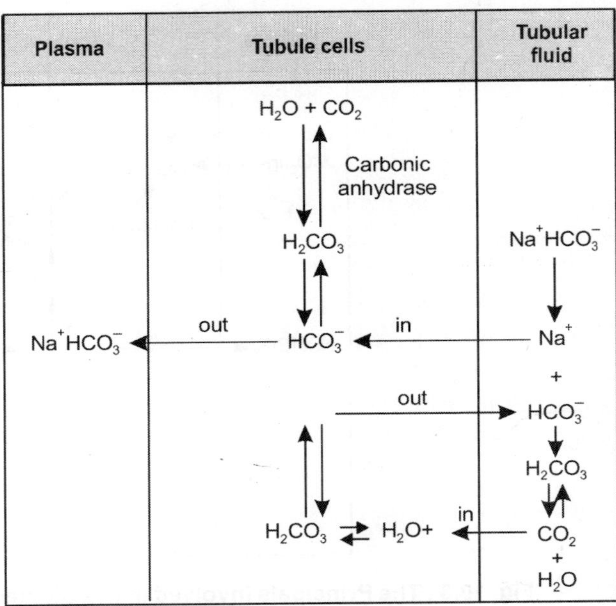

Fig. 10.4 : The Carbonate Ion Exchange

Not all of the Na^+ which is reabsorbed in the kidney tubules is reabsorbed through the H^+-Na^+ exchange mechanism. A part of the glomerular filtrate load of Na^+ is taken up, along with Cl^- and other anions, such as $HPO_4^{=}$ and SO_4^-, by the proximal tubular cells and returned to the plasma by an active Na^+ transport mechanism. The anions appear to be absorbed passively. They are carried along by the attraction of the Na^+ ions and maintain electrical neutrality. Most of the water also is reabsorbed along with the electrolytes in the proximal tubules.

The $Na^+HCO_3^-$ of glomerular fluid normally disappear as the fluid passes through the tubules, and in effect it is reabsorbed into the plasma. Until recently it was believed that this $Na^+HCO_3^-$ is taken up directly by the tubular cells and passed back into the plasma. However, showed that by altering the alveolar pCO_2 (and extracellular CO_2) the amount of $Na^+HCO_3^-$ reabsorbed by the renal tubules can be changed markedly. When respiratory alkalosis was produced (by hyperventilation) with low extracellular pCO_2 and low pCO_2 in renal tubular cells, less $Na^+HCO_3^-$ pCO_2 $Na^+HCO_3^-$ was reabsorbed by the tubules and more excreted in the urine. Conversely, in respiratory acidosis, with high extracellular and tubular cell pCO_2 the tubular reabsorption of $Na^+HCO_3^-$ was greatly increased with a proportionate decrease in urinary excretion. The sulphonamide drug acetazoleamide (Diamox) inhibits the action of the enzyme carbonic-anhydrase. When given to an animal, it accordingly inhibits the conversion of CO_2 to H_2CO_3 in the kidney tubules and limits the supply of H^+ and HCO_3^- for the H^+-Na^+ exchange. Berliner and associates found that experimental animals given acetazoleamide ceased excreting titratable acid in the urine and excreted as much as 50 per cent of the Na^+ HCO^- of the glomerular filtrate in the urine, indicating the operation of the H^+-Na^+ exchange mechanism in the absorption of Na^+ HCO_3^- from the glomerular fluid by the tubules. When respiratory acidosis is produced in an animal which has been given acetazoleamide, the elevated pCO_2 of the tubular cells forms sufficient H_2CO_3 to supply the necessary amounts of H^+ and HCO_3^- to offset the decreased carbonic anhydrase activity and permit normal H^+-Na^+ exchange. The above evidence shows then that Na^+ HCO_3^- absorption from the glomerular filtrate by the tubular cells is dependent upon the H^+-Na^+ exchange mechanism. The process is diagrammed in Figure 10.4. From this illustration it is seen that when Na^+ HCO_3^- of glomerular fluid enters the tubules, Na^+ enters the tubule cells and, with HCO_3^- formed in the cells, is passed on into the plasma as Na^+ HCO_3^-. The HCO_3^- in the tubular fluid reacts with H^+ from the tubular cells to form H_2CO_3, which breaks up into H_2O and CO_2. The CO_2 then enters the tubular cells and forms an equivalent amount of H_2CO_3, which provides H^+ and HCO_3^- for the H^+-Na^+ exchange. Researchers found that all of the Na^+ HCO_3^- normally is reabsorbed in the tubules from glomerular fluid at plasma bicarbonate concentrations from 13 to 24 meq per litre. At plasma concentrations above 28 meq per liter absorption is incomplete, a maximum of 2.8 meq per 100 ml of glomerular filtrate being absorbed and all in excess being excreted in the urine. The renal plasma threshold for bicarbonate excretion normally lies between 25 and 27 meq per litre. This means that the plasma bicarbonate concentration tends to be balanced around 25 to 27 meq per litre. When plasma bicarbonate is normal or reduced, all of it in glomerular filtrate is absorbed in the tubules. When the bicarbonate of plasma is elevated by ingestion of bicarbonate that in excess of the normal amount is excreted.

This apparently fixed bicarbonate threshold varies under certain conditions. It is reduced in hyperventilation (respiratory alkalosis), and bicarbonate is excreted in the urine as compensation. The threshold is raised when the plasma Na^+Cl^- concentration is low, which retains more $Na^+HCO_3^-$ in the plasma (and other extracellular fluids) to help maintain the total electrolyte concentration and osmotic pressure. The H^+-Na^+ exchange mechanism is controlled by the steroid hormones of the adrenal cortex (such as aldosterone and deoxycorticosterone). In Addison's disease the secretion of these hormones is deficient, and the H^+-Na^+ exchange may be greatly decreased, causing both decreased H^+ excretion and Na^+ and HCO_3^- reabsorption in the tubules. As a result, plasma HCO_3^- and, pH fall, producing an acidosis, and plasma Na^+ and also Cl^- are depleted. While large amounts of sodium are lost in the urine in adrenal cortical hypofunction (Addison's disease), the excretion of potassium is decreased and the concentration of potassium in the plasma accordingly increases (hyperkalemia). Also, in Addison's disease there is hypotonic contraction of extracellular fluids, and water passes into the intracellular compartment, causing hypotonic expansion of intracellular fluid. Loeb found that giving large amounts of NaCl to patients with Addison's disease causes plasma electrolytes to gradually return toward normal, with striking clinical improvement.

In hyperfunction of the adrenal cortex the H^+-Na^+ exchange mechanism is stimulated to increased activity, leading to increased tubular excretion of acid and reabsorption of Na^+ and HCO_3^-, which causes a rise in plasma pH and alkalosis. Excessive K^+ and Cl^- losses also occur in adrenal cortical hyperfunction.

ACID-BASE BALANCE CONTROL BY AMMONIA FORMATION IN THE KIDNEYS

Nash and Benedict first demonstrated that renal tubular cells form ammonia from a precursor in renal arterial blood and secrete it into tubular fluid. This ammonia formation takes place in the distal tubules.

It has been found that about two-thirds of the ammonia formed in the kidney is produced by deamidation of glutamine and one-third by oxidative de-amination of amino acids:

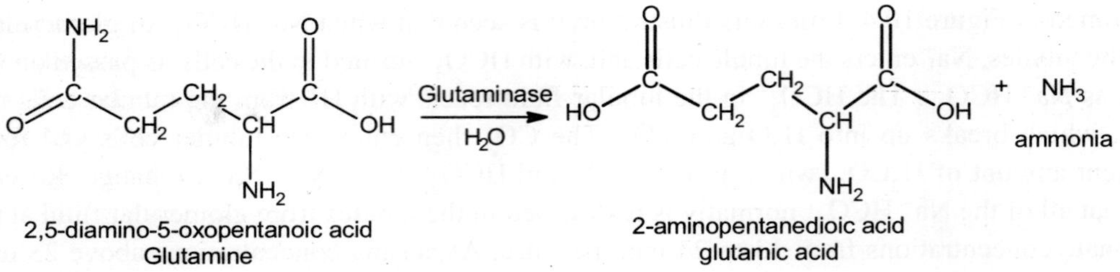

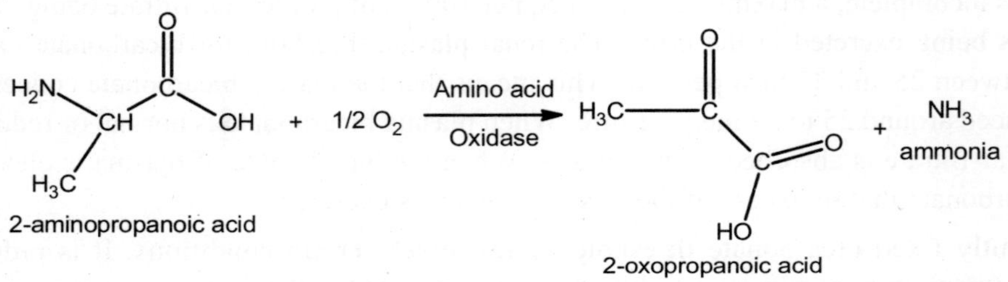

2,5-diamino-5-oxopentanoic acid (Glutamine) $\xrightarrow[H_2O]{\text{Glutaminase}}$ 2-aminopentanedioic acid (glutamic acid) + NH_3 (ammonia)

2-aminopropanoic acid + 1/2 O_2 $\xrightarrow[\text{Oxidase}]{\text{Amino acid}}$ 2-oxopropanoic acid + NH_3 (ammonia)

The NHa formed in the distal tubular cells enters the tubular fluid, where it combines with H^+ from the tubular cells to form NH_4^+. This NH_4^+ then replaces Na^+ in a sodium salt of the tubular fluid, such as Na^+Cl^-. The Na^+ is reabsorbed by the $H^+ - Na^+$ exchange mechanism and re-enters the plasma as $Na{+}HCO_3^-$. The NH_4^+ is excreted in the urine as NH_4Cl. The diagram of Figure 10.5 shows the mechanisms involved.

That the H^+-Na^+ exchange mechanism is involved in the conservation of $Na^+HCO_3^-$ by NH_3 formation in the kidney is shown by giving acetazoleamide to a severely acidotic animal excreting much acid and NH_4^+ in the urine. The administration of acetazoleamide causes the urine to become alkaline, showing that the H^+-Na^+ exchange mechanism has been inactivated, and the NH_4^+ excretion in the urine drops markedly. In such a condition NHa continues to be produced in the kidney, but very little can pass into the tubular fluid and be retained unless at the same time H^+ is added to the fluid to form NH_4^+ and thus maintain a concentration gradient for continued diffusion of NH_3 from the tubular cells to the fluid. Vice versa, the presence of NH_a in the tubular fluid

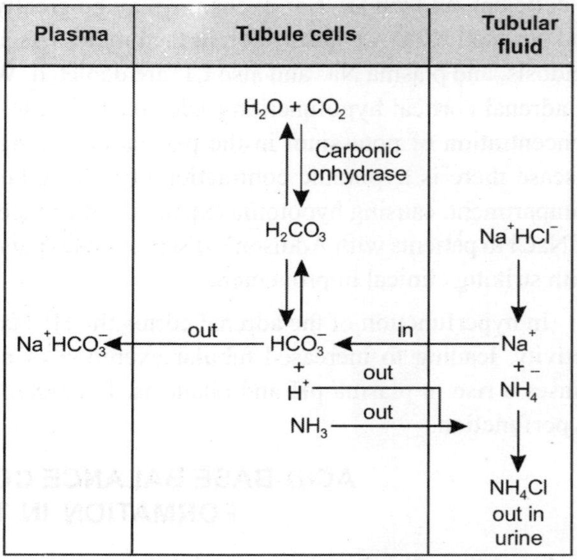

Plasma | Tubule cells | Tubular fluid

$H_2O + CO_2$ ⇅ Carbonic onhydrase ⇅ H_2CO_3 ⇅ HCO_3^- + H^+; NH_3

$Na^+HCO_3^-$ ← out ← HCO_3^- ← in ← Na^+ (from Na^+HCl^-); H^+ out, NH_3 out → NH_4^+ → NH_4Cl out in urine

Fig. 10.5 : The Ion exchange 1

to act as an acceptor for H^+ provides a concentration gradient for H^+ diffusion from the tubular cells into the fluid.

The total protection against acidosis provided by the kidneys is determined by the amount of $Na^+HCO_a^-$ returned to the plasma by the tubules. This bicarbonate equivalence is equal to the equivalents of alkali required to titrate urine to the plasma pH plus the ammonia equivalents in the urine.

Acidosis and dehydration may result from failure of the ammonia-secreting mechanisms of the kidneys, as in the syndromes of Fanconi and of lower nephron nephrosis. From the above discussion it can be seen that the kidneys provide the major control of the electrolyte and acid-base balance of the body, and this is achieved principally through the $H^+.Na^+$ exchange mechanism, NH_3 formation, and the kidney threshold for bicarbonate. While the chief cation concerned in regulation of electrolyte concentration and acid-base balance is Na^+, in severe acidosis the K^+, Ca^{++}, and Mg^{++} concentrations in plasma are elevated, which leads to excessive excretion in the urine.

The operation of the H^+-Na^+ exchange mechanism and the absorption of water from the more concentrated tubular fluid in the kidneys require the expenditure of energy, which must be provided by metabolism of the tubular cells. A minimum of about 850 calories must be expended by the kidneys in the formation of a liter of normal urine. Conditions which limit the supply of O_2 and food materials to the kidneys, such as shock and impaired heart function, result in decreased energy production and kidney function.

CONTROL OF THE ACID-BASE BALANCE BY BUFFER SYSTEMS

The buffer systems of the various body fluid compartments qualitatively are as follows:

Plasma:

$$\frac{B^+HCO_3^-}{H_2CO_3}, \frac{B^+\ Protein^-}{H\ Protein}, \frac{B_2.HPO_4^-}{B^+H_2PO_{4-}}, \frac{B^+org.Acid^-}{H.Org\ Acid}$$

Erythrocyte:

$$\frac{B^+HCO_3^-}{H_2CO_3}, \frac{B^+\ Hb^-}{H\ Hb}, \frac{B_2.HPO_4^-}{B^+H_2PO_{4-}}, \frac{B^+org.\ Acid^-}{H.Org\ Acid}$$

Plasma:

$$\frac{B^+HCO_3^-}{H_2CO_3}, \frac{B^+\ Protein^-}{H\ Protein}, \frac{B_2.HPO_4^-}{B^+H_2PO_{4-}}, \frac{B^+org.Acid^-}{H.Org\ Acid}$$

The balancing cation B^+ is chiefly K^+.

Interstitial fluid. As in plasma, except less protein buffers.

Intracellular. Very much as in plasma, except the protein buffer system and the buffer systems composed of various organic phosphates predominate. There is very little $BHCO_3/H_2CO_3$ buffer within tissue cells. Also, the balancing cation B^+ is chiefly K^+. From the quantitative viewpoint the important buffers in plasma are the bicarbonate-carbonic acid and proteinate-protein buffers. In the erythrocytes the hemoglobin buffer systems dominate, with much less bicarbonate-carbonic acid buffer. While the quantitative buffer relations within cells are much less well known than they are in extracellular fluids, protein and phosphate buffer systems appear to represent most of the buffer capacity. The approximate quantities of the chief buffer components of plasma and erythrocytes are shown in Table 10.12.

Table 10.12: The Approximate Quantities of Chief Buffer Components in Plasma and Erythrocytes

	$B^+HCO_3^-$	H_2CO_3	Protein*
Plasma	25	1.25	13
Erythrocytes	11	1.0	55

*The B^+ protein–/H. protein buffer systems are complex, and only the total protein-milliequivalents can be given.

The pH values of different body compartments vary. On the average they are somewhat as follows: intracellular fluid 7 or a bit below, erythrocytes 7.25, plasma 7.4, interstitial fluid 7.4, cerebrospinal fluid 7.4.

All the values vary within certain ranges under different physiologic and pathologic conditions. Since tissue membranes generally are freely permeable to OH^- or H^+ ions or both, the pH values of different fluids are proportional if not equal. Disturbances in the pH of one fluid thus are reflected in pH disturbances in other fluids, and the buffering capacities of plasma, eryth-rocytes, interstitial fluid, and intracellular fluid are mutually supportive.

Since blood represents a large mass of rapidly circulating fluid carrying a very efficient system of buffers, it is ideally suited to help regulate the pH of interstitial and intracellular fluids and to prevent extreme local variations in pH. Also, since the condition of the blood buffers is indicative of the situation relative to the buffer systems of the body in general, determination of the condition of blood buffers is used clinically to assay the general state of body buffers.

1. *Buffer action against CO_2:* Since very great quantities of CO_2 (H_2CO_3) are being continually formed by the tissues, its buffering is of much importance and is effected chiefly by the extracellular fluids, particularly by the hemoglobin buffer systems of the erythrocytes.

In plasma the following reversible reactions occur:

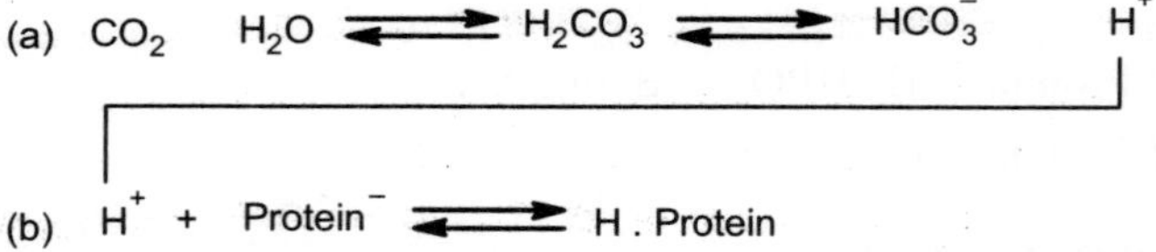

Reaction *(a),* uncatalyzed by carbonic anhydrase, is slow, and the amount of protein- to react with H^+ is relatively small, with the result that only a small amount of CO_2 entering the blood is converted to HCO_3- in the plasma. Interstitial fluid has less capacity (less protein-) to buffer CO_2 than does plasma.

Most of the buffering action against CO_2 is provided by the hemoglobin buffer systems of the erythrocytes, where the following reactions take place:

(c) $CO_2 + H_2O \underset{\text{anhydrase}}{\overset{\text{carbonic}}{\rightleftharpoons}} H_2CO_3 \rightleftharpoons HCO_3^- + H^+$

(d) $H^+ + Hb^- \rightleftharpoons HHb$

(e) $H^+ + HbO_2^- \rightleftharpoons HHbO_2$

Reaction *(c)* is very fast due to the presence of the enzyme carbonic anhydrase, and there is a very high concentration of hemoglobin-ions (Hb^- and HbO_2^-) to combine with the H^+ ions from H_2CO_3 and form HCO_3^-. Consequently, the buffering capacity of the blood against CO_2 is, for practical purposes, determined by the amount of hemoglobin in blood. In reactions *(d)* and *(e)* Hb^- is more effective than HbO_2^- in buffering against increased CO_2, because HHb dissociates less readily than does $HHbO_2$, causing reaction *(d)* to shift further to the right than reaction *(e)*. Conversely, reaction *(e)* is more effective than reaction *(d)* in buffering against decreased CO_2.

As pCO_2 increases, all of the above reactions shift to the right with the formation of more HCO_3^- as Hb- and HbO_2^- combine with the H^+ from H_2CO_3. The ratios of the buffer systems change:

$$\frac{HCO_3^-\text{ increse}}{H_2CO_3\text{ increase a bit more}} \quad \frac{\text{Hb decrease}}{\text{HHb increase}} \quad \frac{HbO_2\text{ decrease}}{HHbO_2\text{ increase}}$$

As pCO_2 decreases, all of the above reactions shift to the left, and the hemoglobin acids donate H^+ ions to HCO_3^- to form H_2CO_3, which in turn produces H_2O and CO_2. This is associated with changes in the ratios of the buffer systems opposite to those taking place when pCO_2 increases, with a proportionate small rise in pH (few hundredths of a pH unit). Thus, because of the operations of and large capacity of the hemo~n buffer systems, much CO_2 can be taken up by blood as HCO_3^-, or can be given up by blood through decomposition of HCO_3^-, with very little change in pH. In fact, the buffer capacity of the hemoglobin buffers is so great that when the pCO_2 of blood is reduced to zero, the hemoglobin buffer acids decompose all of the HCO_3^- and all of the CO_2 is removed from blood. If the pCO_2 of plasma is reduced to zero, a part of the HCO_3^- is decomposed by the protein acid (reactions (*a*) and (*b*), and the remaining HCO_3^- is converted to $CO_3^=$ with loss of half of its CO_2:

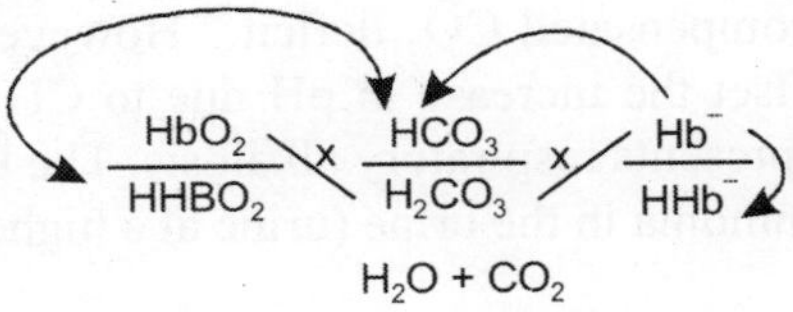

$$\frac{HbO_2}{HHBO_2} \times \frac{HCO_3}{H_2CO_3} \times \frac{Hb^-}{HHb^-}$$

$H_2O + CO_2$

Increasing of pCO_2

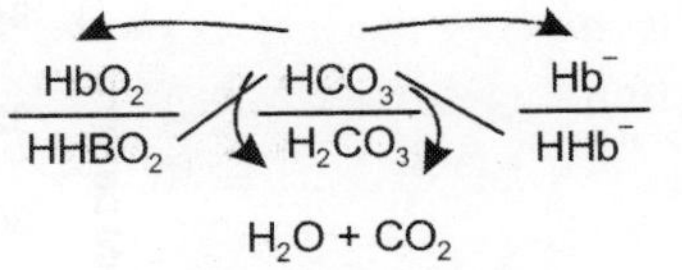

$$\frac{HbO_2}{HHBO_2} \quad \frac{HCO_3}{H_2CO_3} \quad \frac{Hb^-}{HHb^-}$$

$H_2O + CO_2$

decreasing of pCO_2

$$2B^+HCO_3^- \rightarrow B_2^+CO_3^- + H_2O + CO_2$$

The diagram in the right shows the relations of the buffer systems of the eryth-rocytes to changing CO_2 tensions:

As HCO_3^- ions are formed in the cells, many of them diffuse into the plasma and Cl^- ions diffuse from the plasma into the cells (chloride-bicarbonate shift). Thus, while the HCO_3^- is formed from the hemoglobin buffer systems in the cells, it is distributed between cells and plasma. While plasma buffers form little HCO_3, the supply of HCO_3^- in the plasma is maintained indirectly by the hemoglobin buffers.

The effect of varying pCO_2 upon the erythrocyte reactions:

$$CO_2 + H_2O \leftrightarrows H_2CO_3 + Hb^- \leftrightarrows HCO_3^- + HHb$$

$$CO_2 + H_2O \leftrightarrows H_2CO_3 + HbO_2^- \leftrightarrows + HCO_3^- + HHbO_2$$

is of much importance. In conditions in which respiratory exchange is deficient, such as lobar pneumonia, respiratory paralysis, and morphine narcosis, CO_2 continues to be poured into the blood by tissue metabolism, but lung ventilation is below normal. This means that the CO_2 and H_2CO_3 contents of tissue fluids and blood rise above normal. In the B blood the chief compensatory mechanism is the mass action shift of the above reactions to the right, causing an increase in HCO_3 so that the total HCO_3 contents of the cells and the plasma increase. Under these conditions both HCO_3 and H_2CO_3 are above normal, and we have a condition which is called "CO_2 excess." If the pH of the blood remains in the normal range, the condition is compensated CO_2 excess; if the pH is below normal, it is uncompensated CO_2 excess. At times a condition of severe respiratory CO_2 acidosis may develop. The kidneys help compensate the condition by excreting more acid and ammonia in the urine.

Under certain conditions the respiration may be increased above normal with the blowing off of abnormally large quantities of CO_2. This occurs at high altitudes as a result of the stimulating effect of hypoxia upon the respiratory mechanism, at times in cases of hysteria, and may be observed in high fevers induced by infections or prolonged hot baths. As excessive amounts of CO_2 are removed from the blood and pCO_2 falls below normal, the above reactions shift abnormally far to the left with the hæmoglobin buffer acids reacting with HCO_3 to form H_2CO_3 which is eliminated as CO_2. Under these conditions of low pCO_2 in blood (CO_2 deficit) the HCO_3 is decreased. If the HCO_3 and H_2CO_3 are decreased in proportion so that the pH remains within the normal range, the condition is referred to as "compensated CO_2 deficit." However, if the condition is severe and the decrease in HCO_3 does not offset the increase in pH due to CO_2 loss, we have a condition of uncompensated CO_2 deficit; this represents respiratory alkalosis. The kidneys help compensate the condition by excreting less acid and ammonia in the urine (urine at a higher pH).

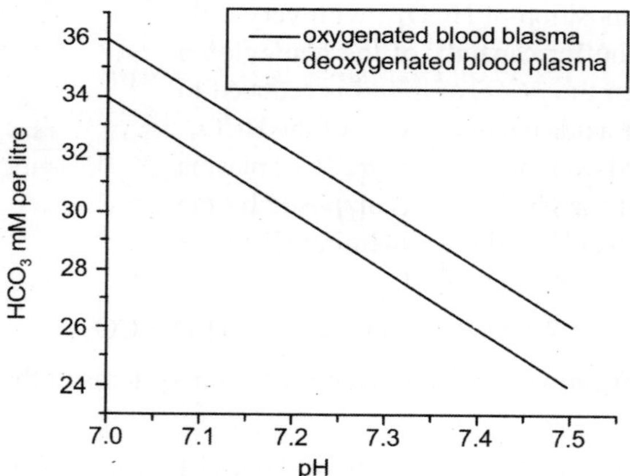

Fig. 10.6: The CO_2 Titration Curves of the Blood; Changing pH by Changing pCO_2 (1) true plasma oxygenated blood and (2) true plasma of deoxygenated blood. True plasma is that plasma separated after whole blood has been equibrated with CO_2 at 10, 20, 40 60, and 80 mm

A condition of CO_2 excess and respiratory acidosis may be readily induced by breathing air or O_2 containing CO_2 (5 to 7 per cent). Such a condition is difficult to cause by holding the breath because of the stimulating effect of CO_2 upon the respiratory center. However, severe CO_2 deficit and respiratory alkalosis may result from prolonged forced breathing.

Since the components of the hæmoglobin and oxyhæmoglobin buffer systems interact with the components of the bicarbonate-carbonic acid buffer system as shown above, when other variables are constant, the total CO_2 and HCO_3 of whole blood vary inversely as the hemoglobin content.

It should be kept in mind that deoxygenated blood is much more efficient than oxygenated in buffering against CO_2, because Hb^- combines with H^+ more firmly than does HbO_2, and consequently Hb^- reacts with H_2CO_3 to form more HCO_3 than does HbO_2 at a given physiologic pH.

Erythrocytes per unit volume normally contain about 60 per cent as much HCO_3 as plasma does. In polycythemic blood, in which the cell volume may be very large compared to plasma volume, the whole blood HCO_3 and CO_2 are low even though the concentrations in cells and plasma are normal. Conversely, HCO_3 and total CO_2 may be above normal in anemic blood simply because of the relatively large plasma volume.

Carbon dioxide titration curves of blood If blood is equilibrated with increasing CO_2 tensions and the total CO_2 and pH of the true plasma determined, the quantities of HCO_3 present for each pH may be calculated by procedures outlined previously. This constitutes a titration of blood with CO_2 and indicates the capacity of the blood to form HCO_3 with variations in pCO_2 and pH. As is to be expected, true plasma of deoxygenated blood forms more HCO_3 per increment of pH and pCO_2 change than does the true plasma of oxygenated blood. Henderson (24) carried out such experiments on oxygenated and deoxygenated normal blood. Table 10.13 gives his values for pCO_2 pH, and total CO_2 in the true plasmas of oxygenated and deoxygenated blood. The values for [HCO_3] per liter were calculated from his data.

Table 10.13: The Variations of Total CO_2 and HCO_3 in True Plasma of Blood with Variations in pCO_2 and pH

pCO_2	Oxygenated			Deoxygenated		
	pH	Total CO_2 mM/1	CO_3^- mM/1	pH	Total CO_2 mM/1	CO_3^- mM/1
10	7.819	15.5	15.2	7.847	16.6	16.3
20	7.621	20.1	19.5	7.652	21.6	21.0
40	7.413	25.4	24.2	7.445	27.2	26.0
60	7.287	28.8	27.0	7.318	30.8	29.0
80	7.192	31.5	29.1	7.224	33.6	31.2

Using data from the table, we may plot millimols of HCO_3 per litre against pH for the true plasma of oxygenated and deoxygenated blood. Such a plot is shown in Figure 10.6. It will be noted that straight lines are obtained. These represent the normal buffer lines against CO_2.

Buffer Action against Nonvolatile Acids

When nonvolatile acids enter the blood, they react with the anions of the buffer systems and are neutralized. However, it is the HCO_3 of the HCO_3/H_2CO_3 buffer system which is primarily concerned with such neutralization. This is true for two reasons. In the first place, the concentration of HCO_3 is highest of any of the buffer anions. In the second place, HCO_3 is peculiarly efficient in neutralizing acids because in the process H_2CO_3 is formed which breaks up into CO_2, and this being exhaled makes the reaction irreversible:

$$HCO_3 + HA \leftrightarrows A^{-} + H_2CO_3 \rightarrow CO_2 \text{ (exhaled)}$$

A reaction such' as the following would be much less efficient in neutralizing HA:

$$HPO_4^{-} + HA \leftrightarrows A^{-} + H_2PO_4$$

Here we have only exchanged a stronger acid, HA, for a mixture of it and a weaker acid, H_2PO_4. In case of neutralization with HCO_3 the H_2CO_3 is removed, leaving a relatively neutral salt and a decreased HCO_3 concentration. The efficiency of HCO_3 in neutralizing nonvolatile acids is such that as much as 80 per cent of the blood bicarbonate may be used up before symptoms of extreme acidosis appear (increased respiration or hyperpnea, and acidotic coma if sufficiently severe). Researchers estimate that when the pH of blood falls from 7.4 to 7.0, as may happen in severe diabetic acidosis, about 18 mM of acid may be neutralized by HCO_3, 8 mM by Hb^{-} and HbO_2, and 2 mM by the anions of other buffer systems. Plasma protein anions neutralize only about 1.7 mM of acid under these conditions. Because blood HCO_3 is readily available for the neutralization of acids and quantitatively is by far the most important, the blood HCO_3 is referred to as the "alkali reserve". Since the quantity of HCO_3 present in blood reflects the condition of the blood buffers in general, and, since it may be determined readily, estimation of blood HCO_3 or alkali reserve is a routine clinical procedure. Excessive amounts of HCO_3 in the body are buffered by the hemoglobin and other buffer systems to some extent. However, as the pH of the blood rises, the respiration is depressed, and this permits the accumulation of excessive amounts of CO_2 (H_2CO_3), which tends to bring the ratio HCO_3/H_2CO_3 and the pH back to normal. At the same time the kidneys excrete more alkaline urine. When kidney excretion has lowered HCO_3 to normal, the lungs have blown off the excess CO_2 and normal conditions are re-established. Alkalies introduced into the blood react quickly with H_2CO_3 to form HCO_3^{-}:

In previous discussion it has been pointed out that respiratory acidosis and alkalosis may be produced by CO_2 retention (CO_2 excess) and excessive CO_2 excretion (CO_2 deficit), respectively. The so-called metabolic types of acidosis and alkalosis are preferable to decreases or increases in total buffer anions, chiefly HCO_3, not related to respiratory changes in blood CO_2. Examples of metabolic acidosis are observed in diabetic and nephritic acidosis when the quantities of nonvolatile acids to be neutralized react with abnormal quantities of buffer anions, particularly HCO_3. A similar condition may be caused by the ingestion or injection of HCl, NH_4Cl, or other acids. The ingestion of alkaline earth chlorides, such as $CaCl_2$ and $MgCl_2$, produces an effect about equal to that produced by an equivalent amount of HCl. This is due to the fact that very little of the metal ion is absorbed, while most or all of the Cl^{-} passes into the blood. While the exact mechanism is unknown, the effect is equivalent to membrane hydrolysis of the"metal chloride with absorption of HCl, which, of course, is quickly neutralized by blood buffers.

In cases of metabolic acidosis the chief effects noted are a decrease in the alkali reserve HCO_3 and total CO_2 of the blood and lowered blood pH. The lowered pH stimulates respiration to blow off CO_2 in an attempt to bring the ratio HCO_3/H_2CO_3 and the pH to normal. Also, the kidneys excrete a more acid urine and more ammonıa to conserve HCO_3. If the condition is mild and the pH is within normal range, it is'referred to as "compensated metabolic acidosis" ("compensated base or alkali deficit"). If the pH is below the normal range, the condition is "uncompensated metabolic acidosis" ("uncompensated base or alkali deficit").

Metabolic alkalosis may be caused by the ingestion or injection of Na HCO_3 or other basic substances, or of substances, such as sodium lactate and sodium citrate, which give $NaHCO_3$ when metabolized. As the pH rises, the respiration is depressed to retain CO_2 and lower the pH, and the kidneys excrete urine at a higher pH to remove excess HCO_3 from the blood. The blood HCO_3, H_2CO_3, and total CO_2 are high. If the condition is mild and the pH within normal ranges due to a normal HCO_3/H_2CO_3 ratio, the con-dition is "compensated metabolic alkalosis" ("compensated base or alkali excess"). If the pH is, above normal, the condition is "uncompensated metabolic alkalosis."

THE ROLE OF TISSUE BUFFERS IN THE REGULATION OF ACID-BASE BALANCE OF EXTRACELLULAR FLUIDS

Researchers have estimated the total buffer capacity of the adult human body available for neutralization of acid before a fatal pH is reached as about six times that of blood, equivalent to the neutralization of about 900 ml of 1 *N* acid (9000 meq). This means that the tissue buffers (including bone) must be capable of neutralizing about 750 ml of this 1 *N* acid.

The tissue cells aid in regulating the acid-base balance of the body through exchanging H^+, K^+, and Na^+ between the extracellular and intracellular fluids.

Figures 10.7 and 10.7A show the exchanges taking place between extra-cellular and intracellular fluids in conditions of metabolic alkalosis and acidosis.

In metabolic alkalosis H^+ and K^+ shift from intracellular to extracellular fluid in exchange for Na^+, which enters the cells. The H^+ reacts with extra-cellular HCO_3^- to form H_2CO_3. The kidneys excrete K^+ and Na^+ with HCO_3^-, and respiration is decreased to retain CO_2 (H_2CO_3). Each of these processes helps compensate the alkalosis.

In metabolic acidosis the ionic shifts between extracellular and intracellular fluids are reversed. The movement of H^+ (and K^+) from extracellular fluid into the cells in exchange for Na^+ leaves HCO_3- in the extracellular fluid:

$$H_2CO_3 \leftrightarrows HCO_3^- + H^+$$

Into cells

to help raise the pH. The H^+ entering the cells is taken up by the intracellular buffer systems. The kidneys increase excretion of H^+ and decrease excretion of Na^+, K^+, and HCO_3. Respiration is stimulated to increase excretion of CO_2. Each of these processes helps compensate the acidosis.

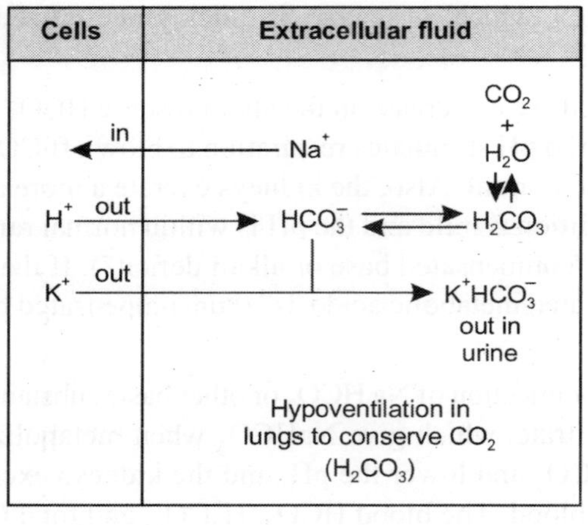

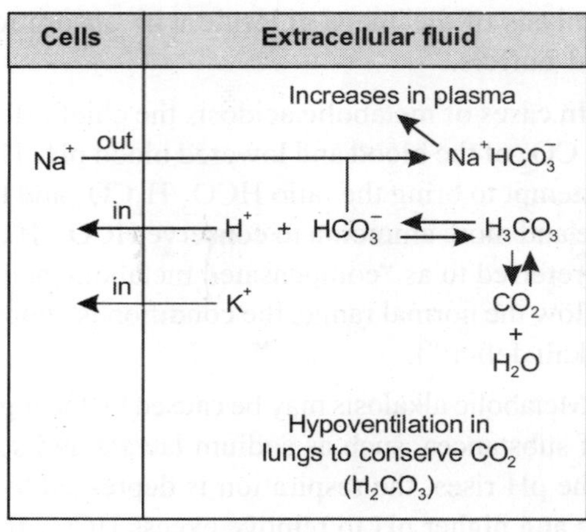

Fig. 10.7 and 10.7A. The Acidisis and Alaklosis in the E.C.F.

In respiratory alkalosis due to hyperventilation H+ and K+ pass from intracellular to extracellular fluid in exchange for Na+, which enters the cells.

The H^+ reacts with H HCO_3 to form H_2CO_3 to compensate the alkalosis. The kidneys excrete less H^+ and more HCO_3, along with K^+ and Na^+.

In respiratory acidosis the ionic shifts between extracellular and intracellular compartments are reversed to those in respiratory alkalosis, with H^+ and K^+ entering the cells and Na^+ passing from the cells into extracellular fluid. The shift of H^+ into the cells leaves HCO_3^- in the extracellular fluid to raise the pH:

$$H_2CO_3 \leftrightharpoons HCO_3^- + \underset{\text{Into cells}}{H^+}$$

and the H^+ is taken up by intracellular buffers. The kidneys compensate by increased excretion of H^+ and of NH_4^+ with anions such as Cl^-, and by decreased excretion of Na^+ and K^+ (NH_4^+) is substituted for these).

In prolonged severe metabolic acidosis the $Ca_3(PO_4)_2$ of bone may be drawn upon to neutralize acid:

$$Ca_3(PO_4)_2 + 4HA \rightarrow 3Ca^{++} + 2H_2PO_4- + 4A^-$$

The Ca^{++}, $H_2PO_4^-$, and A^- ions are excreted in the urine. This process may result in serious demineralization of the skeleton.

Effect of Digestion upon the Acid-base Balance. The Alkaline Tide

It has been found that shortly after a meal the alveolar CO_2 is increased, the urine pH rises, the plasma bicarbonate rises, and the chloride falls. This so-called alkaline tide after a meal is due to formation of gastric HCl from Cl^- ions and H^+ ions in plasma, leaving HCO_3 and OH^- ions to increase pH. The condition is very mild and disappears after the gastric HCl is neutralized in the intestine to form NaCl, which is absorbed into the blood and re-establishes the *status quo*. The alkaline tide is most pronounced after meals which stimulate much gastric secretion, such as heavy protein meals. It may be greatly diminished or disappear in conditions of achlorhydria, in which little gastric HCl is secreted. This may be true after meals consisting chiefly of fat and carbohydrate.

The bicarbonate~carbonic acid buffer system of blood as an index of the acid-base balance of the body. Since the various changes in the acid-base balance of the body are reflected in the buffer systems of the blood, and since the state of the blood buffers is mirrored in the bicarbonate-carbonic acid buffer system, and finally because the bicarbonate buffer system is easily assayed in the laboratory, practically all estimations of body acid-base balance are based upon analysis of the bicarbonate buffer system. In practice it is customary to use plasma for such analyses, since its buffer composition reflects the condition of the blood buffers, and because it represents a more constant material than whole blood where the volume of cells relative to the volume of plasma may vary considerably. Heparin is used as anticoagulant instead of oxalate or other salts which may affect the buffer systems of the blood.

The variables relating to a bicarbonate-carbonic acid buffer system are pCO_2 total CO_2, HCO_3, H_2CO_3, and pH. If two of these are known, the others may be calculated.

Total CO_2 may be obtained readily by acidification in the Van Slyke gas apparatus, and pH may be determined with the glass electrode. In order that results obtained from determinations of total CO_2 and pH may be compared and properly interpreted, it is necessary to use oxygenated (arterial) blood, since the total CO_2, HCO_3, H_2CO_3, and pH all change from arterial to venous blood. The equivalent of arterial blood may be obtained by soaking the arm in hot water (46°–47°C) for ten minutes or longer and withdrawing the sample from the antecubital vein or a vein on the back of the hand. Special equipment and techniques are required in drawing the samples and separating and analyzing the plasma.

Blood may be equilibrated at 380°C and at a definite pCO_2, generally 40 mm in the presence of excess O_2, and the total CO_2 determined on plasma separated from the equilibrated blood (so-called true plasma). Under these conditions the plasma total CO_2 essentially represents the alkali reserve of arterial blood at the normal pCO_2 of 40 mm. If the pCO_2 of the arterial blood is higher than 40 mm in the subject, the values at 40 mm CO2 are too low, and if the pCO_2 of the subject's blood is lower than 40 mm, the values ob-tained are too high. However, the results at 40 mm indicate the capacity of the blood to take up CO_2 under normal conditions. Since nearly all the CO_2 taken up represents HCO_3, the CO_2 capacity at 40 mm pCO2 is defined specifically as the alkali reserve.

Since the $[H_2CO_3]$ varies directly as pCO_2, the total quantity of H_2CO_3 at any pCO_2 is readily calculated ($0.03 \times pCO_2$), and under all physiologic and pathologic conditions is small relative to $[HCO_3]$. When total CO_2 is high, HCO_3 is high and vice versa. When HCO_3 or total CO_2 is high and pH is below normal, pCO_2 is high. If pC02 is normal and pH above normal, HCO_3 and total CO_2 are high. When the ratio $HCO_3/\ H_2CO_3$ is high (above 20:1), the pH is above normal; and when it is low, the pH is below normal.

The true plasma' of oxygenated normal blood may be considered to be characterized by the following relations:

pH	pCO_2mmHg	[total CO_2]		[HCO_3]		[H_2CO_3]		[HCO_3] / [H_2CO_3]
		vols%	mM/L	vols%	mM/L	vols%	mM/L	
7.4	40	56	25.2	53.4	24	2.57	1.2	20

[HCO_3^-] may be calculated. Suppose the total plasma CO_2 is 15mM per liter and the plasma pH is 7.2. Then from the Henderson-Hasselbalch equation:

$$7.2 = 6.1 + \log\frac{[HCO_3^-]}{[H_2CO_3]}$$

and

$$\log\frac{[HCO_3^-]}{[H_2CO_3]} = 7.2 - 6.1 = 1.1 \text{ and } \frac{[HCO_3^-]}{[H_2CO_3]} = \frac{12.6}{1}$$

which means that the total CO_2 present as HCO_3^- is

$$\frac{12.6}{13.6}\times 100 = 92.6\% = 15\times 0.926 = \text{ mM/litre}.$$

REPRESENTATION OF THE ACID-BASE BALANCE OF BLOOD ON CHARTS

As has been explained previously, if either pCO_2 and total CO_2, or pH and total CO_2 are determined on blood plasma, then all the other variables of the bicarbonate-carbonic acid buffer system may be calculated. In order to visualize better the whole picture of the buffer system, it is customary to prepare charts upon which a number of these variables are shown and from which the others may be calculated. One of the best systems is based upon plotting pH against millimols of HCO_3^- per liter in true plasma of oxygenated blood. Another system is to plot pCO_2 against total CO_2.

The nature of the HCO_3/ H_2CO_3 buffer system of any blood may be quickly seen by plotting the pH and [HCO_3] of its true plasma separated from an oxygenated sample.

Table 10.15 shows the values calculated for pCO_2, [H_2CO_3], and [HCO_3] when the pH and [total CO_2] values indicated are found by analysis of the true plasma of oxygenated blood.

The points A, B, C, etc., corresponding to the pH and HCO_3^- values of Table 10.13 are plotted on the chart. It will be noted that B represents un-compensated CO_2 excess or respiratory acidosis, D and E represent uncompensated base excess or metabolic alkalosis, G represents uncompensated CO_2 deficit or respiratory alkalosis, and K and I represent uncompensated base deficit or metabolic acidosis. Point C may represent either compensated CO_2 excess or base excess-that is, either compensated respiratory acidosis or metabolic alkalosis-while H may represent either compensated CO_2 defi-cit or base deficit-that is, either compensated respiratory alkalosis or metabolic acidosis. Point A, lying on the normal buffer line, represents uncom-pensated CO_2 excess or respiratory acidosis; while point F, also lying on the buffer line, represents uncompensated CO_2 deficit or respiratory alkalosis.

Table 10.14 : The Relation of pH, pCO_2 [total CO_2][H_2CO_3] and Ration[HCO_3/ H_2CO_3] in plasma at 38°C Values given in milimoles-multiplied by 2.226 = volumes per cent

pCO_2→	10 mm			20 mm			30 mm			
pH	[Total CO_2]* mM	[H_2CO_3]@ mM	[HCO_3^-]# mM	[Total CO_2]$ mM	[H_2CO_3] mM	[HCO_3^-] mM	[Total CO_2] mM	[H_2CO_3] mM	[HCO_3^-] mM	[HCO_3^-]% =R / [H_2CO_3] =R
7.0	2.7	0.3	2.4	5.4	0.6	4.8	8.1	0.9	7.2	8.0
7.2	4.1	0.3	3.8	8.2	0.6	7.6	12.2	0.9	11.3	12.6
7.4	6.3	0.3	6.0	12.6	0.6	12.0	18.9	0.9	18.0	20.0
7.6	9.8	0.3	9.5	19.6	0.6	19.0	29.4	0.9	28.5	31.7
7.8	15.4	0.3	15.1	30.8	0.6	30.2	46.2	0.9	45.3	50.2
8.0	24.1	0.3	23.8	48.2	0.6	47.6	72.3	0.9	71.4	79.5
pCO_2→	40 mm			50 mm			60 mm			
7.0	10.8	1.2	9.6	13.5	1.5	12.0	16.2	1.8	14.4	8.0
7.2	16.3	1.2	15.1	20.4	1.5	18.9	24.5	1.8	22.7	12.6
7.4	25.2	1.2	24.0	31.5	1.5	30.0	37.8	1.8	36.0	20.0
7.6	39.2	1.2	38.0	49.0	1.5	47.5	58.9	1.8	57.1	31.7
7.8	61.6	1.2	60.4	77.0	1.5	75.5	92.4	1.8	90.6	50.2
8.0	96.4	1.2	95.2	120.5	1.5	119.0	144.6	1.8	142.8	79.5

* $\log\dfrac{[\text{total } CO_2]-(0.03\times pCO_2)}{(0.03\times pCO_2)} = pH - pK_1 = pH - 6.1$, or $[\text{total } CO_2] = (R+1)\times 0.03\times pCO_2$.

@ $[H_2CO_3] = 0.03\times pCO_2$; $[H_2CO_3]$ in mM × 2.226 = volumes per cent CO_2 as H_2CO_3.

\# $[HCO_3] = [\text{total } CO_2] - [H_2CO_3]$.

$ $pH = pK_1 + \log\dfrac{[HCO_3]}{[H_2CO_3]}$ and $\log\dfrac{[HCO_3]}{[H_2CO_3]} - pH - pK_1 = pH - 6.1$.

The analysis of plasma buffer systems as outlined above obviously is too complicated and time-consuming to be used routinely in clinical laboratories. For clinical purposes it is rather generally customary to separate the plasma from venous blood, equilibrate it with CO_2 at 40 mm, and determine the total CO_2 content. This is expressed as the alkali reserve of the plasma. It represents the capacity of the plasma alone to take up CO_2 at normal pCO_2, which is different from the capacity in the presence of cells (true plasma). Such a procedure gives some idea of the quantitative value of the blood buffer systems but does not show the whole picture and cause of the condition. Since the clinical history of the patient generally suggests the cause of plasma alkali reserve abnormalities, this enables the clinician to assess approximately the significance of results obtained in this simplified manner. The CO_2 capacity of plasma at 40 mm pCO_2 (alkali reserve) as commonly determined in clinical laboratories correlates roughly as follows with unspecified conditions of acidosis and alkalosis

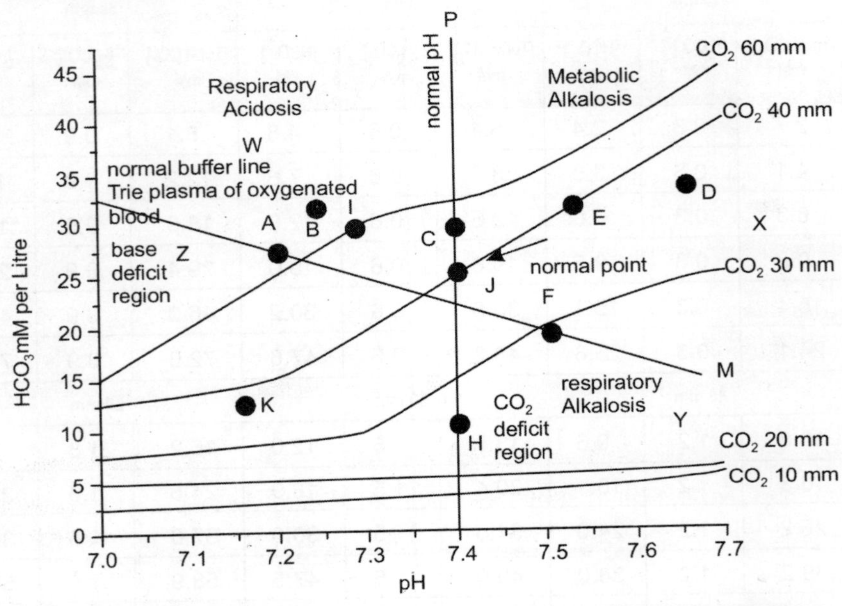

Fig. 10.8 : The Respiration of the acid-base Balance of Blood on Chart.

Table 10.15 : If total CO_2 and pH Values of True Plasma of Oxygenated Blood Shown in Colums 1 and 2 are Obtained by Analysis. The other values are obtained by calculation

	pH	Total CO_2 mM/1	pCO_2 mm	H_2CO_2 mM/1	HCO_2 mM/1	R
A	7.25	28.05	61.5	1.85	26.2	14.2
B	7.30	33.0	65.3	1.96	31.04	15.8
C	7.40	36.0	57.1	1.71	34.29	20.0
D	7.60	34.0	34.7	1.04	32.96	31.7
E	7.50	31.32	40.0	1.2	30.12	25.1
F	7.50	22.46	28.7	0.86	21.60	25.1
G	7.50	17.23	22.0	0.6	16.57	25.1
H	7.40	15.75	25.0	0.75	15.0	20.1
I	7.20	24.48	60.0	1.80	22.68	12.6
J	7.40	25.2	40.0	1.20	24.0	20.0
K	7.25	15.0	32.9	0.99	14.01	14.2

	Plama CO_2 Capacity at 40 mm pCO_2	
	Volumes%	mM/1
Normal extremes	53-78	23.8-35
Alkalosis	> 78	> 35
Mild acidosis (no apparent symptoms)	40-53	17.9-23.8
Moderate acidosis (symptoms)	30-40	13.5-17.9
Severe aciodis (severe symptoms)	< 30	< 13.5
Terminal acidosis (diabetic), H <7.1	< 15	< 6.7

Extreme Acidosis	Extreme Alkalosis
1. pH around 4.5	1. pH around 7.8
2. No HCO_3	2. Much HCO_2
3. Increased chloride	3. Decreased chloride
4. Increased acid, up to 150 ml of 1N	4. Urine becomes alkaline, alkali titratable
5. Increased NH3, up to 600 ml of 1 N	5. No ammonia
6. Decreased cations (Na^+, K^+, etc.)	6. Increased cations
7. Large volume (diuresis)	7. Some diuresis

Table 10.16: The Variation of Plasma Electrolytes in Several Pathological States. The Cations Variations Chiefly Involve Na^+

Condition of Subject		Anions			
	Cations, Total	HCO_3^-	Cl^-	Keto-acids	Other anions
1	2	3	4	5	6
Normal	155	27	103		25
Diabetic ketosis (acidosis)	142	5	80	24	33
Fasting ketosis (acidosis)	155	15	101	13	26
Diarrhea	148	11	100		37
Addison's disease	126	22	73		31
NH_4Cl acidosis	155	12	117		26
Nephrosis	150	20	113		17
Chronic nephritis	157	21	101		35
Chronic nephritis (terminal)	146	6	87		53
Pyloric obstruction (vomiting)	146	59	42		45
Duodenal obstruction (vomiting)	150	43	48	32	27
Habitual vomiting	155	33	96		27

POTASSIUM METABOLISMS

Maintenance of the proper K^+ concentration of extracellular fluids is essential, particularly for proper function of the heart. High concentrations of K^+ cause widespread intra-cardiac block, while low concentrations impair the contractility of heart muscle. Maintenance of proper intracellular K^+ concentrations is essential for various enzymatic reactions, for the normal contractility of muscle, and for impulse transmission and other functions of the nervous system. The widespread distribution of potassium in foods normally precludes the development of potassium deficiency. Experimental potassium deficiency in rats causes diminished growth, kidney hypertrophy, necrosis of heart muscle, loss of hair, and finally death.

EXCRETION OF POTASSIUM BY THE KIDNEYS.

Control of the K^+ balance of the body by the kidneys is not as efficient as the control of Na^+ balance. Whereas an animal on a Na^+-deficient diet may show practically no Na^+ excretion in the urine, there is always an obligatory excretion of potassium, regardless of the potassium intake which continues even during prolonged starvation, the K^+ arising from cellular breakdown. This obligatory excretion (30-60 meq per day) is apparently related to the $K^+H^+–Na^+$ exchange in the distal tubules, where the reabsorption of Na^+ is dependent upon exchange for H^+ and K^+. However, it has been found that in both animals and humans deprived of potassium but maintained without stress to minimize adrenal cortical activity K^+ is conserved by the kidneys.

The mechanisms by which the kidneys excrete K^+ differ from those for the excretion of Na^+. It appears that normally most or all of the K^+ of glomerular filtrate is reabsorbed along with anions such as Cl^- in the proximal tubules, and the K^+ which appears in the urine is secreted by the distal tubular cells in the $H^+K^+–Na^+$ exchange mechanism. The operation of the $H^+K^+–Na^+$ exchange mechanism in the distal tubules is shown in Figure 361. Anions other than $HPO_4^=$ may serve as acceptors of K^+ as well as of H^+.

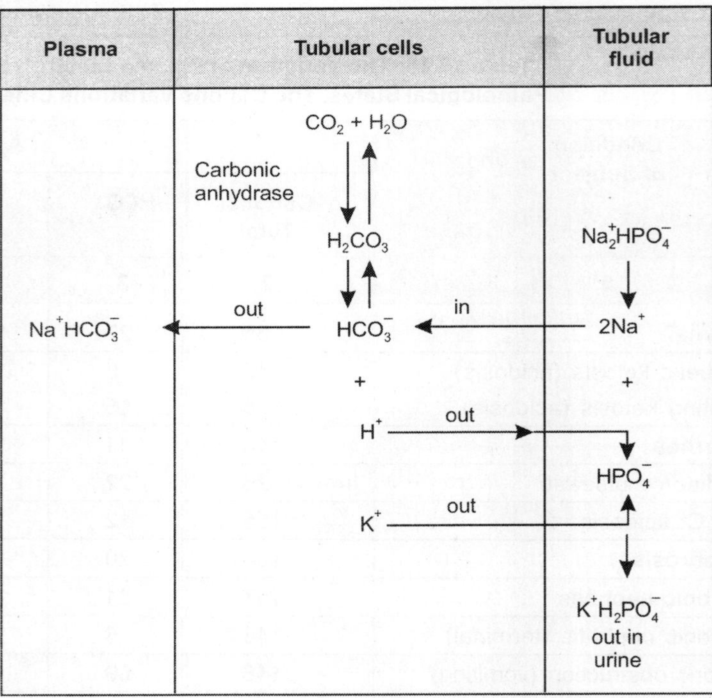

Fig. 10.9 : The Operation of the H^+K^+-Na^+ Exchange Mechanism in the Distal Tubules

Since K^+ and H^+ are competitive with each other in the H^+K^+–Na^+ exchange mechanism, conditions which increase H^+ excretion decrease K^+ excretion and vice versa. However, H^+ has much greater affinity for the exchange mechanism than does K^+.

When extracellular K^+ is elevated (also tubular cell K^+) by administration of potassium salts, tubular H^+ excretion is decreased, K^+ and HCO_3 excretions are increased, and acidosis results. Conversely, when extracellular K^+ is depleted, H^+ excretion is increased, and K^+ and HCO_3^- excretions are decreased, causing alkalosis.

In respiratory alkalosis, with decreased pCO_2 and H_2CO_3 in the tubular cells, H^+ excretion is decreased and K^+ and HCO_3^- excretions are increased. In respiratory acidosis, with increased pCO_2 and H_2CO_3 in tubular cells, H^+ excretion is increased and the excretions of K^+ and HCO_3^- are decreased. The drug acetazoleamide (Diamox) and maleate and Li^+, which inhibit the carbonic anhydrase enzyme, decrease H^+ excretion and increase Na^+, K^+, and HCO_3^- excretion. The operation of the H^+K^+–Na^+ exchange is under the control of adrenal corticosteroid hormones (aldosterone, deoxycorticosterone). Administration or increased production of these steroids (in hyperadrenocorticism), increases the exchange, with the retention of Na^+ and excretion of increased amounts of H^+, K^+, and NH_4^+ in the urine. An alkalosis and K^+ depletion result. In a deficiency of adrenal corticosteroids, as seen in Addison's disease, the exchange is decreased, with decreased urinary excretion of both H^+ and K^+, and increased Na^+ and HCO_3^- excretion, leading to an acidosis and K^+ elevation in extracellular and intracellular fluids. Distribution of potassium in the body. The potassium intake of the adult generally is between 4 and 8 g (100 to 200 meq) per day. Total body potassium of the normal male adult determined by K 42 dilution appears to be about 47 milliequivalents per kilogram of body weight. In females the average value found was about 41 meq per kilogram. Researchers estimate the following amounts of K^+ to be present in the major depots of the body of a healthy 60 kg adult.

Muscle (intracellular)	2600 meq
Erythrocytes (intracellular)	200 meq
Liver (intracellular)	160 meq
Extracellular fluid	54 meq

About 100 meq of K^+ are secreted in the gastrointestinal fluids and 100 meq excreted in the urine per day.

While most of the K^+ excretion is through the kidneys, more K^+ than Na^+ is excreted in the feces.

The unequal distribution of K^+ between extracellular and intracellular fluid exists only in the living organism and entirely disappears when the animal dies.

The maintenance of high concentrations of K^+ in intracellular and low concentrations in extracellular fluid, and of high concentrations of Na^+ in extracellular and low concentrations in intracellular fluid, is the result of active transport of Na^+ from intracellular to extracellular fluid. The concept is that K^+, Na^+, and H^+ pass across the cell membrane rather freely, as occurs when the cell dies. However, the sodium pump mechanism, which requires metabolic energy, in the living cell continually moves Na^+ from the intracellular to the extracellular fluid at a rate which maintains a high $[Na^+]$ in extracellular fluid and low $[Na^+]$ in intracellular fluid. Electrical neutrality in the cell is maintained by the diffusion of sufficient K^+ and H^+ into the cell to balance the deficit of cations caused by the sodium pump mechanism. Most of this balancing is done by K^+, accounting for the high concentration of K^+ in intracellular fluid. Thus, we see that the $K^+H^+–Na^+$ exchange mechanism appears to operate not only in the renal tubules, but probably also in tissues generally.

ALTERATIONS IN THE POTASSIUM CONTENTS OF EXTRACELLULAR AND INTRA-CELLULAR FLUIDS

There are a number of conditions which lead to changes in the K^+ concentrations of extracellular and intracellular fluids and the distributions of K^+ between these fluids. The administration and retention of excessive amounts of water or glucose solution (as in parenteral therapy) dilutes both the extracellular and intracellular fluids, with an attendant drop in $[K^+]$ in both fluids, though the total body K^+ is unchanged. This situation occurs in patients unable to excrete their water loads, as in cases of kidney disease with acute tubular damage. There are a number of processes which lead to increased cellular K^+. Since Na^+, the movement of K^+ from the cells is balanced by the movement of Na^+ and H^+ into the cells.

Severe diarrhea, particularly as seen in infants, causes acidosis because of excessive loss of HCO_3^- over Cl^- and depletion of total electrolytes and dehydration. The K^+ stores are depleted, but because of the dehydration the extracellular $[K^+]$ may be elevated.

The factors causing the exchange of K^+ between extracellular and intracellular fluids are shown in Figure 10.10.

Cause of K^+ shift	Extracellular fluid	Cells
Growth and tissue repair	K^+ →	
Testosterone administration	K^+ →	
Dehydration	←	K^+
Rehydration after dehydration	K^+ →	
Glycogen deposition	K^+ →	
Deglycogenation	←	K^+
Acidosis	K^+ →	
Alkalosis	←	K^+
Periodic paralysis	K^+ →	
Anoxia	←	K^+
Excess adrenal steroids	←	K^+
Deficit of adrenal steroids	K^+ →	
Diarrhea	←	K^+
Vomiting	←	K^+
Carbonic anhydrase inhibitors	←	K^+
Excessive H_2O administration	←	K^+

Fig. 10.10 : Shift of K^+ between Intracellular and Extracellular Compartments

While extracellular [K^+] is increased in a number of clinical conditions, generally excess K^+ is readily disposed of by the kidneys. Persistent increase in extracellular K^+ generally indicates impaired excretion, as seen in oliguricor anuric patients with lower nephron nephrosis, severe diarrhea, and diabetic coma. The most significant consequence of increased extracellular K^+, (hyperkalemia) is its effects upon cardiac function. Increases in cellular K^+ which ordinarily occur appear not to be deleterious.

Serious depletion of total body and intracellular K^+ leads to a number of : disturbances: histologic kidney lesions, diminished growth, impaired glucose 1 metabolism (tolerance), alterations in gastric secretion, impaired neuromus-1 cular function, alterations in the electrocardiogram, cardiac arrest, and death.

PARENTERAL FLUID THERAPY

The clinical correction of disturbances in fluid, electrolyte, and acid-base balances is based upon the application of principles discussed above and constitutes a major problem in medical practice. Generally extracellular fluid and electrolyte deficits are repaired by oral administration of water, sodium chloride, and potassium compounds, or potassium-rich foods when this is possible. Otherwise, isotonic NaCl solution, isotonic NaCl solution plus KCl, or such electrolyte solutions containing glucose (for energy) are given intravenously. Generally, if kidney function is adequate, fluid volume and electrolyte repair may be achieved with isotonic NaCl solution, and by this solution with RCl and glucose if there is potassium deficiency. The metabolism of glucose provides energy to the tissues, and its deposition as glycogen in cells increases the transfer of K^+ from extracellular fluid into the cells.

In the correction of hypertonic fluid contraction with electrolyte excess where there is a water deficit, isotonic (5%) glucose solution may be used to advantage. The provision of water is generally accomplished by giving isotonic glucose, since the glucose is taken up by the cells and metabolized.

In isotonic fluid contraction where there is proportionate loss of water and electrolytes, as in severe vomiting and diarrhea, if kidney function is adequate, the administration of isotonic NaCl and glucose restores extracellular volume. The kidneys excrete the proper ions to establish the correct acid-base and electrolyte balance.

In conditions of moderate acidosis, as found in diarrhea and diabetes, the administration of NaCl solution initially increases the severity of the acidosis by diluting the already deplenished extracellular HCO_3^-. However, it restores fluid volume, overcomes dehydration, and promotes kidney function. If kidney function is adequate, acidic urine is formed, and the excess Cl^- of the Na^+Cl^- in the glomerular filtrate is combined with NH_4^+ in the tubules and excreted in the urine. The Na^+ is reabsorbed with HCO_3^- back into the plasma to compensate the acidosis. The over-all process in the kidneys amounts to the following reaction, in which Na^+Cl^- of plasma (extra-cellular fluid) is converted to $Na^+ HCO_3^-$:

$$Na^+ Cl^- + NH_4^+ + HCO_3^- \rightarrow Na^+ HCO_3^- + NH_4^+ Cl^-$$

When the acidosis and dehydration are severe, treatment usually consists in administering a solution containing NaCl, $NaHCO_3$, and glucose (with insulin in diabetes) to insure more prompt correction of the acidosis. Often in such a situation there is a K^+ deficiency, and KCl is also added to the solution. Sodium lactate may be substituted for $NaHCO_3$, since it is metabolized to yield bicarbonate mol for mol.

Conditions of alkalosis are seldom treated by the administration of acids, though occasionally NH_4Cl solutions are used, the ammonia being converted to neutral urea in the liver with the release of HCl, which reacts with body buffers. The administration of isotonic NaCl solution dilutes the extracellular HCO_3^-, to ameliorate the alkalosis somewhat, and expands the extra-cellular fluid volume, which may be desirable. Often in such cases as the alkalosis of vomiting there is K^+ depletion and this deficit must be corrected by administering potassium before the kidneys can compensate the alkalosis by excreting bicarbonate and retaining Cl^-.

Each patient with fluid and electrolyte disturbances presents an individual problem to the physician, who, to apply the proper corrective measures, must take many facts into consideration.

Introduction

To start writing something about antibiotic we have to long back in the time of Egyptians (1550) used to treat the human wounds which they called *'Katuh'* – the cuts by a mixture of *'mrht'* *'byt'* and *'ftt'* which is lard honey and lint.

They also used the extraction of herbs one of them was myrrh for the treatment of fevers and other physical disorders. Antibiotics are not definitely proved on the wounds to cure them but honey and butter cures the wound easily. Many vegetable extracts such as onion, radish garlic are used in the mummies by the ancient Egyptian experts, which now reveal that after 5000 years time has unable to putrefy the mummies.

Egyptians were excellent healers and master of medicines, these are found in the Ebres papyrus. This papyrus also provides us information about the possible documentation of tumors. After defeat of Egypt by the Romans, the study of medicine flourished in Rome. The Romans invented many surgical instruments unique to women. Romans were pioneer in the cataract surgery.

Medieval medicine was an evolving mixture of scientific and unani. In the early middle ages following the fall of the Roman Empire, standard medical knowledge was based chiefly Greek and Roman texts which survived.

Persia's position at the crossroads of the East and West frequently placed it in the midst of the development in both ancient Greek and Indian medicine. Muhammad Ibn Zakariya Al Razi became the first physician who systematically used alcohol in medicine in A.D. 1200. He also wrote books giving the introduction on measles and small pox which also became very influential in Europe.

In India the Vedas were the main source and reference of knowledge and custom. Ayurveda was the epitome of the medicine. From these Charak and Susshruta (300BC) was the eminent physicians and surgeon, who used different techniques now the modern medicine is not fully aware of. Nagarjuna the famous Alchemist in the A.D. 900 used metals in the medicine and he named it as *Kajjali*, the combination of sulphur and mercury, the precursors of modern sulpha drugs.

In the modern medical era in the year 1942 Selman Watsmen used the term Antibiotic when he used microorganism extract in high dilution. The word antibiotic means the building up a force against some biological organism (Greek *anti*-against and *Bios*-life).

Biochemically antibiotics are those groups of medicines which develop sensitivity against a specific microorganism and on the other hand also develop a resistivity against other microorganisms. For example Gatifloxacin as well known modern antibiotic, creates sensitivity against *E coli* but creates resistivity against H. Pyllori. *E coli* causes urinary tract infection and H Pyllori causes Gastro-intestinal infection, both are gram negative bacteria.

This we are going to read in this chapter in details and also we are going to read about the classification, synthesis and uses of the antibiotics in detail.

THE INTRODUCTION AND DISCOVERY OF PENICILIN

Sir Alexander Fleming (6 August 1881-11 March 1955) was a Scottish biologist and pharmacologist. Fleming published many articles on bacteriology, immunology and chemotherapy. His best-known achievements are the discovery of the enzyme lysozyme in 1923 and the antibiotic substance penicillin from the fungus *Penicillium notatum* in 1928, for which he shared the Nobel Prize in Physiology or Medicine in 1945 with Howard Walter Florey and Ernst Boris Chain.

In 1999, *Time Magazine* named Fleming one of the 100 Most Important People of the 20th Century for his discovery of penicillin, and stated; "It was a discovery that would change the course of history. The active ingredient in that mold, which Fleming named penicillin, turned out to be an infection-fighting agent of enormous potency. When it was finally recognized for what it was—the most efficacious life-saving drug in the world—penicillin would alter forever the treatment of bacterial infections. By the middle of the century, Fleming's discovery had spawned a huge pharmaceutical industry, churning out

Fig. 11.1 : The Penicilium Notatum Fungus

synthetic penicillins that would conquer some of mankind's most ancient scourges, including syphilis, gangrene and tuberculosis". After the war Fleming actively searched for anti-bacterial agents, having witnessed the death of many soldiers from septicemia resulting from infected wounds. Antiseptics killed the patients' immunological defences more effectively than they killed the invading bacteria. In an article he submitted for the medical journal *The Lancet* during World War I, Fleming described an ingenious experiment, which he was able to conduct as a result of his own glass blowing skills, in which he explained why antiseptics were killing more soldiers than infection itself during World War I. Antiseptics worked well on the surface, but deep wounds tended to shelter anaerobic bacteria from the antiseptic agent, and antiseptics seemed to remove beneficial agents produced that protected the patients in these cases at least as well as they removed bacteria, and did nothing to remove the bacteria that were out of reach. Sir Almroth Wright strongly supported Fleming's findings, but despite this, most army physicians over the course of WWI continued to use antiseptics even in cases where this worsened the condition of the patients. "When I woke up just after dawn on September 28, 1928, I certainly didn't plan to revolutionize all medicine by discovering the world's first antibiotic, or bacteria killer," Fleming would later say, "But I guess that was exactly what I did".

By 1928, Fleming was investigating the properties of staphylococci. He was already well-known from his earlier work, and had developed a reputation as a brilliant researcher, but his laboratory was often untidy. On 3 September 1928, Fleming returned to his laboratory having spent August on holiday with his family. Before leaving he had stacked all his cultures of staphylococci on a bench in a corner of his laboratory. On returning, Fleming noticed that one culture was contaminated with a fungus, and that the colonies of staphylococci that had immediately surrounded it had been destroyed, whereas other colonies further away were normal. Fleming showed the contaminated culture to his former assistant Merlin Price who said "that's how you discovered lysozyme." Fleming identified the mould that had contaminated his culture plates as being from the *Penicillium* genus, and—after some months' of calling it "mould juice"—named the substance it released *penicillin* on 7 March 1929.

He investigated its positive anti-bacterial effect on many organisms, and noticed that it affected bacteria such as staphylococci, and many other Gram-positive pathogens that cause scarlet fever, pneumonia, meningitis and diphtheria, but not typhoid fever or paratyphoid fever—which are caused by Gram-negative bacteria—for which he was seeking a cure at the time. It also affected *Neisseria gonorrhoeae*, which causes gonorrhoea although this bacterium is Gram-negative.

Fleming published his discovery in 1929, in the *British Journal of Experimental Pathology*, but little attention was paid to his article. Fleming continued his investigations, but found that cultivating penicillium was quite difficult, and that after having grown the mould, it was even more difficult to isolate the antibiotic agent. Fleming's impression was that because of the problem of producing it in quantity, and because its action appeared to be rather slow, penicillin would not be important in treating infection. Fleming also became convinced that penicillin would not last long enough in the human body (*in vivo*) to kill bacteria effectively. Many clinical tests were inconclusive, probably because it had been used as a surface antiseptic. In the 1930s, Fleming's trials occasionally showed more promise, and he continued, until 1940, to try to interest a chemist skilled enough to further refine usable penicillin.

Fleming soon abandoned penicillin, and not long after Florey and Chain took up researching and mass producing it with funds from the U.S and British governments. They started mass production after the bombing of Pearl Harbor. When D-day arrived they had made enough penicillin to treat all the wounded allied forces. Ernst Chain worked out how to isolate and concentrate penicillin. He also correctly theorised the structure of penicillin. Shortly after the team published its first results in 1940, Fleming telephoned Howard Florey, Chain's head of department to say that he would be visiting within the next few days. When Chain heard that he was coming he remarked "Good God! I thought he was dead".

Norman Heatley suggested transferring the active ingredient of penicillin back into water by changing its acidity. This produced enough of the drug to begin testing on animals. There were many more people involved in the Oxford team, and at one point the entire Dunn School was involved in its production.

After the team had developed a method of purifying penicillin to an effective first stable form in 1940, several clinical trials ensued, and their amazing success inspired the team to develop methods for mass production and mass distribution in 1945. Fleming was modest about his part in the development of penicillin, describing his fame as the "Fleming Myth" and he praised Florey and Chain for transforming the laboratory curiosity into a practical drug. Fleming was the first to discover the properties of the active substance, giving him the privilege of naming it: penicillin. He also kept, grew and distributed the original mould for twelve years, and continued until 1940 to try to get help from any chemist who had enough skill to make penicillin. Sir Henry Harris said in 1998: "Without Fleming, no Chain; without Chain, no Florey; without Florey, no Heatley; without Heatley, no penicillin." Fleming's accidental discovery and isolation of penicillin in September 1928 marks the start of modern antibiotics. Fleming also discovered very early that bacteria developed antibiotic resistance whenever too little penicillin was used or when it was used for too short a period. Almroth Wright had predicted antibiotic resistance even before it was noticed during experiments. Fleming cautioned about the use of penicillin in his many speeches around the world. He cautioned not to use penicillin unless there was a properly diagnosed reason for it to be used, and that if it were used, never to use too little, or for too short a period, since these are the circumstances under which bacterial resistance to antibiotics develops.

CLASSIFICATION OF PENICILLINS

There are four classes of penicillins, based upon their ability to kill various types of bacteria. From narrow to broad range of effectiveness they include:

- **Natural Penicillins** (Penicillin G, Procaine, Penicillin G, Penicillin V, Benzathine). The natural penicillins were the first agents in the penicillin family to be introduced for clinical use. The natural penicillins are based on the original penicillin-G structure. They are effective against gram-positive strains of streptococci, staphylococci, and some gram-negative bacteria such as meningococcus. Penicillin V is the drug of choice for the treatment of streptococcal pharyngitis. It is also useful for anaerobic coverage in patients with oral cavity infections.
- **Penicillinase-Resistant Penicillins** (Cloxacillin, Dicloxacillin, Methicillin, Nafcillin, Oxacillin). Methicillin was the first member of this group, followed by oxacillin, nafcillin, cloxacillin and dicloxacillin. The penicillinase-resistant penicillins have a more narrow spectrum of activity than the natural penicillins. Their antimicrobial efficacy is aimed directly against penicillinase-

producing strains of gram-positive cocci, particularly Staphylococcal species and these drugs are sometimes called anti-staphylococcal penicillins.

- **Aminopenicillins** (Ampicillin, Amoxicillin, Bacampicillin). The aminopenicillins were the first penicillins discovered to be active against gram-negative bacteria (such as E. coli and H. influenzae). Aminopenicillins are acid-resistant so administered orally. Orally administered amoxicillin and ampicillin are used primarily to treat mild infections such as otitis media, sinusitis, bronchitis, urinary tract infections and bacterial diarrhea. Amoxicillin is the agent of choice for the treatment of otitis media.
- **Extended Spectrum Penicillins** (sometimes called anti-pseudomonal penicillins). Extended Spectrum Penicillins include both alpha-carboxypenicillins (carbenicillin and ticarcillin) and acylaminopenicillins (piperacillin, azlocillin, and mezlocillin). These agents have similar spectrums of activity as the aminopenicillins but with additional activity against several gram negative organisms of the family Enterobacteriaceae, including many strains of Pseudomonas aeruginosa. Like the aminopenicillins, these agents are susceptible to inactivation by beta-lactamases. These agents may be used alone or in combination with Aminoglycosides.

MODE OF ACTION

All penicillin derivatives produce their bacteriocidal effects by inhibition of bacterial cell wall synthesis. Penicillins prevent bacteria from using a substance that is necessary for the maintenance of the bacterias outer cell wall. Unable to use this substance for cell wall maintenance, the bacteria swell, rupture, assume unusual shapes, and finally die. The penicillins may be bactericidal (kill the bacteria) or bacteriostatic (stop the growth of bacteria). They are bactericidal against sensitive microorganisms provided there is an adequate concentration of penicillin in the body. An inadequate concentration of penicillin may produce bacteriostatic activity, which may or may not control the infection.

CONDITIONS TREATED WITH PENICILLINS, INDICATIONS & USES

Penicillins may be used to treat infections such as urinary tract infections, septicemia, meningitis, intra-abdominal infection, gonorrhea, syphilis, pneumonia, respiratory infections, ear, nose and throat infections, skin and soft tissue infections. Examples of infectious microorganisms (bacteria) that may respond to penicillin therapy include gonococci, staphylococci, streptococci, and pneumococci.

Penicillins are used to prevent bacterial infection before, during and after surgery and to prevent Group A streptococcus ("strep") infections in people with a history of rheumatic heart disease.Most penicillins work best when taken on an empty stomach (either one hour before meals or two hours after) with an 8-ounce glass of water. The water helps prevent the medicine from irritating the delicate lining of the esophagus and stomach. However, some types of penicillin can be taken on either a full or empty stomach. These include amoxicillin, penicillin V, and the tablet form of bacampicillin.Penicillins work best when there is a constant amount circulating in the body. So it's important not to miss a dose. Also, it's best to take doses at evenly spaced intervals, both day and night.

SIDE EFFECTS

Although most penicillins are safe for the majority of people, some people may experience side effects. Allergic or hypersensitivity reactions are thought to be the most frequently occurring side effect. An estimated 3-10% of the general population are allergic to penicillin. Once an individual is allergic to one penicillin, he or she is most likely allergic to all of the penicillins. Those allergic to penicillin also have a higher incidence of allergy to the cephalosporin antibiotics.The most serious allergic reaction is anaphylaxis, a severe allergic reaction that can cause skin rash, hives, itching, difficulty breathing, shock, and unconsciousness. An early sign of anaphylaxis is a feeling of warmth and flushing. If any of these occurs, the medicine should be stopped and emergency help sought immediately. Anaphylactic shock occurs more frequently after parenteral administration but can occur with oral use. Other most common side effects are mild diarrhea, vomiting, headache, vaginal itching and discharge, sore mouth or tongue, or white patches in the mouth or on the tongue. These problems usually go away as the body adjusts to the drug and do not require medical treatment unless they continue or they are bothersome. Occasionally, certain types of penicillin may cause the tongue to darken or discolor. This condition is temporary and will go away when the medicine is stopped. On rare occasions some types of penicillin may cause severe abdominal or stomach cramps, pain, or bloating or severe or bloody diarrhea. Other rare side effects include fever, increased thirst, severe nausea or vomiting, unusual tiredness or weakness, weight loss, seizures, or unusual bleeding or bruising.

THE ADVANTAGES AND DISADVANTAGES OF PENICILLIN

Penicillins advantages:

- bactericidal against sensitive strains
- relatively nontoxic
- have excellent tissue penetration
- efficacious in the treatment of infections
- relatively inexpensive in comparison with other antibiotics.

Newer penicillins are resistant to stomach acid, such as penicillin V, or have a broader spectrum, such as ampicillin and amoxicillin.

Penicillins disadvantages:

- acid lability - most of these drugs are destroyed by gastric acid
- short duaration of action - because of this short half-life, the penicillins must be administered at short intervals, usually every 4 hours
- lack of activity against most Gram-negative organisms
- drug hypersensivity - about 10% of population has allergy
- many patients experience GI upset
- painful if given intramuscularly

THE CLASSIFICATION OF ANTIBIOTICS

The antibiotics are classified according to three criteria and although that each category contains several drugs but each one of them is unique in some features and effects. The first classification is according to the spectrum. The spectrum means the number of the organisms affected by the same drug. There are narrow and wide spectrum antibiotics. The wide spectrum antibiotics affect several types of bacteria and fungi and it is usually used where the specific type of the microorganism is unknown. For example, when we are treating an bacterial caused inflammation, we know that we are dealing with a staphylococcus or streptococcus microorganism so the doctor can proceed with the treatment without asking for more lab tests to identify the specific type of the microorganism using the broad spectrum antibiotics but in other cases, where we know the specific type of the microorganism, we can use the narrow spectrum antibiotics that are more effective on specific microorganism but less effective on others. The second classification is according to the type of the action of antibiotics. It could be bactericidal or bacteriostatic. The bactericidal antibiotics kill the harmful microorganism while the baceriostatic ones tend to slow down their growth and give the body the chance to use its immune system against the microorganisms. In case of virulent microorganisms or in case of weak immunity, bactericidal antibiotics are preferred because they will omit the problem from its roots but they will affect the normal microorganisms in the body. In mild cases, bacteriostatic antibiotics could be used because of their minor side effects. The third classification of antibiotics is according to the route of administration of the drug. The prevalent route of administration is the oral route but, there are other routes of administration that are more effective in certain cases like injection or topical applications.The most common method classifies them according to their chemical structure as antibiotics sharing the same or similar chemical structure will generally show similar patterns of antibacterial activity, effectiveness, toxicity and allergic potential.

Class (chemical structure)	Mechanism of action	Examples
1	2	3
B-lactam antibiotics • Penicillins • Cephalosporins • Carbapenems	Inhibit bacterial cell wall synthesis	Penicillins • Penicillin G • Amoxicillin • Flucloxacillin • Cloxacillin • Cefazolin • Ampicilin Cephalosporins • Cefoxitin • Cefotaxime • Ceftriaxone

1	2	3
		Carbapenem • Imipenem
Macrolides	Inhibit bacterial protein synthesis	• Erythromycin • Roxithromycin • Azithromycin • Clarithromycin • Fhosphomycin
Tetracyclines	Inhibit bacterial protein synthesis	• Tetracycline • Minocycline • Doxycycline • Lymecycline
Fluoroquinolones	Inhibit bacterial DNA synthesis	• Norfloxacin • Ciprofloxacin • Enoxacin • Ofloxacin • Gatifloxacin
Sulphonamides	Blocks bacterial cell metabolism by inhibitingenzymes	• Co-trimoxazole • Trimethoprim
Aminoglycosides	Inhibit bacterial protein synthesis	• Gentamicin • Amikacin • Tobramycin • Neomycin • Streptomycin
Imidazoles	Inhibit bacterial DNA synthesis	• Metronidazole
Peptides	Inhibit bacterial cell wall synthesis	• Bacitracin
Lincosamides	Inhibit bacterial protein synthesis	• Clindamycin • Lincomycin
Other	Inhibit bacterial protein synthesis	• Fusidic acid • Mupirocin
Benzoquinine	Inhibit Tumer cell DNA	• Geldana mycin

THE SYNTHESIS OF SOME OF THE IMPORTANT ANTIBIOTICS

In this section we shall know about the synthesis of the important antibiotics among them the neomycin and fosphomycin from their respective sources, it showed be kept in mind that the process of antibiotic synthesis is more or less same but the difference is the growth medium the pH and the source material from which the antibiotic is going to me synthesized.

NEOMYCIN SYNTHESIS

Source: **Streptomyces fradiae** Incubation Temp: **28°C**

pH: **7.2** Killing agnet: **Helico bactor Pylori**

The neomycin-producing culture of *S. fradiae* was maintained on a potato-yeast extract agar slant at 28°C and was subcultured at monthly intervals. The effects of different carbon and nitrogen sources were studied in the basal medium consisting of K_2HPO_4, 1.0 g; $MgSO_4*7H_20$, 0.5 g; $CaCl_2$-$2H_20$, 0.04 g; $FeSO_4$-$7H_20$, 0.005 g; $ZnSO_4$ $7H_20$, 0.0005 g; water, 1,000 ml; pH was 7.5 1t 0.1.

Table 11.1 : Effect of Carbohydrates as Carbon Sources on the Growth of Streptomyces Fradiae and Neomycin Formation in a Synthetic Medium[a]

Carbohydrate (1%)	Incubation of 3 days		5 days		7 days	
	Growth (mg/ 100 ml)	Neomycin (µg/ml)	Growth (mg/	Neomycin (µg/ml)	Growth (mg/	Neomycin (µg/ml)
Control (No crbohydrate)	13.5	0	18.5	0	28.5	0.69
Dextrin	135.5	3.8	130	40.7	120	70.4
Starch (soluble)	163	4.0	138	34.8	111	62.0
Reffinose	15	0	20	0	26	7.9
Maltose	111	2.3	101	24.1	93.5	48.1
Sucrose	27.5	1.3	54	9.4	83.5	18.9
Lactose	36	0.52	60.5	12.5	73	14.6
D-Galactose	22.5	2.3	49	17.6	77.5	32.8
D-Glucose	80	3.9	119	19.2	124	23
L-Rhamnose	40.5	0.26	49	4.4	33	11.3
D-Fructose	25.5	0	55	1.95	53.5	8.2
D-Mannose	104	5.2	147.5	2.6	140.5	0
D(+)-Xylose	99	0.75	111	9.5	127	20.8
L(+)-Acrabinose	88	4.3	123.5	9.6	142.5	13

[a]Medium (30 ml per 100-ml Erlenmeyer flsk) contained basal mineral salts and 0.2% glutamic acid and was incubated on a rotary shaker at 250 rev/min at 28°C.

The carbon sources and phosphate were sterilizedseparately and added just prior to inoculation. The nitrogen source was always included in the basal medium, and the medium was adjusted to pH 7.2 and sterilized.A well-sporulated slant culture (10 days old) was washed with 5 ml of sterile water, and 0.2 ml wasused to inoculate each 100-mi Erlenmeyer flask containing 30 ml of medium. After a night of stationary rest, the flasks were placed on a rotary shaker [250 rev/min; eccentricity 0.5 inch (1.27 cm)]. Incubation temperature was 28°C. Occasional checking of the flask to drop adhering cells into the medium was necessary during the first 48 hr. Determination of neomycin potency. The beer and neomycin B bases were diluted with potassium phosphate buffer (pH 8) and the estimation of neomycin was made by a modified cupplate method, with *Bacillus subtilis* (strain B3) as the test organism. The results are expressedin terms of micrograms per millilitre of neomycinB base (U-6682, lot 6846 HKJ-78; The Upjohn Co., Kalamazoo, Mich.), which was used as standard. Estimation of components in a neomycin complex. The components of neomycin in the fermentation broth were determined. Growth was determined asthe dry weight of cells. The mycelium was separated by filtration through a sintered-glass filter and was washed with water. It was quantitatively transferred to a dry aluminum dish and was then dried at 70 ± 5°C. Determination of residual sugar. Since the fermentation broth did not contain any reducing substance other than the supplied sugar.

Results and Discussion

Effect of carbohydrates. A number of carbohydrates were investigated for their effect on growth of *S. fradiae* and neomycin production. Dextrin, starch, and maltose were excellent carbon sources for neomycin formation and also for growth (Table 11.1). Although glucose, mannose, xylose, and arabinose supported abundant growth of the organism, these sugars were poor carbon sources for neomycin production when used alone. Mannose failed to produce the antibiotic after the 7th day of

Table 11.2 : Effect of Alcohols, Sugar Alcohols and Sodium Salts of Organic Acids as Carbon Sources on the Growth of Streptomyces Fradiae and Neomycin Formationin a Synthetic Medium[a]

Nitrogen sources	Incubation of 3 days		5 days		7 days	
	Growth (mg/ 100 ml)	Neomycin (μg/ml)	Growth (mg/ 100 ml)	Neomycin (μg/ml)	Growth (mg/ 100 ml)	Neomycin (μg/ml)
Ethylalcohol	16	0	15	0	39.5	0
Butanol	0.5	0	1.0	0	2.0	0
Glyceron	53.5	0	49	0	38.5	0
Dulcitol	39.5	0.33	47	5.2	30.5	17.5
Sorbitol	56.5	0.26	28.5	4.4	21.5	12.2
Mannitol	40	0.26	59	4.0	31	10.1
Inositol	29	0	26.5	0.33	19.5	1.3
Na lactate	15	0	17	6.6	22	9.2
Na succinate	61	1.52	107	14.2	98.5	8.35
Na acetate	31	0	30.5	0.65	34.5	1.8
Na citrate	0.5	0	1.0	0	1.0	0

Table 11.3 : Effect of Carbohydrates as Carbon Sources on the Growth of Streptomyces Fradiae and Neomycin Formation in a Synthetic Medium[a]

Nitrogen sources	Incubation of 3 days		5 days		7 days	
	pH	Growth (mg/100 ml)	Neomycin (µg/ml)	pH	Growth (mg/100 ml)	Neomycin (µg/ml)
L-Aspartic acid	7.5	170.6	45.7	7.5	133.3	100.4
L-Glutamic acid	8.0	125	28.4	8.0	103	55
DL-Serine	7.2	36.6	9.3	7.5	93	50
Glycine	7.1	163.3	22.9	7.2	127	46
DL-Alanine	7.0	142.6	29	7.0	139	40.8
DL-Threonine	7.0	34.3	10	7.0	85.6	25
L-Tyrosine	7.2	55.3	0	7.2	90	12.5
L-Asparagine	7.5	77	4.25	7.2	140.6	13.5
L-Lysine	6.4	101.3	2.9	6.2	89.6	4.45
DL-Valine	6.6	65	5.3	6.4	72	3
L-Histidine	6.4	153.3	0	6.8	135.6	6
DL-Methionine	7.0	80.6	0	6.2	73.3	0
L-Proline	7.2	13.3	0	6.8	34.3	0
DL-Phenylalanine	7.5	11	0	7.2	20	0
DL-Tryptophan	7.5	12.3	0	7.5	22	0
L-Cystine	7.2	59.3	0	7.2	68.3	0
L-Leucine	7.2	10	0	7.0	14	0
L-Arginine	6.4	96	0	6.2	73	0
Na nitrate	7.5	177.6	57.5	7.5	165.3	77
Ammonium chloride	5.8	59.3	0	4.9	59	0

[a]Cultivation conditions as in Table 11.1 without glutamate). N sources were added at 19 mg of Nper 100 ml.

fermentation in spite of good growth of the organism. This might be due to the acidity of the medium (pH not shown in the table) during rapid utilization of the sugar. It was reported earlier that little or almost no neomycin was produced at a pH below 7. It is, however, interesting to note that neomycin production is not related solely to the amount of cellular growth. Studies on penicillin and actinomycin biosynthesis show that a slowly utilizable carbohydrate gave a higher yield of the antibiotic because of its availability in adequate amount during both the phases of cellular growth and antibiotic production. The influence of different carbohydrates on neomycin formation may be explained in a similar manner, although the pH of the fermentation broth during the phase of neomycin synthesis may be critical. Effect of alcohols, sugar alcohols, and organic acids. Results in Table 11.2 show that alcohols, sugar alcohols, and organic acids, with the exception of succinic acid, were poor carbon sources for growth of the organism. Dulcitol, sorbitol, mannitol, and sodium salts of succinic and lactic acids gave moderate yields of neomycin. Although, among the different carbon sources studied, dextrin and starch gave higher neomycin yields than did maltose, maltose was selected as the carbon source in subsequent investigations, as it is a simpler compound and also available inhighly purified form. A maltose concentration of 1.5 g/100 ml gave maximal neomycin titer; a lower or higher dose decreased the yield.

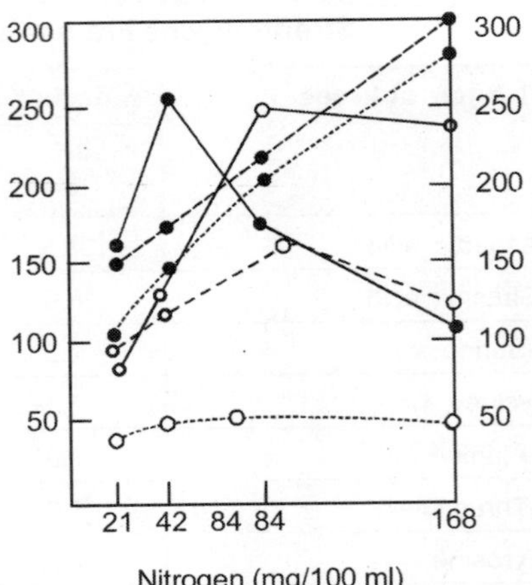

Fig. 11.2 : Effect of nitrogen concentration on growth and neomycin production in a synthetic medium (1.5% maltose, plus basal mineral salts). Symbols: O= neomycin production, mg/ml; ● = growth, mg/100 ml; solid line = sodium nitrate; dashed line, aspartic acid; dotted line, glutamic acid. Analysis made on 7th day of fermentation

Effect of different nitrogen sources. The medium for testing different nitrogen sources contained the basal mineral salts plus 1.5% maltose. Aminoacids and inorganic nitrogen compounds were employed at a concentration equivalent to 19 mg of nitrogen per 100 ml. The data in Table 11.3 show that the highest neomycin yields were obtained with aspartic acid and sodium nitrate as the single source of nitrogen. DL-Methionine, L-proline, DL-phenylalanine, DL-tryptophan, L–cystine, L-leucine, L-arginine yielded no antibiotic, and some of them were very unsuitable for growth. The inhibition of neomycin biosynthesis with many amino acids and ammoniumchloride may be attributed to the acidity of the medium. Since sodium nitrate, L-aspartic acid, and L–glutamicacid proved to be excellent nitrogen sources for neomycin biosynthesis and growth,their optimal levels for neomycin biosynthesis were determined. Figure 11.2 shows that optimal nitrogen concentration for neomycin production was the same (84 mg/100 ml) with each of the nitrogen compounds tested. However, the pattern of the growth curve with different levels of each nitrogenous compound was different, with sodium nitrate showing maximal growth at lower nitrogen concentration (42 mg/100 ml). Moreover, it is evident (Fig. 11.2) that there is no direct relation between the growth of the organism and neomycin formation. Four typical media were selected for studyingbiochemical changes during neomycin fermentation. Medium 1, giving very good growth but low neomycin yield, consisted of basal medium plus glucose (1.5 %) and glutamic acid (84 mg of N_2/100 ml); for medium 2, glucose was replaced by maltose (1.5%), a slowly utilizable sugar. Media 3 and 4 contained aspartic acid (the maximal neomycin-yielding organic nitrogensource) and sodium nitrate (an inorganic nitrogen source), respectively, as the sole nitrogen source with carbohydrate. Mineral salts were the sameas those employed in the preparation of medium (Fig. 11.3 and 11.4).Glucose was more rapidly utilized than maltose,but the rate of consumption of the latter dependedon the nitrogen source supplied. The utilization of of maltose in the presence of either glutamic or aspartic acid was much slower than in the presence of sodium nitrate. The utilization of the carbon from amino acids might be responsible for the less rapid utilization of maltose. The rate of consumption of nitrogen from glutamic acidor aspartic acid was almost similar in the presence of either sugar. Nitrate nitrogen was consumedvery slowly, which might

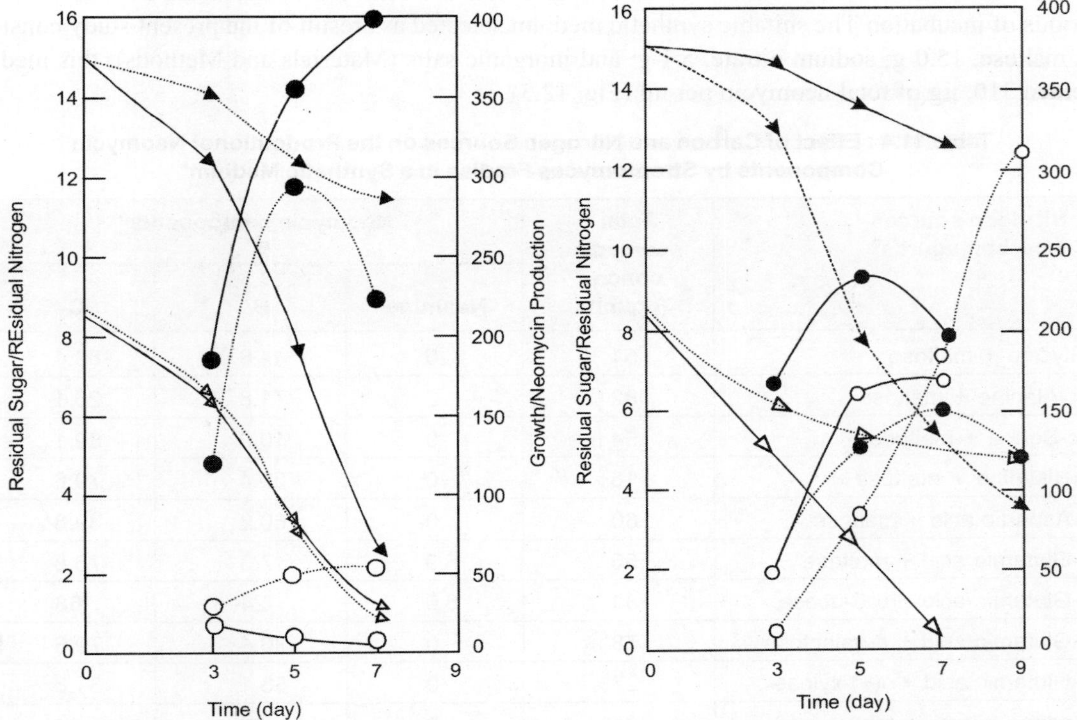

Fig. 11.3 : Biochemical Changes During Neomycin Fermentation. Symbols: O = neomycin, µg/ml; ● = growth, mg/100 ml; ∆ = residual nitrogen mg/10 ml; ▲ = residual sugar, mg/ml; solid line = medium 1 (glucose + glutamic acid + basal mineral salts); and broken line = medium 2 (maltose + glutamic acid + basal mineral salts).

Fig. 11.4 : Biochemical Changes During Neomycin Fermentation. Symbols: O = neomycin, µg/ml; ● = growth, mg/100 ml; ∆ = residual nitrogen mg/10 ml; ▲ = residual sugar, mg/ml; solid line = medium 3 (maltose + aspartic acid + basal mineral salts); and broken line = medium 4 (maltose + sodium nitrate + basal mineral salts).

result in less growth of the organism. There was inhibition of neomycinformation after the 3rd day of fermentation in the medium containing 1.5% glucose as thecarbon source, probably owing to the high acidity of the medium (pH not shown in the figure). The influence of the variation of carbon and nitrogen sources on the composition of neomycincomplex was next studied, and the results are shown in Table 11.3. The quantitative formation of neomycin components depended on the nature carbohydrates and nitrogenous compounds supplied, and the suitability of different sugars for neomycin production varied with the nitrogen source. A combination of glutamic acid with glucose, galactose, or maltose favoured mainly neomycin C, and xylose favored neomycin B production, the most biologically active component. The lower activity of the fermentation broth with glutamic acid, glycine, serine, histidine, and asparagine might be due to the formation of a higher percentage of neomycin C. Among the nitrogen sources tested, alanine, aspartic acid, and sodium nitrate increased mainly the synthesis of neomycin B. Small amounts of neamine were formed in the medium with alanine and maltose or with glutamic acid and maltose or glucose. The composition of neomycin complex did not vary with the periods of incubation The suitable synthetic medium selected as aresult of the present study consisted of: msltose, 15.0 g; sodium nitrate, 5.1 g; and inorganic salts (Materials and Methods); this medium yielded 310, μg of total neomycin per ml (Fig. 12.3).

Table 11.4 : Effect of Carbon and Nitrogen Sources on the Prodcutionof Neomycin Components by Streptomyces Fradiae in a Synthetic Medium*

Nitrogen sources + carbon source[b]	Total neomycin concn (μg/ml)	Neomycin components %		
		Neamine	B	C
Glycine + maltose	51	0	11.6	88.4
DL-Alanine + maltose	82	2	71.8	26.1
DL-Serine + maltose	54	0	10.9	89.1
L-Histidine + maltose	16	0	20.4	79.6
L-Aspartic acid + maltose	160	0	60.2	39.8
L-Glutamic acid + maltose	56	6.9	17.3	75.8
L-Glutamic acid + D-Glucose	31	8.0	24	68
L-Glutamic acid + D-galactose	78	0	20.4	79.6
L-Glutamic acid + D(+)-xylose	27	0	68.1	31.6
Sodium nitrate + maltose	301	0	89	11
Sodium nitrate + D-glucose	75	0	88	12
Sodium nitrate + D-galactose[c]	—	—	—	—
Sodium nitrate + D(+)-xylose[c]	—	—	—	—

[a]Cultivation conditions as in Table 11.1 without glutamate). N sources were added at 19 mg of Nper 100 ml.
[b]The nitrogen sources provide nitrogen at a concentration of 84 mg/100 ml, and the concentration of the carbon sources was 1%.
[c]Inadequate growth.

ACTINOMYCINS SYNTHESIS

Source: **Streptomyces parvulus**
pH: **7.0**
Incubation Period: **30°C for 5 days**
Killing Agent:

Introduction

The actinomycins are a family of chromopeptide antibiotics that present antitumoral properties, being employed in the treatment of several human neoplasies. Structurally, they have a chromophorous group, identical in all actinomycins, and two pentapeptide chains with a variable composition of amino acids. They are synthesized by *Streptomyces* as mixtures of different actinomycins, however, the

S. parvulus species produces actinomycin D almost exclusively (>95%). Over the last ten years, research into actinomycin D has been directed mainly towards clinical applications and no work related to production was found. In a previous study the authors optimized a complex medium for the production of actinomycin D by *S. parvulus* DAUFPE 3124 strain, increasing the final antibiotic concentration from 245 mg/L to 530 mg/L. Few papers have been found that focus on the nutritional requirements of *Streptomyces* species for achieving a high actinomycin yield. The maximum antibiotic synthesis ocurred in a chemically defined medium containing L(-) galactose (10 g/L), L(-) glutamic acid (2 g/L), phosphate (1 g/L) and mineral salts. Although this composition proved to be satisfactory for production of actinomycin mixtures by *S. antibioticus,* the synthesis of actinomycin D by *S. parvulus* was rather poor and extremely variable. Proposed a chemically defined medium for the synthesis of high yields (500 to 600 mg/L) of actinomycin D by *S. parvulus* ATCC 12434. Working with the same medium and the same strain immobilized in calcium alginate, using an air lift column operating under discontinuous, fed batch and continuous conditions, obtained 50, 73 and 80 mg/L, respectively. However, since nutritional requirements for growth and product formation vary from strain to strain, the medium resulted in a low antibiotic production by the *S. parvulus* DAUFPE 3124 strain. This work presents a set of experiments that culminated in the development of a chemically defined medium in which higher concentration of actinomycin D was achieved using the *S. parvulus* DAUFPE 3124 strain.

Materials and Methods

Microorganism: *Streptomyces parvulus* DAUFPE 3124 strain was used in all assays. This strain was maintained on slants of 4 g/L glucose, 4 g/L yeast extract and 10 g/L malt extract agar medium. Incubation was carried out for 3 to 5 days at 30° C until gray spores developed and the slants were stored at 4°C.

Preparation of Spore Suspension: The spores were scraped from the surface of heavily sporulated slants with 3 ml of 9 g/L NaCl solution containing 1 ml/L of Tween 80, following the Hopwood *et al.,* (1985) technique. The spore concentration was quantified in the suspension and stored in the freezer at -4°C.

Growth and Production Media: The medium used during the biomass growth had the following composition: 5 g/L tryptone and 3 g/L yeast extract. In the antibiotic production several media were used with pH adjusted to 7.0 and in all of them, the following basal mineral salts medium K_2HPO_4 1.0 g/L; $MgSO_4.7H_2O$ 25 mg/L; $ZnSO_4.7H_2O$ 25 mg/L; $CaCl_2.2H_2O$ 25 mg/L; $FeSO_4.7H_2O$ 25 mg/L was used. A number of amino acids and carbohydrates were tested as carbon and nitrogen sources so that the C/N ratio was held around 41.7, as recommended by Williams & Katz (1977), and several C/N ratios were employed when the best carbon and nitrogen sources were selected. The amino acids, carbohydrates and salts were sterilized separately and added just prior to inoculation.

Experimental Procedure: The growth was performed in two steps. In the first one, the inoculum was prepared in a 250 ml flask containing 25 ml of growth medium inoculated with 50/1 of spore suspension (8.0×10^7 cfu/ml).The flasks were placed on a rotary shaker at 250 rpm at 30ºC for 48 h. In the second step, 20 ml of the mycelial suspension was used to inoculate a 2000 ml flask containing 200 ml of growth medium which was subjected to the same conditions as described above for 24 h.

The antibiotic production was conducted in 500 ml shaken flasks, containing 50 ml of the fermentation medium. These flasks were inoculated with 3 ml of mycelial suspensions and subjected to the same conditions as cited above. Samples were taken at 24 hour intervals and analysed in terms of actinomycin D concentration by spectrophotometric method. The data reported in all experiments represent the yield of actinomycin D for 144 hours of cultivation except for the last one.

Results and Discussion

As observed, the association of L(-)glutamic acid with another amino acid potentialized the synthesis of actinomycin D by *S. parvulus* ATCC 12434. This fact was tested using the *S. parvulus* DAUFPE 3124 and the results are shown in Table 11.4.

This table shows that a maximum concentration of actinomycin D was obtained when the L(-) glutamic acid was combined with L(-)histidine, L(-)ornithine or L(-)threonine (70.8 mg/L).Assays were carried out with several L(-) amino acids to determine the one more appropriated for antibiotic synthesis (Table 11.2). The best result (87.7 mg/L) was achieved when L(-) threonine was the sole nitrogen source. Surprisingly, this result was higher than the one obtained (70.8 mg/L) by the combination of L(-)glutamic acid and L(-) threonine. In the literature there is no explanation for this effect and, furthermore, it results in a cheaper chemically defined medium composition.

Table 11.5 : Effect of L(-) amino acids provdied in combination with L(-) glutamic acid for actionomycin production by *S. parvulus* DAUFPE 3124 grown in a 40 gL (D(+) fructose, 2.2 g/l (L(-) glutamic acid basal mineral salts medium for 144 h at 30°C (C/N ratios between 41.2 and 42.8)

L (-) glutamic acid + L (-) amino acids (g/L)	Actionomycin D (mg/L)
Histidine (0.775)	70.8
Ornithine (0.99)	70.8
Threonine (1.787)	70.8
Alanine (1.336)	65.0
Leucine (1.967)	62.7
Glutamine (1.096)	38.6
Glicine (1.126)	47.3
Valine (1.757)	53.7
Proline (1.727)	68.5
Serine (1.576)	22.1
Asparagine (0.99)	53.7
Tryptophan (1.53)	42.1
Methionine (2.387)	18.4

L(-) threonine is one of the amino acids constituents of the actinomycin molecules. Researchers employed labeled L(-) threonine-^{14}C in experiments of short incubation and verified that *S. antibioticus* strains metabolizes this amino acid in both protein and actinomycin synthesis. Due to this evidence, these authors came to the conclusion that there is an intracellular pool of threonine commom to the two synthesizer systems although these systems were differents.A number of carbohydrates were investigated as carbon sources for actinomycin-D production employing the basal mineral salts medium with 1.8 g/L of threonine and the results can be seen in Table 11.5. This table shows that D(+) fructose was superior to the others carbohydrates tested in the same concentration. This result agrees with the findings of Williams & Katz (1977), who affirmed that D fructose was the most effective carbon source for antibiotic synthesis by *S. parvulus*. No actinomycin-D production was detected when glucose and galactose were employed as sole carbon sources. The glucose supports a rapid growth rate of *S. antibioticus*, but it exerts severe catabolite repression in antibiotic production, particulary in the synthesis of phenoxazinone synthase, an essential enzyme for the formation of the actinomycin chromophorous group. Researchers working with *S. parvulus* ATCC 12434, also showed the repression caused by glucose additions to a chemically defined medium containing fructose (40g/L) as the main substrate. The inhibitory effect of galactose on phenoxazinone synthetase specific activity in cells of *S. antibioticus* harvested after six hours of incubation.

Table 11.6 : Influence of several carbon sources on actinomycin production by S. parvulus DAUFPE 3124 grown in a 1.8 g/L L(-) threonine basal mineral salts medium for 144h at 30°C (C/N = 41.5)

Carbon sources (20g/L)	Actinomycin D (mg/L)
D(-) arabinose	5.3
D(+) fructose	55.8
D(-) galactose	ND*
D(+) glucose	ND*
Myo-inositol	43.2
D(-) manytol	36.2
D(+) manose	18.2
Sucrose	21.6
D(+) xylose	44.8

*ND-not detected.

Table 11.7 : Effect of Several C/N Ratios on Actinomycin D Production by *S. parvulus* DAUFPE 3124 grown in a basal mineral salts medium for 144h at 30°C

D(+) Fructose (g/L)	L(-) Threonine (g/L)	C/N	Actinomycin D (mg/L)
80	3.57	22.5	67.9
20	2.38	32.0	57.5
20	1.78	41.6	58.9
20	1.19	60.6	53.8

To evaluate the quantitative influence of C/N ratios upon antibiotic synthesis by *S. parvulus* DAUFPE 3124, the concentration of D(+) fructose (20 g/L) was held constant while the amount of L(-) threonine was varied to provide different C/N ratios in the medium. The results of these experiments are shown in Table 11.7.

Higher production of actinomycin-D (67.9mg/L) was achieved at C/N ratio of 22.5 which corresponded to 20 g/L D(+) fructose and 3.57 g/L L(-) threonine.

Comparative study of the medium reported by Williams & Katz (1977) and the one developed in this work was done employing *S. parvulus* DAUFPE 3124 strain. The concentration curves of actinomycin D production are shown in Figure 1 where can see that the antibiotic concentration reached the maximum value of 133 mg/L in the proposed medium and in the medium suggested by cited authors the maximum value was 43 mg/L.

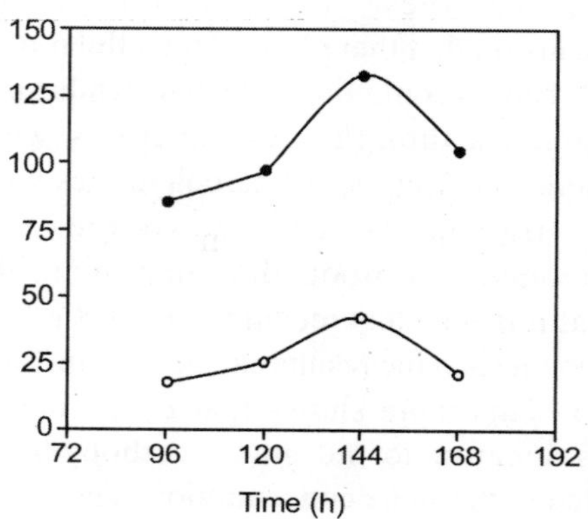

Fig. 11.5. Actinomycin D Concentrations obtained in Shaken Flasks by *S. parvulus* DAUFPE 3124, using chemically defined media (O).

The results clearly show the dependence of the antibiotic synthesis on medium constituents. The medium developed in this study to obtain high yield of actinomycin D by *S. parvulus* DAUFPE 3124 has the following composition: 20 g/L D(+) fructose; 3.57 g/L L(-) threonine; 1.0 g/L K_2HPO_4; 25 mg/L $MgSO_4.7H_2O$, 25 mg/L $ZnSO_4.7H_2O$, 25 mg/L $CaCl_2.2H_2O$ and 25 mg/L $FeSO_4.7H_2O$. This composition allowed an improvement of over 200% in the concentration of actinomycin D.

THE DERIVATIVES AND REACTION OF ANTI BIOTICS

Benzylpenicillins and phenoxymethylpenicillins (penicillins G and V, respectively) are produced by fermentation and are the basic precursors of a wide range of semi-synthetic antibiotics, e.g. ampicillin. The amide link may be hydrolysed conventionally but the conditions necessary for its specific hydrolysis, whilst causing no hydrolysis of the intrinsically more labile but pharmacologically essential β-lactam ring, are difficult to attain. Such specific hydrolysis may be simply achieved by use of penicillin amidases (also called penicillin acylases). Different enzyme preparations are generally used for the hydrolysis of the penicillins G and V, pencillin-V-amidase being much more specific than pencillin-G-amidase.Penicillin amidase may be obtained from *E. coli* and has been immobilised on a number of supports including cyanogen bromide-activated Sephadex G200. It represents one of the earliest successful processes involving immobilised enzymes and is generally used in batch or semicontinuous STR processes (40,000 Ukg^{-1}penicillin G, 35°C, pH 7.8, 2 h) where it may be reused over 100 times. It has also been used in PBRs, where it has an active life of over 100 days, producing about two tonnes of 6-aminopenicillanic acid $kg^{-1}$of immobolised enzyme.

benzyl penicillin + H_2O → phenylacetic acid + 6-aminopenicillanic acid

phenoxymethyl penicillin + H_2O → phenoxyacetic acid + 6-aminopenicillanic acid

The penicillin-G-amidases may be used 'in reverse' to synthesise penicillin and cephalosporin antibiotics by non -equilibrium kinetically controlled reactions (. Ampicillin has been produced by the use of penicillin-G-amidase immobilised by adsorption to DEAE -cellulose in a packed bed column:

6-aminopenicillanic acid + D-phenylglycine methyl ester → ampicillin + methanol

Many other potential and proven antibiotics have been synthesised in this manner, using a variety of synthetic b-lactams and activated carboxylic acids.

β-LACTAM ANTIBIOTICS

β-Lactam antibiotics are a broad class of antibiotics that include penicillin derivatives (penams), cephalosporins (cephems), monobactams, and carbapenems, that is, any antibiotic agent that contains a β-lactam nucleus in its molecular structure. They are the most widely-used group of antibiotics.While not true antibiotics, the β-lactamase inhibitors are often included in this group.

β-Lactams are classified according to their core ring structures.

- β-Lactams fused to saturated five-membered rings:
  - β-Lactams containing thiazolidine rings are named penams.
  - β-Lactams containing pyrrolidine rings are named carbapenams.
  - β-Lactams fused to oxazolidine rings are named oxapenams or clavams.
- β-Lactams fused to unsaturated five-membered rings:
  - β-Lactams containing 2,3-dihydrothiazole rings are named penems.
  - β-Lactams containing 2,3-dihydro-1H-pyrrole rings are named carbapenems.
- β-Lactams fused to unsaturated six-membered rings:
  - β-Lactams containing 3,6-dihydro-2H-1,3-thiazine rings are named cephems.
  - β-Lactams containing 1,2,3,4-tetrahydropyridine rings are named carbacephems.

  - β-Lactams containing 3,6-dihydro-2H-1,3-oxazine rings are named oxacephems.
- β-Lactams not fused to any other ring are named monobactams.

Common β-lactam antibiotics

Penicillins (Penams)

Main article: penicillin

Semisynthetic penicillins are prepared starting from the penicillin nucleus 6-APA.

Narrow-spectrum

- β-lactamase sensitive
  - benzathine penicillin
  - benzylpenicillin (penicillin G)
  - phenoxymethylpenicillin (penicillin V)
  - procaine penicillin
- Penicillinase-resistant penicillins
  - methicillin
  - oxacillin
  - nafcillin
  - cloxacillin
  - dicloxacillin
  - flucloxacillin
- β-lactamase-resistant penicillins
  - temocillin

Moderate-spectrum

- amoxicillin
- ampicillin

Broad-spectrum

- co-amoxiclav (amoxicillin+clavulanic acid)

Extended-spectrum

- azlocillin
- carbenicillin
- ticarcillin
- mezlocillin
- piperacillin

Cephalosporins (Cephems)

Main article: cephalosporin

First generation

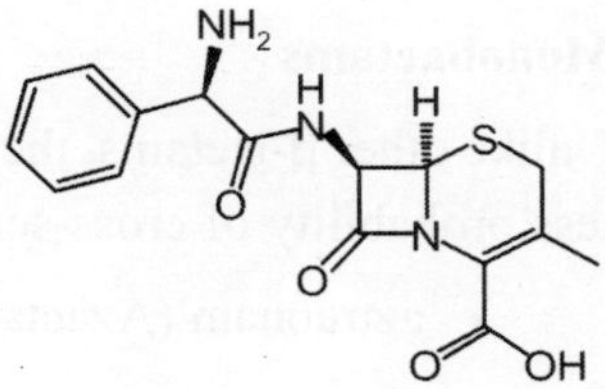

Skeletal formula of cefalexin, a first-generation cephalosporin

Moderate spectrum.

- cephalexin
- cephalothin
- cefazolin

Second generation

Moderate spectrum with anti-*Haemophilus* activity.

- cefaclor
- cefuroxime
- cefamandole

Second generation cephamycins

Moderate spectrum with anti-anaerobic activity.

- cefotetan
- cefoxitin

Third generation

Broad spectrum.

- ceftriaxone
- cefotaxime
- cefpodoxime

Broad spectrum with anti-*Pseudomonas* activity.

- ceftazidime

Fourth generation

Broad spectrum with enhanced activity against Gram positive bacteria and β-lactamase stability.

- cefepime
- cefpirome

Carbapenems and Penems

Main article: carbapenem

Main article: penem

Structural Formula of Imipenem 1

Broadest spectrum of β-lactam antibiotics.

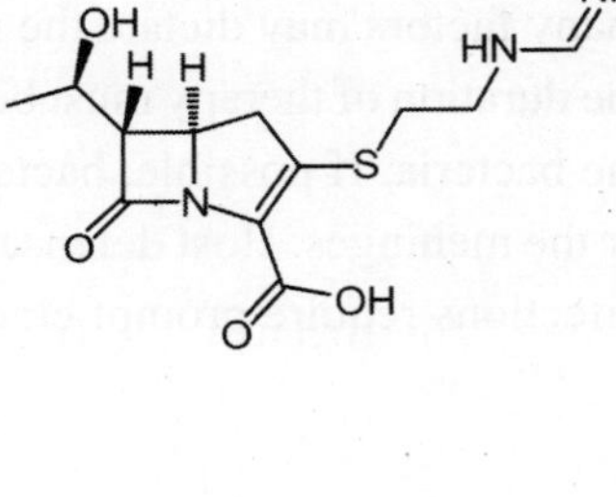

- imipenem (with cilastatin)
- meropenem
- ertapenem
- faropenem
- doripenem

Monobactams

Unlike other β-lactams, the monobactam contains a nucleus with no fused ring attached. Thus, there is less probability of cross-sensitivity reactions.

- aztreonam (Azactam)
- tigemonam
- nocardicin A
- tabtoxinine-β-lactam

β-lactamase inhibitors

Although they exhibit negligible antimicrobial activity, they contain the β-lactam ring. Their sole purpose is to prevent the inactivation of β-lactam antibiotics by binding the β-lactamases, and, as such, they are co-administered with β-lactam antibiotics.

- clavulanic acid
- tazobactam
- sulbactam

THE ACTION OF ANTIBIOTICS

Action of Antibiotics

Clinically effective antimicrobial agents all exhibit selective toxicity toward the bacterium rather than the host. It is this characteristic that distinguishes antibiotics from disinfectants. The basis for selectivity will vary depending on the particular antibiotic. When selectivity is high the antibiotics are normally not toxic. However, even highly selective antibiotics can have side effects.

The Theraputic Index

The therapeutic index is defined as the ratio of the dose toxic to the host to the effective therapeutic dose. The higher the therapeutic index the better the antibiotic.

The Category of Antibiotics

Antibiotics are categorized as bactericidal if they kill the susceptible bacteria or bacteriostatic if they reversibly inhibit the growth of bacteria. In general the use of bactericidal antibiotics is preferred but many factors may dictate the use of a bacteriostatic antibiotic. When a bacteriostatic antibiotic is used the duration of therapy must be sufficient to allow cellular and humoral defense mechanisms to eradicate the bacteria. If possible, bactericidal antibiotics should be used to treat infections of the endocardium or the meninges. Host defenses are relatively ineffective at these sites and the dangers imposed by such infections require prompt eradication of the organisms.

Antibiotic Susceptibility Testing

The basic quantitative measures of the *in vitro* activity of antibiotics are the minimum inhibitory concentration (MIC) and the minimum bactericidal concentration (MBC). The MIC is the lowest concentration of the antibiotic that results in inhibition of visible growth (*i.e.* colonies on a plate or turbidity in broth culture) under standard conditions. The MBC is the lowest concentration of the antibiotic that kills 99.9% of the original inoculum in a given time. The pharmacological absorption and distribution of the antibiotic will influence the dose, route and frequency of administration of the antibiotic in order to achieve an effective dose at the site of infection. In clinical laboratories, a more common test for antibiotic susceptibility is a disk diffusion test (Fig. 11.6). In this test the bacterial isolate is inoculated uniformly onto the surface of an agar plate. A filter disk impregnated with a standard amount of an antibiotic is applied to the surface of the plate and the antibiotic is allowed to diffuse into the adjacent medium.. The result is a gradient of antibiotic surrounding the disk. Following incubation, a bacterial lawn appears on the plate. Zones of inhibition of bacterial growth may be present around the antibiotic disk. The size of the zone of inhibition is dependent on the diffusion rate of the antibiotic, the degree of sensitivity of the microorganism, and the growth rate of the bacterium. The zone of inhibition in the disk diffusion test is inversely related to the MIC. The test is performed under standardized conditions and standard zones of inhibition have been established for each antibiotic. If the zone of inhibition is equal to or greater than the standard, the organism is considered to be sensitive to the antibiotic. If the zone of inhibition is less than the standard, the organism is considered to be resistant. Figure 1 also illustrates how the disk diffusion test is done and Figure 2 illustrates some of the standard zones of inhibition for several antibiotics.

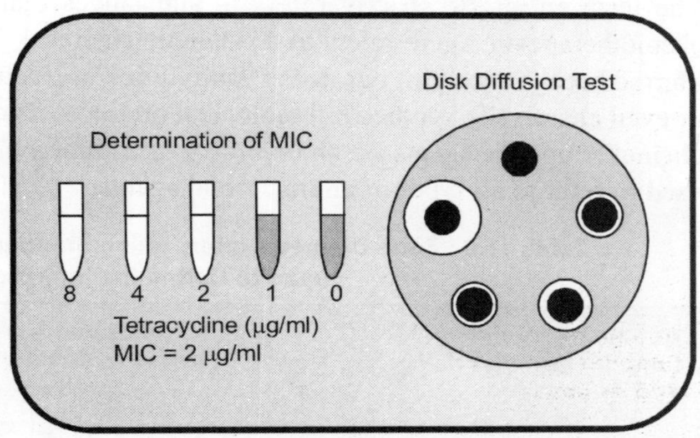

Fig. 11.6 : Antibiotic susceptibility testing

Combination Therapy

Combination therapy with two or more antibiotics is used in special cases:

- To prevent the emergence of resistant strains
- To treat emergency cases during the period when an etiological diagnosis is still in progress
- To take advantage of antibiotic synergism.

Antibiotic synergism occurs when the effects of a combination of antibiotics is greater than the sum of the effects of the individual antibiotics. Antibiotic antagonism occurs when one antibiotic, usually the one with the least effect, interferes with the effects of another antibiotic.

Antibiotics and Chemotherapeutic Agents

The term antibiotic strictly refers to substances that are of biological origin whereas the term chemotherapeutic agent refers to a synthetic chemical. The distinction between these terms has been blurred because many of our newer "antibiotics" are actually chemically modified biological products or even chemically synthesized biological products. The generic terms to refer to either antibiotics or chemotherapeutic agents are antimicrobic or antimicrobial agent. However, the term antibiotic is often used to refer to all types of antimicrobial agents.

Table 11.8 : Zone Diameter Interpretive Standards and Approximate MIC Correlates used to Define the Interpretive Categories

Antimicrobial agent (amount per disk) and organism	Zone diameter (nearest whole millimeter) for each interpretive category				Approximate MIC correlates (micro gm/ml) for	
	R	I	MS	S	R	S
Ampicillin (10 micro gm)						
Enterobacteriaceae	≤11	12-13		≥14	≥32	≤8
Staphylococcus spp.	≤28			≥29	beta-Lactamase	≤0.25
Haemophilus spp.	≤19			≥20	≥4	≤2
Enterococci	≤16		≥17		≥16	
Other streptococci	≤21		22-29	≥30	≥4	≤0.12
Chloramphenicol (30 micro g)	≤12	13-17		≥18	≥25	≤12.5
Erythromycin (15 micro g)	≤13	14-17		≥18	≥8	≤2
Nalidixic acid (30 micro g)	≤13	14-18		≥19	≥32	≤12
Streptomycin (10 micro g)	≤11	12-14		≥15		
Tetracycline (30 micro g)	≤14	15-18		≥19	≥16	≤4
Trimethoprim (5 micro g)	≤10	11-15		≥16	≥16	≤4

Note: R, Resistant; I, intermediate; MS, moderately susceptible; S, susceptible. An I result should be reported since it indicates an equivocal test result that may require further testing. When designated in the table, an MS result should be reported to indicate a level of susceptibility that should require the maximal safe dosage for therapy. Strains in the MS category are susceptible and not intermediate.

Approximate MIC correlates used for the definition of the resistant and susceptible categories. Theses correlates should not be used for the interpretation of antimicrobial dilution test results.

Protein Synthesis and Site of Action of Antimicrobials that Inhibit Protein Synthesis

Initiation of Protein Synthesis

The initiation of protein synthesis and the site of action of antimicrobials that inhibit this process.

Elongation

Figure 11.7 illustrates the process of elongation and the site of action of antimicrobials that inhibit this process.

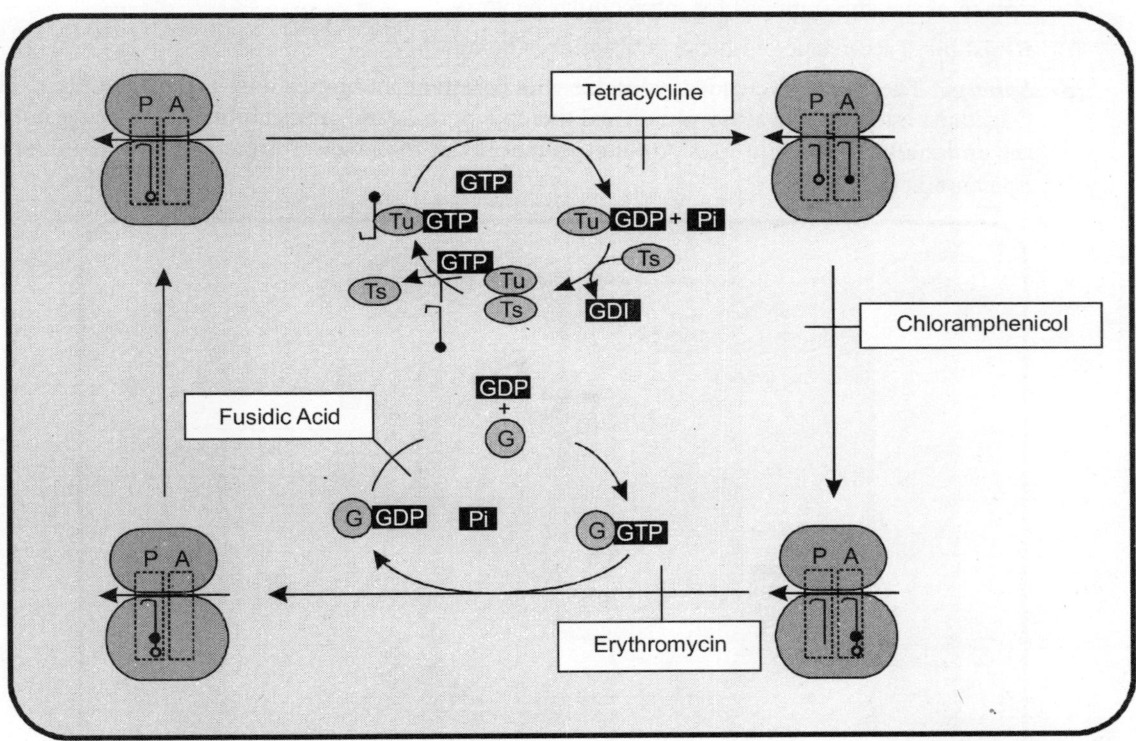

Fig. 11.7 : Antibiotics that Act at the Level of the Elongation Phase of Protein Synthesis

Inhibitors of Protein Synthesis

The selectivity of these agents is a result of differences in the prokaryotic 70S ribosome and the 80S eukaryotic ribosome. Since mitochondrial ribosomes are similar to prokaryotic ribosomes, these antimetabolites can have some toxicity. They are mostly bacteriostatic.

Antimicrobials that Bind to the 30s Ribosomal Subunit

Aminoglycosides(bactericidal)

Streptomycin, kanamycin, gentamicin, tobramycin, amikacin, netilmicin and neomycin (topical)

(a) ***Mode of action***: The aminoglycosides irreversibly bind to the 30S ribosome and freeze the 30S initiation complex (30S-mRNA-tRNA), so that no further initiation can occur. The aminoglycosides also slow down protein synthesis that has already initiated and induce misreading of the mRNA.

(b) ***Spectrum of Activity***: Aminoglycosides are active against many gram-negative and some gram-positive bacteria. They are not useful for anaerobic bacteria, since oxygen is required for uptake of the antibiotic, or for intracellular bacteria.

(c) ***Resistance***: Resistance to these antibiotics is common.

(d) ***Synergy***: The aminoglycosides synergize with β-lactam antibiotics such as the penicillins. The β-lactams inhibit cell wall synthesis and thereby increase the permeability of the bacterium to the aminoglycosides. Fig 11.8 Antibiotics that act at the level of protein synthesis initiation Spectinolycin.

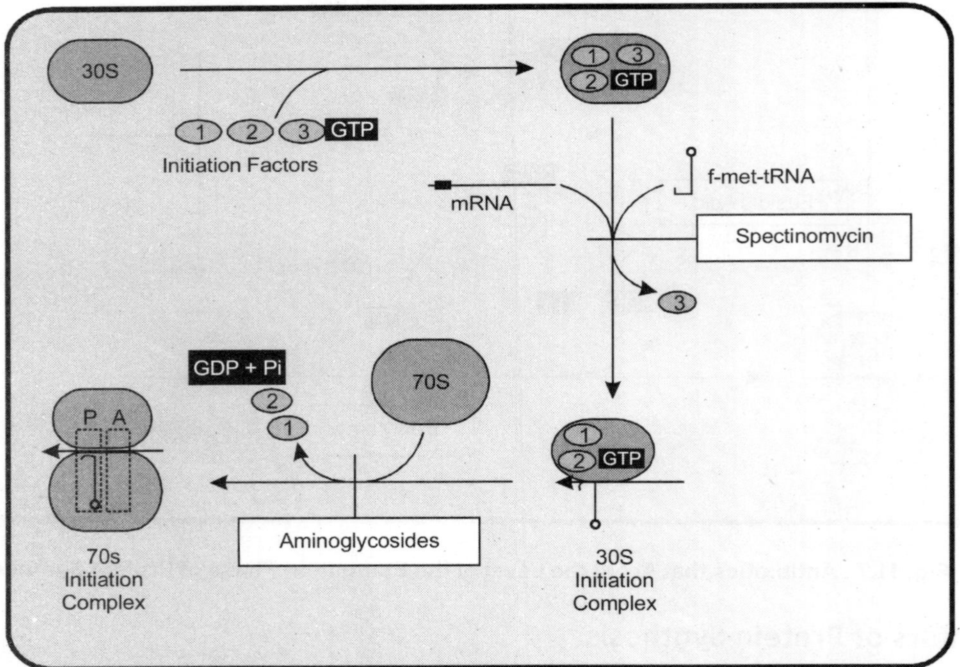

Fig. 11.8 : Antibiotics that act at the Level of Protein Synthesis Initiation

Tetracyclines (bacteriostatic)**:** Tetracycline, minocycline and doxycycline

(a) ***Mode of action***: The tetracyclines reversibly bind to the 30S ribosome and inhibit binding of aminoacyl-t-RNA to the acceptor site on the 70S ribosome.

(b) ***Spectrumofactivity*:** These are broad spectrum antibiotics and are useful against intracellular bacteria.

(c) ***Resistance*:** Resistance to these antibiotics is common.

(d) ***Adverseeffects*:** Destruction of normal intestinal flora often occurs, resulting in increased secondary infections. There can also be staining and impairment of the structure of bone and teeth.

3. Spectinomycin (bacteriostatic)

(a) ***Modeofaction*** Spectinomycin reversibly interferes with mRNA interaction with the 30S ribosome. It is structurally similar to aminoglycosides but does not cause misreading of mRNA.

(b) ***Spectrum of activity*** Spectinomycin is used in the treatment of penicillin-resistant *Neisseria gonorrhoeae.*

(c) ***Resistance*** This is rare in *Neisseria gonorrhoeae.*

Antimicrobials that Bind to the 50s Ribosomal Subunit

Chloramphenicol, lincomycin, clindamycin (bacteriostatic)

(a) ***Mode of action*** These antimicrobials bind to the 50S ribosome and inhibit peptidyl transferase activity.

(b) Spectrum of activity

- Chloramphenicol - Broad range
- Lincomycin and clindamycin - Restricted range

(c) ***Resistance*** Resistance to these antibiotics is common

(d) ***Adverse effects*** Chloramphenicol is toxic (bone marrow suppression) but it is used in the treatment of bacterial meningitis.

2. Macrolides (bacteriostatic): Erythromycin (also azithromycin, clarithromycin)

(a) ***Mode of action*** The macrolides inhibit translocation of the peptidyl tRNA from the A to the P site on the ribosome by binding to the 50S ribosomal 23S RNA.

(b) ***Spectrum of activity*** Gram-positive bacteria, *Mycoplasma, Legionella*

(c) ***Resistance*** Resistance to these antibiotics is common. Most gram-negative antibiotics are resistant to macrolides.

Antimicrobials that Interfere with Elongation Factors

Fusidic acid (bacteriostatic)

(a) ***Mode of action*** Fusidic acid binds to elongation factor G (EF-G) and inhibits release of EF-G from the EF-G/GDP complex.

(b) ***Spectrum of activity*** Fusidic acid is only effective against gram-positive bacteria such as *Streptococcus, Staphylococcus aureus* and *Corynebacterium minutissimum.*

Inhibitors of Nucleic Acid Synthesis and Function

The selectivity of these agents is a result of differences in prokaryotic and eukaryotic enzymes affected by the antimicrobial agent.

Inhibitors of RNA Synthesis And Function

Rifampin, rifamycin, rifampicin (bactericidal)

(a) ***Mode of action***: These antimicrobials bind to DNA-dependent RNA polymerase and inhibit initiation of RNA synthesis.

(b) ***Spectrum of activity***: They are wide spectrum antibiotics but are used most commonly in the treatment of tuberculosis

(c) ***Resistance***: Resistance to these antibiotic is common.

(d) ***Combination therapy***: Since resistance is common, rifampin is usually used in combination therapy

Inhibitors of DNA Synthesis and Function

Quinolones: nalidixic acid, ciprofloxacin, oxolinic acid (bactericidal)

(*a*) Mode of action These antimicrobials bind to the A subunit of DNA gyrase (topoisomerase) and prevent supercoiling of DNA, thereby inhibiting DNA synthesis.

(*b*) Spectrum of activity - These antibiotics are active against Gram-positive cocci and are used in urinary tract infections

(*c*) Resistance This is common for nalidixic acid and is developing for ciprofloxacin

Antimetabolite Antimicrobials

Inhibitors of Folic Acid Synthesis

The selectivity of these antimicrobials is a consequence of the fact that bacteria cannot use preformed folic acid and must synthesize their own folic acid. In contrast, mammalian cells use folic acid obtained from food.

Figure 11.9 summarizes the pathway of folic acid metabolism and indicates the sites at which antimetabolites act.

Sulphonamides, sulfones (bacteriostatic)

(a) ***Mode of action***: These antimicrobials are analogues of para-aminobenzoic acid and competitively inhibit formation of dihydropteric acid.

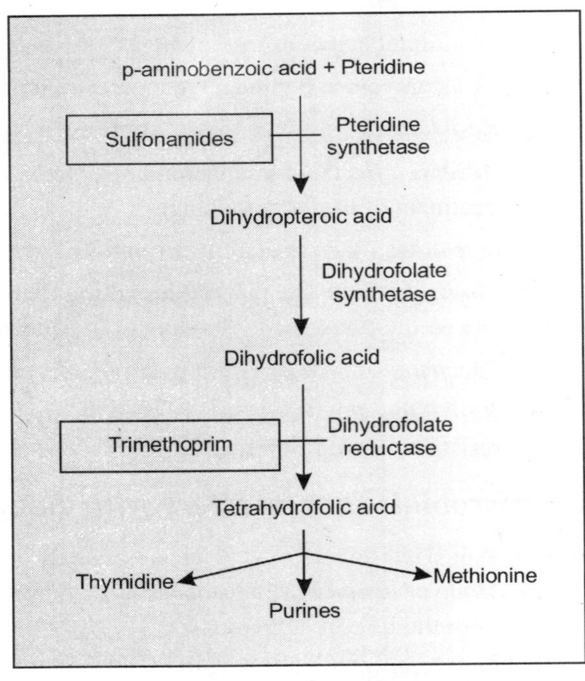

Fig. 11.9 : The Folic Acid Metabolism

(b) ***Spectrum of activity***: They have a broad range activity against gram-positive and gram-negative bacteria and are used primarily in urinary tract infections and in *Nocardia* infections.

(c) ***Resistance***: Resistance to these antibiotics is common

(d) ***Combination therapy***: The sulphonamides are used in combination with trimethoprim. This combination blocks two distinct steps in folic acid metabolism and prevents the emergence of resistant strains.

2. Trimethoprim, methotrexate, pyrimethamine (bacteriostatic)

(a) ***Mode of action***: These antimicrobials bind to dihydrofolate reductase and inhibit formation of tetrahydrofolic acid.

(b) ***Spectrum of activity***: They have a broad range activity against gram-positive and gram-negative bacteria and are used primarily in urinary tract infections and in *Nocardia* infections.

(c) ***Resistance***: Resistance to these antibiotics is common

(d) ***Combination therapy***: These antimicrobials are used in combination with the sulphonamides. This combination blocks two distinct steps in folic acid metabolism and prevents the emergence of resistant strains.

Anti-mycobacterial Agents

Anti-mycobacterial agents are generally used in combination with other antimicrobials since treatment is prolonged and resistance develops readily to individual agents.

Para-aminosalicylic acid (PSA) (bacteriostatic)

(a) ***Mode of action*** This is similar to sulphonamides

(b) ***Spectrum of activity*** PSA is specific for *Mycobacterium tuberculosis*

Dapsone (bacteriostatic)

(*a*) Mode of action Similar to sulphonamides

(*b*) Spectrum of activity Dapsone is used in treatment of leprosy

Isoniazid (INH) (bacteriostatic)

(a) ***Mode of action*** Isoniazid inhibit synthesis of mycolic acids.

(b) ***Spectrum of activity*** INH is used in treatment of tuberculosis

(c) ***Resistance Resistance*** has developed

ANTIBIOTIC RESISTANCES

Antibiotic resistance is a specific type of drug resistance when a microorganism has the ability of withstanding the effects of antibiotics. Antibiotic resistance evolves via natural selection acting upon random mutation, but it can also be engineered by applying an evolutionary stress on a population. Once such a gene is generated, bacteria can then transfer the genetic information in a horizontal fashion (between individuals) by conjugation, transduction, or transformation. Many antibiotic resistance genes reside on plasmids, facilitating their transfer. If a bacterium carries several resistance genes, it is called multiresistant or, informally, a superbug. The term antimicrobial resistance is sometimes used to explicitly encompass organisms other than bacteria. Antibiotic resistance can also be introduced artificially into a microorganism through laboratory protocols, sometimes used as a selectable marker to examine the mechanisms of gene transfer or to identify individuals that absorbed a piece of DNA that included the resistance gene and another gene of interest.

THE CAUSES

The widespread use of antibiotics both inside and outside of medicine is playing a significant role in the emergence of resistant bacteria. They are often used in animals but also in other industries which at least in the case of agricultural use lead to the spread of resistant strains to human populations. In some countries antibiotics are sold over the counter without a prescription which compounds the problem. In human medicine the major problem of the emergence of resistant bacteria is due to misuse and overuse of antibiotics by doctors as well as patients. Other practices contributing towards resistance include the addition of antibiotics to the feed of livestock. Household use of antibacterials in soaps and other products, although not clearly contributing to resistance, is also discouraged (as not being effective at infection control). Also unsound practices in the pharmaceutical manufacturing industry can contribute towards the likelihood of creating antibiotic resistant strains. Certain antibiotic classes are highly associated with colonisation with superbugs compared to other antibiotic classes. The risk for colonisation increases if there is a lack of sensitivity (resistance) of the superbugs to the antibiotic used and high tissue penetration as well as broad spectrum activity against “good bacteria”. In the case of MRSA, increased rates of MRSA infections are seen with glycopeptides, cephalosporins and especially quinolones. In the case of colonisation with C difficile the high risk antibiotics include cephalosporins and in particular quinolones and clindamycin. The volume of antibiotic prescribed is the major factor in increasing rates of bacterial resistance rather than compliance with antibiotics. Inappropriate prescribing of antibiotics has been attributed to a number of causes including: people who insist on antibiotics, physicians simply prescribe them as they feel they do not have time to explain why they are not necessary, physicians who do not know when to prescribe antibiotics or else are overly cautious for medical legal reasons. A third of people for example believe that antibiotics are effective for the common cold and 22% of people do not finish a course of antibiotics primarily due to that fact that they feel better (varying from 10% to 44% depending on the country).Compliance with once daily antibiotics is better than with twice daily antibiotics. Sub optimum antibiotic concentrations in critically ill people increase the frequency of antibiotic resistance organisms. While taking antibiotics doses less than those recommended may increase rates of resistance, shortening the course of antibiotics may actually decrease rates of resistance. Poor hand hygiene by hospital staff has been associated with the spread of resistant organisms and an increase in hand washing compliance results in decreased rates of these organisms.

THE MECHANISM

Antibiotic resistance can be a result of horizontal gene transfer,[30] and also of unlinked point mutations in the pathogen genome and a rate of about 1 in 10^8 per chromosomal replication. The antibiotic action against the pathogen can be seen as an environmental pressure; those bacteria which have a mutation allowing them to survive will live on to reproduce. They will then pass this trait to their offspring, which will result in a fully resistant colony.

The four main mechanisms by which microorganisms exhibit resistance to antimicrobials are:

1. Drug inactivation or modification: e.g. enzymatic deactivation of *Penicillin* G in some penicillin-resistant bacteria through the production of β-lactamases.
2. Alteration of target site: e.g. alteration of PBP—the binding target site of penicillins—in MRSA and other penicillin-resistant bacteria.
3. Alteration of metabolic pathway: e.g. some sulfonamide-resistant bacteria do not require para-aminobenzoic acid (PABA), an important precursor for the synthesis of folic acid and nucleic acids in bacteria inhibited by sulfonamides. Instead, like mammalian cells, they turn to utilizing preformed folic acid.
4. Reduced drug accumulation: by decreasing drug permeability and/or increasing active efflux (pumping out) of the drugs across the cell surface.

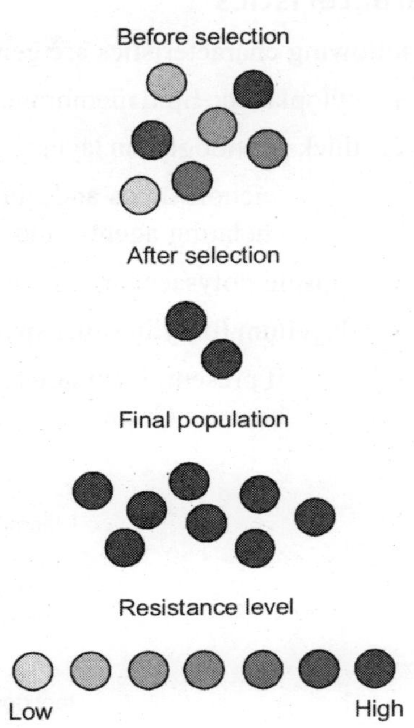

Fig. 11.10 : The Resistive Mechanism of Antibiotics

There are three known mechanisms of fluoroquinolone resistance. Some types of efflux pumps can act to decrease intracellular quinolone concentration. In gram-negative bacteria, plasmid-mediated resistance genes produce proteins that can bind to DNA gyrase, protecting it from the action of quinolones. Finally, mutations at key sites in DNA gyrase or Topoisomerase IV can decrease their binding affinity to quinolones, decreasing the drug's effectiveness. Research has shown that the bacterial protein LexA may play a key role in the acquisition of bacterial mutations giving resistance to quinolones and rifampicin.

THE GRAM POSITIVE AND GRAM NEGATIVE BACTERIA

Gram-positive bacteria are those that are stained dark blue or violet by Gram staining. This is in contrast to Gram-negative bacteria, which cannot retain the crystal violet stain, instead taking up the counterstain (safranin or fuchsin) and appearing red or pink. Gram-positive organisms are able to retain the crystal violet stain because of the high amount of peptidoglycan in the cell wall. Gram-positive cell walls typically lack the outer membrane found in Gram-negative bacteria.

When treated as a clade, the term "Posibacteria" is sometimes used.

Characteristics

The following characteristics are generally present in a Gram-positive bacterium:

1. cytoplasmic lipid membrane
2. thick peptidoglycan layer
   - teichoic acids and lipoids are present, forming lipoteichoic acids which serve to act as chelating agents, and also for certain types of adherence.
3. capsule polysaccharides (only in some species)
4. flagellum (only in some species)
   - if present, it contains two rings for support as opposed to four in Gram-negative bacteria because Gram-positive bacteria have only one membrane layer.

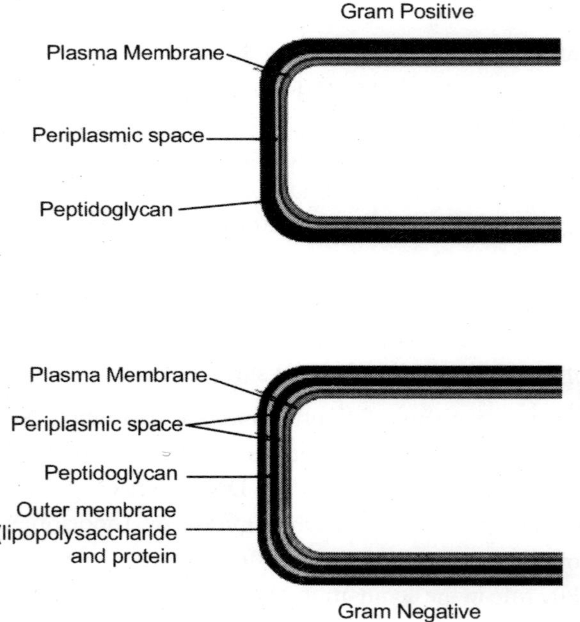

Fig. 11.11 : The Gram-Positive and Negative Cell Wall Structure

Classification

In the original bacterial phyla, the Gram-positive organisms made up the phylum Firmicutes, a name now used for the largest group. It includes many well-known genera such as *Staphylococcus, Streptococcus, Enterococcus,* (which are cocci) and *Bacillus, Corynebacterium, Nocardia, Clostridium, Actinobacteria,* and *Listeria* (which are rods and can be remembered by the mnemonic obconical). It has also been expanded to include the Mollicutes, bacteria-like *Mycoplasma* that lack cell walls and cannot be Gram stained, but are derived from such forms. Actinobacteria are the other major group of

Gram-positive bacteria, which have a high guanine and cytosine content in their genomes (high G+C group). This contrasts with the Firmicutes, which have a low G+C content.

Both Gram-positive and Gram-negative bacteria may have a membrane called an S-layer. In Gram-negative bacteria, the S-layer is directly attached to the outer membrane. In Gram-positive bacteria, the S-layer is attached to the peptidoglycan layer. Unique to Gram-positive bacteria is the presence of teichoic acids in the cell wall. Some particular teichoic acids, lipoteichoic acids, have a lipid component and can assist in anchoring peptidoglycan, as the lipid component is embedded in the membrane.

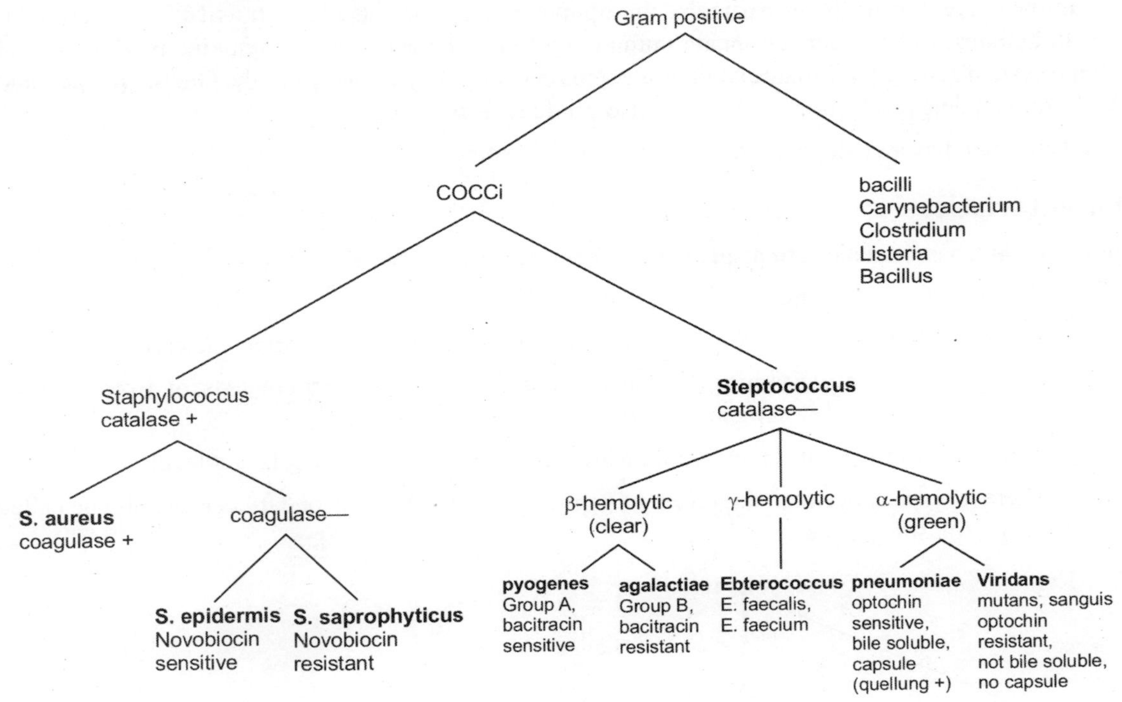

Exceptions

The Deinococcus-Thermus bacteria have Gram-positive stains, although they are structurally similar to Gram-negative bacteria.

Pathogenesis

Most pathogenic bacteria in humans are gram-positive organisms. Classically, six gram-positive genera are typically pathogenic in humans. Two of these, *Streptococcus* and *Staphylococcus*, are cocci (sphere-shaped bacteria). The remaining organisms are bacilli (rod-shaped bacteria) and can be subdivided based on their ability to form spores. The non-spore formers are *Corynebacterium* and *Listeria* (a coccobacillus), while *Bacillus* and *Clostridium* produce spores. The spore-forming bacteria can again be divided based on their respiration: *Bacillus* is a facultative anaerobe, while *Clostridium* is an obligate anaerobe.

Gram Negative Bacteria

Gram-negative bacteria are those bacteria that do not retain crystal violet dye in the Gram staining protocol.[1] In a Gram stain test, a counterstain (commonly safranin) is added after the crystal violet, coloring all Gram-negative bacteria with a red or pink color. The test itself is useful in classifying two distinct types of bacteria based on the structural differences of their cell walls. On the other hand, Gram-positive bacteria will retain the crystal violet dye when washed in a decolorizing solution.

The pathogenic capability of Gram-negative bacteria is often associated with certain components of Gram-negative cell walls, in particular the lipopolysaccharide (also known as LPS or endotoxin) layer. In humans, LPS triggers an innate immune response characterized by cytokine production and immune system activation. Inflammation is a common result of cytokine (from the Greek *cyto*, cell and *kinesis*, movement) production, which can also produce host toxicity.

When treated as a clade, the term "negibacteria" is sometimes used.

Characteristics

The following characteristics are displayed by Gram-negative bacteria:

1. Cytoplasmic membrane.
2. Thin peptidoglycan layer (which is much thinner than in Gram-positive bacteria).
3. Outer membrane containing lipopolysaccharide (LPS, which consists of lipid A, core polysaccharide, and O antigen) outside the peptidoglycan layer.
4. Porins exist in the outer membrane, which act like pores for particular molecules.
5. There is a space between the layers of peptidoglycan and the secondary cell membrane called the periplasmic space.

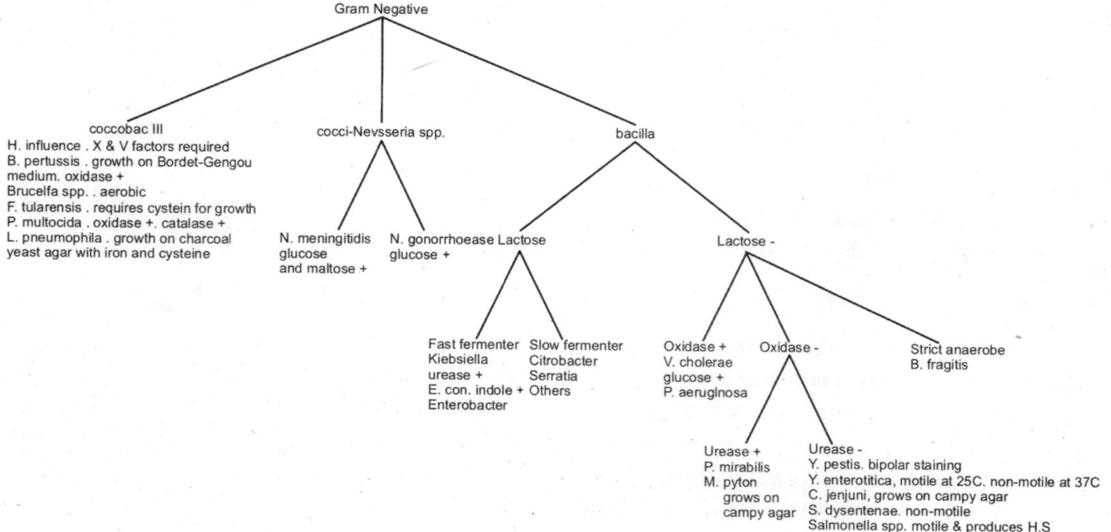

6. The S-layer is directly attached to the outer membrane, rather than the peptidoglycan.
7. If present, flagella have four supporting rings instead of two.
8. No teichoic acids or lipoteichoic acids are present.
9. Lipoproteins are attached to the polysaccharide backbone.
10. Most do not sporulate (*Coxiella burnetii*, which produces spore-like structures, is a notable exception).

Example Species

The proteobacteria are a major group of Gram-negative bacteria, including *Escherichia coli*, *Salmonella*, *Shigella*, and other Enterobacteriaceae, *Pseudomonas*, *Moraxella*, *Helicobacter*, *Stenotrophomonas*, *Bdellovibrio*, acetic acid bacteria, *Legionella* and alpha-proteobacteria as *Wolbachia* and numerous others. Other notable groups of Gram-negative bacteria include the cyanobacteria, spirochaetes, green sulfur and green non-sulfur bacteria.

Medically relevant Gram-negative cocci include three organisms, which cause a sexually transmitted disease (*Neisseria gonorrhoeae*), a meningitis (*Neisseria meningitidis*), and respiratory symptoms (*Moraxella catarrhalis*).

Medically relevant Gram-negative bacilli include a multitude of species. Some of them primarily cause respiratory problems (*Hemophilus influenzae*, *Klebsiella pneumoniae*, *Legionella pneumophila*, *Pseudomonas aeruginosa*), primarily urinary problems (*Escherichia coli*, *Proteus mirabilis*, *Enterobacter cloacae*, *Serratia marcescens*), and primarily gastrointestinal problems (*Helicobacter pylori*, *Salmonella enteritidis*, *Salmonella typhi*).

Gram-negative bacteria associated with nosocomial infections include *Acinetobacter baumannii*, which cause bacteremia, secondary meningitis, and ventilator-associated pneumonia in intensive care units of hospital establishments.

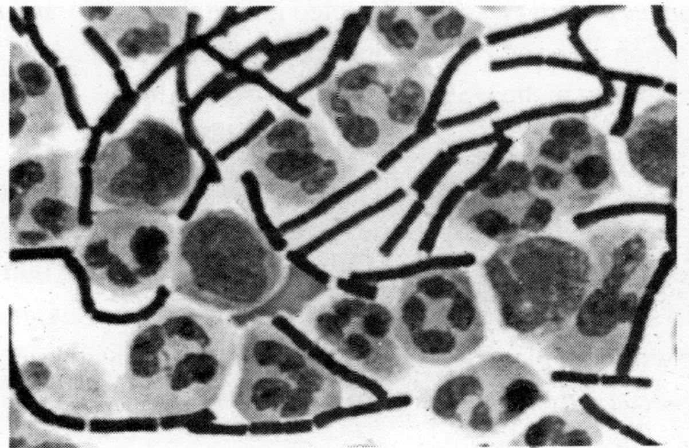

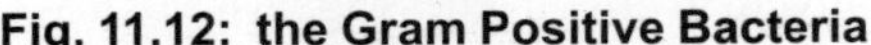

Fig. 11.12: the Gram Positive Bacteria

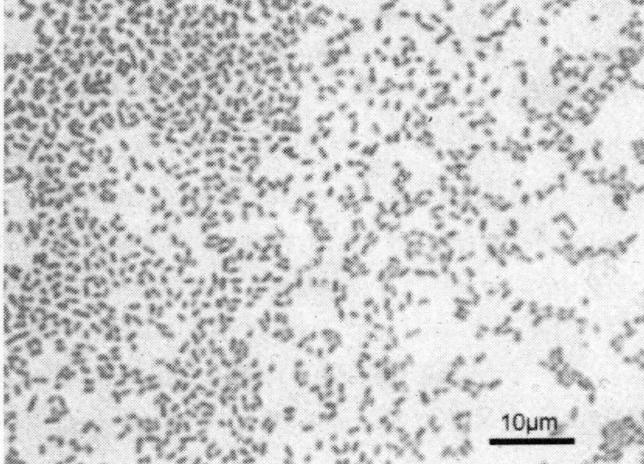

Fig. 11.13 : Gram Negative Bacteria

Gram Staining Method

1. The Specimen Smear is dried at first.
2. The Specimen is fixed with methanol.
3. Then the Specimen is covered with Crystal violet stain.
4. Then the stain is washed away and the specimen is allowed to dry.
5. Then the specimen is again stained with Lugol iodine.
6. Then the specimen is again washed and finally.
7. The specimen is again decolorized with acetone alcohol stain.
8. The specimen after final dry is treated with neutral red solution for 30 seconds:

The gram positive bacteria:	violet of blue colour
The gram Negative Bacteria:	red or orange colour
The agar:	red or orange colour
The pus cell :	red or orange colour

RESISTANT PATHOGENS

Staphylococcus Aureus

Staphylococcus aureus (colloquially known as "Staph aureus" or a *Staph infection*) is one of the major resistant pathogens. Found on the mucous membranes and the human skin of around a third of the population, it is extremely adaptable to antibiotic pressure. It was the first bacterium in which penicillin resistance was found—in 1947, just four years after the drug started being mass-produced. Methicillin was then the antibiotic of choice, but has since been replaced by oxacillin due to significant kidney toxicity. MRSA (methicillin-resistant *Staphylococcus aureus*) was first detected in Britain in 1961 and is now "quite common" in hospitals. MRSA was responsible for 37% of fatal cases of sepsis in the UK in 1999, up from 4% in 1991. Half of all *S. aureus* infections in the US are resistant to penicillin, methicillin, tetracycline and erythromycin.

This left vancomycin as the only effective agent available at the time. However, strains with intermediate (4-8 ug/ml) levels of resistance, termed GISA (glycopeptide intermediate *Staphylococcus aureus*) or VISA (vancomycin intermediate *Staphylococcus aureus*), began appearing in the late 1990s. The first identified case was in Japan in 1996, and strains have since been found in hospitals in England, France and the US. The first documented strain with complete (>16 ug/ml) resistance to vancomycin, termed VRSA (Vancomycin-resistant *Staphylococcus aureus*) appeared in the United States in 2002.

A *new* class of antibiotics, oxazolidinones, became available in the 1990s, and the first commercially available oxazolidinone, linezolid, is comparable to vancomycin in effectiveness against MRSA. Linezolid-resistance in *Staphylococcus aureus* was reported in 2003.

CA-MRSA (Community-acquired MRSA) has now emerged as an epidemic that is responsible for rapidly progressive, fatal diseases including necrotizing pneumonia, severe sepsis and necrotizing fasciitis. Methicillin-resistant *Staphylococcus aureus* (MRSA) is the most frequently identified antimicrobial drug-resistant pathogen in US hospitals. The epidemiology of infections caused by MRSA is rapidly changing. In the past 10 years, infections caused by this organism have emerged in the community. The 2 MRSA clones in the United States most closely associated with community outbreaks, USA400 (MW2 strain, ST1 lineage) and USA300, often contain Panton-Valentine leukocidin (PVL) genes and, more frequently, have been associated with skin and soft tissue infections. Outbreaks of community-associated (CA)-MRSA infections have been reported in correctional facilities, among athletic teams, among military recruits, in newborn nurseries, and among men who engage in frequent homosexual activities. CA-MRSA infections now appear to be endemic in many urban regions and cause most CA-S. aureus infections.

Streptococcus and Enterococcus

Streptococcus pyogenes (Group A Streptococcus: GAS) infections can usually be treated with many different antibiotics. Early treatment may reduce the risk of death from invasive group A streptococcal disease. However, even the best medical care does not prevent death in every case. For those with very severe illness, supportive care in an intensive care unit may be needed. For persons with necrotizing fasciitis, surgery often is needed to remove damaged tissue. Strains of *S. pyogenes* resistant to macrolide antibiotics have emerged, however all strains remain uniformly sensitive to penicillin.

Resistance of *Streptococcus pneumoniae* to penicillin and other beta-lactams is increasing worldwide. The major mechanism of resistance involves the introduction of mutations in genes encoding penicillin-binding proteins. Selective pressure is thought to play an important role, and use of beta-lactam antibiotics has been implicated as a risk factor for infection and colonization. Streptococcus pneumoniae is responsible for pneumonia, bacteremia, otitis media, meningitis, sinusitis, peritonitis and arthritis.

Penicillin-resistant pneumonia caused by *Streptococcus pneumoniae* (commonly known as *pneumococcus*), was first detected in 1967, as was penicillin-resistant gonorrhea. Resistance to penicillin substitutes is also known as beyond *S. aureus*. By 1993 *Escherichia coli* was resistant to five fluoroquinolone variants. *Mycobacterium tuberculosis* is commonly resistant to isoniazid and rifampin and sometimes universally resistant to the common treatments. Other pathogens showing some resistance include *Salmonella*, *Campylobacter*, and *Streptococci*.

Enterococcus faecium is another superbug found in hospitals. Penicillin-Resistant Enterococcus was seen in 1983, vancomycin-resistant enterococcus (VRE) in 1987, and Linezolid-Resistant Enterococcus (LRE) in the late 1990s.

Pseudomonas Aeruginosa

Pseudomonas aeruginosa is a highly prevalent opportunistic pathogen. One of the most worrisome characteristics of *P. aeruginosa* consists in its low antibiotic susceptibility. This low susceptibility is attributable to a concerted action of multidrug efflux pumps with chromosomally-encoded antibiotic resistance genes (e.g. *mexAB-oprM*, *mexXY* etc.) and the low permeability of the bacterial cellular envelopes. Besides intrinsic resistance, *P. aeruginosa* easily develop acquired resistance either by mutation in chromosomally-encoded genes, or by the horizontal gene transfer of antibiotic resistance determinants. Development of multidrug resistance by *P. aeruginosa* isolates requires several different genetic events that include acquisition of different mutations and/or horizontal transfer of antibiotic resistance genes. Hypermutation favours the selection of mutation-driven antibiotic resistance in *P. aeruginosa* strains producing chronic infections, whereas the clustering of several different antibiotic resistance genes in integrons favours the concerted acquisition of antibiotic resistance determinants. Some recent studies have shown that phenotypic resistance associated to biofilm formation or to the emergence of small-colony-variants may be important in the response of *P. aeruginosa* populations to antibiotics treatment.

Clostridium Difficile

Clostridium difficile is a nosocomial pathogen that causes diarrheal disease in hospitals world wide. Clindamycin-resistant *C. difficile* was reported as the causative agent of large outbreaks of diarrheal disease in hospitals in New York, Arizona, Florida and Massachusetts between 1989 and 1992. Geographically dispersed outbreaks of *C. difficile* strains resistant to fluoroquinolone antibiotics, such as Cipro (ciprofloxacin) and Levaquin (levofloxacin), were also reported in North America in 2005.

Salmonella and *E. Coli*

E. coli and Salmonella come directly from contaminated food. Of the meat that is contaminated with E. coli, eighty percent of the bacteria are resistant to one or more drugs made; it causes bladder infections that are resistant to antibiotics ("HSUS Fact Sheet"). Salmonella was first found in humans in the 1970s and in some cases is resistant to as many as nine different antibiotics ("HSUS Fact Sheet"). When both bacterium are spread, serious health conditions arise. Many people are hospitalized each year after becoming infected, and some die as a result.

Acinetobacter Baumannii

On November 5, 2004, the Centers for Disease Control and Prevention (CDC) reported an increasing number of *Acinetobacter baumannii* bloodstream infections in patients at military medical facilities in which service members injured in the Iraq/Kuwait region during Operation Iraqi Freedom and in Afghanistan during Operation Enduring Freedom were treated. Most of these showed multidrug resistance (MRAB), with a few isolates resistant to all drugs tested.

THE OTHER ANTIBACTERIALS

Nitrofurantoin

Nitrofurantoin is another synthetic antibiotic, used mainly for urinary tract infections. (Since it is excreted in the urine, it concentrates in the bladder very nicely.) Nitrofurantoin stops bacteria from growing, and can kill bacteria with a high enough level, by blocking the bacteria's ability to use energy it makes by "digesting" nutrients like sugar, and by blocking other chemical reactions that use the same system. It is not usually used for infections other than UTIs, and there are several side effects (ranging from stomach upset to (very rarely) malfunctioning nerves) which limit its use.

Aminoglycosides

The aminoglycosides are drugs which stop bacteria from making proteins; they work by attaching permanently to the protein machinery. Since they attach permanently, the bacterial cell will die if it gets enough of the drug. They can be used by themselves, or along with penicillins or cephalosporins to give a two-pronged attack on the bacteria.

Aminoglycosides work quite well, but bacteria can become resistant to them. The drawbacks are large, though. Since aminoglycosides are broken down easily in the stomach, they can't be given by mouth and must be injected or given IV (although we can use them as eyedrops for "pink eye"). When injected, their side effects include possible damage (temporary or permanent) to the ears and to the kidneys; this can be minimized by checking the amount of the drug in the blood and adjusting the dose so that there is enough drug to kill bacteria but not too much of it. Generally, aminoglycosides are given for short time periods, and in the hospital where we can check both the drug levels and the bacteria's sensitivity easily.

Quinolones

The quinolones, of which the best known is ciprofloxacin (Cipro®), interfere with an enzyme called DNA gyrase that is essential for duplication of bacterial DNA. (Bacteria have only one long chromosome (DNA molecule); the chromosome gets twisted during replication, like a telephone cord, and, again like the telephone cord, the chromosome can become so twisted that nothing more can be done with it. DNA gyrase is the "untwisting" enzyme.) This interference is completely different from the interference of other antibiotics with bacterial "machinery", and so bacteria that are resistant to other antibiotics may be sensitive to the quinolones.

However, bacteria can develop resistance to the quinolones, too. Also, researchers have noticed that young animals given quinolones can have damage to their cartilage (the hard but slippery material that connects some bones and covers the slding surfaces of joints). In the past we have avoided using quinilones in children because of this finding, but we sometimes have to give some children quinolones when there is no alternative antibiotic available.

Polymyxin B

Polymyxin B is an antibacterial that is produced by another bacteria. It kills bacteria by damaging the cell wall chemically—just the way soap does. It can't be taken internally, but it's very useful for skin infections (it's part of "Polysporin") and for conjunctivitis ("pink eye").

Tetracyclines

Tetracyclines are yet another family of antibiotics oringinally found in bacteria. They also block the protein-making machinery of certain bacteria. One of the tetracyclines, *doxycycline*, is often used to treat certain sexually transmitted diseases (such as chlamydia and gonorrhea) in older patients. One known side effect of the tetracyclines is that they affect development of bone and of tooth enamel in young children, and because of this we do not usually give tetracyclines to children under age 8 years.

However, tetracycline may be the best antibiotic for some life-threatening infections, such as cholera and anthrax, and in such cases we may use tetracycline to treat a young child (tetracycline often leaves a permanent brown stain on developing teeth, but that's better than death...).

Antifungals

Some microorganisms, known as *fungi* (*fungus* in the singular), are cells that are biologically more similar to animal cells than to bacteria. Since many of the antibacterial antibiotics take advantage of the difference between bacterial cells and animal cells, the fungi's similarlity to animal cells makes them immune to the antibacterial antibiotics. However, there are antibiotics available for fungi such as *Candida*. These include *nystatin*, the *azoles* (including fluconazole, ketoconazole, and similar antibiotics), and *amphotericin B*. These work by disrupting the fungal cells' machinery. Some of these antibiotics may be applied to the skin or taken by mouth, while others must be given IV.

Antivirals

Since viruses can't live outside the person or animal they infect, they are much harder to kill off. Our immune system can find and kill many of the viruses that attack us, but sometimes a virus can multiply and overwhelm the immune system before the immune system "comes up to full speed". We immunize or vaccinate people against diseases—mostly viral, but some bacterial—so that their immune systems do have that head start. That seems to be the most succesful way to kill viruses permanently. An example is smallpox, which has been eradicated due mainly to the use of vaccines against it—without which the virus killed thousands, if not millions, in epidemics. Some viruses, such as HIV (which specifically attacks the immune system), are very hard to become immune to, but a great deal of research is being aimed at producing a working vaccine for those diseases.

Unfortunately, since viruses are completely dormant outside a "host" (an infected human or animal), they can't be attacked biologically unless they infect someone. The immune system can't go after the virus unless it's in the body, and all of the antiviral medicines we have work only when the virus is trying to reproduce in the body. We can destroy viruses in the environment if we know they are there (an example is using household bleach to kill HIV that might be on equipment contaminated with body fluids—but bleach won't kill HIV *in* the body, even if we could get it into the body safely). Once the virus is in the body, however, all we can do is let the immune system do its work, and in very rare cases (perhaps half-a-dozen viruses at most) give drugs that slow down the infection so that the body can clear it out more easily.

Acyclovir

One often-used antiviral medicine is *acyclovir*; *ganciclovir* and *valciclovir* are similar to acyclovir. These medicines slow down infections with viruses of a certain family, which include both varicella (chickenpox and shingles) and the herpes viruses. Acyclovir slows down the virus' multiplication and therefore slows down the infection. The problem is that the varicella and herpes viruses are never actually eradicated—they stay in the body forever, and "reactivate" later (sometimes years later). The recurrent sores of herpes, and the appearance of shingles years after you have chickenpox, are examples of reactivation, and although acyclovir can help you get over the reactivation infection, it can't actually get rid of the viruses.

AZT and other Reverse-Transcriptase Inhibitors

Another very well-known antiviral is *triazidothymidine*, better known as *zidovudine* or *AZT*. This drug, and others like it, are used to inhibit an enzyme called "reverse transcriptase" which HIV uses to "copy" its own genes into the genes of the cells it infects. Once the HIV genes are copied, the infected cell and all its offspring can produce more HIV. (This is why an AIDS patient cannot actually get rid of all of the virus once infected: the virus may lie dormant as inactive genes for months or years, and the anti-AIDS drugs cannot get to the gene copies.) Like bacteria, viruses can mutate, changing their structure so that drugs that used to work no longer help; this explains why AZT and other reverse-transcriptase inhibitors eventually lose their effectiveness in many patients.

Protease Inhibitors

A newer class of anti-AIDS drugs, the *protease inhibitors*, work by blocking a different HIV enzyme. HIV uses reverse transcriptase to copy its genes into the cell it's infecting; it uses "protease" (an enzyme that breaks down protein) to get into the cell in the first place. Many people with AIDS have been able to eliminate the virus from their bloodstream—or almost eliminate it—by using both reverse-transcriptase inhibitors and protease inhibitors at the same time. However, since the virus has copied itself into cells where neither kind of drug can attack it, a patient must keep taking the drugs forever to keep the virus from reactivating.

Note, by the way, that the antiviral drugs, even more than the antibacterials, are tailored to the kind of viruses they are intended to attack. AZT won't do anything for a cold, and neither will acyclovir. In fact, there are—so far—*no* antivirals that will do anything for the common cold. And, since there are many different viruses in several different families that can cause colds, we are not likely to have any anti-common-cold drugs in the near future.